MATERIALS CHEMISTRY

A Multidisciplinary Approach to Innovative Methods

MATERIALS CHEMISTRY

A Multidisciplinary Approach to Innovative Methods

Edited by

Klaus Friedrich, PhD

Gennady E. Zaikov, DSc

A. K. Haghi, PhD

AAP | APPLE ACADEMIC PRESS

Apple Academic Press Inc.
3333 Mistwell Crescent
Oakville, ON L6L 0A2
Canada

Apple Academic Press Inc.
9 Spinnaker Way
Waretown, NJ 08758
USA

Exclusive worldwide distribution by CRC Press, a member of Taylor & Francis Group
No claim to original U.S. Government works
Printed in the United States of America on acid-free paper
International Standard Book Number-13: 978-1-77188-251-4 (Hardcover)
International Standard Book Number-13: 978-1-77188-252-1 (eBook)

Library and Archives Canada Cataloguing in Publication

Materials chemistry : a multidisciplinary approach to innovative methods/edited by Klaus Friedrich, PhD, Gennady E. Zaikov, DSc, A.K. Haghi, PhD.

Includes bibliographical references and index.
Issued in print and electronic formats.
ISBN 978-1-77188-251-4 (hardcover).--ISBN 978-1-77188-252-1 (pdf)
1. Materials. 2. Materials--Analysis. 3. Chemistry, Technical. 4. Materials science. I. Friedrich, Klaus, 1945-, editor II. Zaikov, G. E. (Gennadi™i Efremovich), 1935-, author, editor III. Haghi, A. K., author, editor

TA403.M38 2016 620.1'1 C2016-901114-3 C2016-901115-1

Library of Congress Cataloging-in-Publication Data

Names: Friedrich, Klaus, 1945- editor. | Zaikov, G. E. (Gennadiæi Efremovich), 1935- editor. | Haghi, A. K., editor.
Title: Materials chemistry : a multidisciplinary approach to innovative methods / Klaus Friedrich, PhD, Gennady E. Zaikov, DSc, A.K. Haghi, PhD, editors.
Description: Toronto : Apple Academic Press, 2016. | Includes bibliographical references and index.
Identifiers: LCCN 2016006428 (print) | LCCN 2016009023 (ebook) | ISBN 9781771882514 (hardcover : alk. paper) | ISBN 9781771882521 ()
Subjects: LCSH: Supramolecular chemistry. | Chemistry, Physical and theoretical.
Classification: LCC QD878 .M38 2016 (print) | LCC QD878 (ebook) | DDC 541/.2--dc23
LC record available at http://lccn.loc.gov/2016006428

Apple Academic Press also publishes its books in a variety of electronic formats. Some content that appears in print may not be available in electronic format. For information about Apple Academic Press products, visit our website at **www.appleacademicpress.com** and the CRC Press website at **www.crcpress.com**

CONTENTS

LIST OF CONTRIBUTORS

Tatiana S. Babicheva
Educational and Research Institute of Nanostructures and Biosystems, Saratov State University, 83 Astrakhanskaya St., Saratov 410012, Russian Federation; E-mail: Tatyana.babicheva.1993@mail.ru

M. V. Bazunova
Bashkir State University, 32 Zaki Validi Street, 450076 Ufa, Republic of Bashkortostan, Russia

A. A. Berlin
Semenov Institute of Chemical Physics, Russian Academy of Sciences, ul. Kosygina 4, Moscow, 119991 Russia

N. S. Borisova
Bashkir State University, 32 Z, Validi Street, Ufa, 450076, Republic of Bashkortostan. Russia, E-mail: NSGuskova@rambler.ru

Natalia O. Gegel
Educational and Research Institute of Nanostructures and Biosystems, Saratov State University, 83 Astrakhanskaya St., Saratov 410012, Russian Federation; E-mail: GegelNO@yandex.ru

A. K. Haghi
University of Guilan, Rasht, Iran

A. L. Iordanskii
Semenov Institute of Chemical Physics, Russian Academy of Sciences, ul. Kosygina 4, Moscow, 119991 Russia

O. Ismailov
Institute of General and Inorganic Chemistry, Uzbek Academy of Sciences, 77-a M. Ulugbek Street, 100 170, Tashkent, Uzbekistan; E-mail: ismoilovnmpi@mail.ru

S. G. Karpova
Emanuel Institute of Biochemical Physics, Russian Academy of Sciences, ul. Kosygina 4, Moscow, 119991 Russia

S. V. Kolesov
Bashkir State University, 32 Zaki Validi Street, 450076 Ufa, Republic of Bashkortostan, Russia

G. A. Korablev
Izhevsk State Agricultural Academy, Russian Federation

R. G. Korablev
Izhevsk State Agricultural Academy, Russian Federation

E. V. Koverzanova
Federal State Budgetary Establishment of a Science of Institute of Chemical Physics of N. N. Semenov of Russian Academy of Sciences, Russian Federation

F. H. Kudasheva
Bashkir State University, 9, Ufa, Republic of Bashkortostan, 450076, Russia

E. I. Kulish
Bashkir State University, 32 Zaki Validi Street, 450076 Ufa, Republic of Bashkortostan, Russia

L. N. Kurkovskaja
Federal State Budgetary Establishment of a Science of Institute of Biochemical Physics of N. M. Emanuelja of Russian Academy of Sciences, Russian Federation

I. M. Levina
Federal State Budgetary Establishment of a Science of Institute of Biochemical Physics of N. M. Emanuelja of Russian Academy of Sciences, Russian Federation

Eugene M. Lisitsyn
North-East Agricultural Research Institute of Rosselkhoz Academy, 166-a Lenin St., Kirov, 610007, Russia, E-mail: shikhova-l@mail.ru

S. M. Lomakin
Federal State Budgetary Establishment of a Science of Institute of Biochemical Physics of N. M. Emanuelja of Russian Academy of Sciences, Russian Federation

P. L. Maksimov
Izhevsk State Agricultural Academy, Russian Federation

M. V. Motyakin
Semenov Institute of Chemical Physics, Russian Academy of Sciences, ul. Kosygina 4, Moscow, 119991 Russia

A. A. Ol'khov
Plekhanov Russian University of Economics, Stremyannyi per. 36, Moscow, 117997 Russia

Ilia Valerievich Pankov
Voronezh State University, Russian Federation, 394006, Voronezh, Universitetskaya pl, No.1, Russia

N. G. Petrova
Agency of Informatization and Communication, Udmurt Republic, Russian Federation, E-mail: biakaa@mail.ru

Saeedeh Rafiei
University of Guilan, Rasht, Iran

S. Z. Rogovina
Semenov Institute of Chemical Physics, Russian Academy of Sciences, ul. Kosygina 4, Moscow, 119991 Russia

Z. Salimov
Institute of General and Inorganic Chemistry, Uzbek Academy of Sciences, 77-a M. Ulugbek Street, 100 170, Tashkent, Uzbekistan; E-mail: ismoilovnmpi@mail.ru

Sh. Saydahmedov
Fergansky Refinery, Russian Academy of Sciences, 4 Kosygin Street, 119 334 Moscow, Russia

D. I. Shafigullina
Bashkir State University, 9, Ufa, Republic of Bashkortostan, 450076, Russia

V. G. Shamratova
Bashkir State University, 32 Zaki Validi Street, 450076 Ufa, Republic of Bashkortostan, Russia

L. A. Sharafutdinova
Bashkir State University, 32 Zaki Validi Street, 450076 Ufa, Republic of Bashkortostan, Russia

Lyudmila N. Shikhova
Vyatka State Agricultural Academy, 133 Oktyabrsky Prospect, Kirov, 610017, Russia

N. G. Shilkina
Semenov Institute of Chemical Physics, Russian Academy of Sciences, ul. Kosygina 4, Moscow, 119991 Russia

Anna B. Shipovskaya
Educational and Research Institute of Nanostructures and Biosystems, Saratov State University, 83 Astrakhanskaya St., Saratov 410012, Russian Federation; E-mail: ShipovskayaAB@rambler.ru

R. N. Shiryaeva
Bashkir State University, 9, Ufa, Republic of Bashkortostan, 450076, Russia

O. V. Staroverova
Semenov Institute of Chemical Physics, Russian Academy of Sciences, ul. Kosygina 4, Moscow, 119991 Russia

G. R. Timerbaeva
Branch of Ufa State Aviation Technical University, 5 Youth Md. Tuimazy, 452750, Republic of Bashkortostan, Russia, E-mail: guzel.timerbaeva@mail.ru

R. F. Tukhvatullin
Bashkir State University, 32 Zaki Validi Street, 450076 Ufa, Republic of Bashkortostan, Russia

Viktor Nikolaevich Verezhnikov
Voronezh State University, Russian Federation, 394006, Voronezh, Universitetskaya pl, No.1, Russia

A. A. Volodkin
Federal State Budgetary Establishment of a Science of Institute of Biochemical Physics of N. M. Emanuelja of Russian Academy of Sciences, Russian Federation

Viktor Petrovich Yudin
Voronezh Affiliated Societies of Science Researching Synthetic Rubber, Russian Federation, 394014, Voronezh, Mendeleeva St., No.3, Building B, Russia, E-mail: skilful25@mail.ru

G. E. Zaikov
N.M. Emanuel Institute of Biochemical Physics, Russian Academy of Sciences, 4 Kosygin Street, 119 334 Moscow, Russia; E-mail: chembio@sky.chph.ras.ru

Maryam Ziaei
University of Guilan, Rasht, Iran

Yu. S. Zimin
Bashkir State University, 32 Z, Validi Street, Ufa, 450076, Republic of Bashkortostan. Russia, E-mail: ZiminYuS@mail.ru

LIST OF ABBREVIATIONS

AC	activated carbons
AC	alternate current
ACNF	activated carbon nanofibers
AFC	alkali fuel cells
AFM	atomic force microscope
AIBN	azoisobutyronitrile
AP	apple pectin
BSF	besieged silica filler
CNF	carbon nanofibers
CNG	compressed natural gas
CNT	carbon nanotubes
CP	citrus pectin
CRDCSC	Canadian Research and Development Center of Sciences and Cultures
Ct	total conversion
CTC	charge-transfer complex
CTZ	chitosan
CVD	chemical vapor deposition
DC	direct current
DMFC	direct methanol fuel cell
EM	electron microscope
FLWIP	first layer of the water-insoluble polymer
G	grafting efficiency
GA	glycolic acid
GE	grafting efficiency
GF	grafting frequency
GTP	group transfer polymerization
HIPS	High impact polystyrenes
HM	heavy metals
HP	hydroperoxide
ICE	internal combustion engine

IVW	Institute for Composite Materials
LH2	liquid hydrogen
MAC	maximum allowable concentration
MCFC	molten carbonate fuel cells
MEA	membrane electrode assembly
MWCNT	multi-walled carbon nanotubes
NTP	normal temperature and pressure
PAFC	phosphoric acid fuel cells
PAM	polyacrylamide
PCCs	polymer-colloid complexes
PE	polyethylene
PEEK	poly(ether-ketones)
PEFC	polymer electrolyte fuel cells
PEKKA	poly(ether-ketone-ketone
PEM	polymer electrolyte membrane
PEMFC	proton exchange membrane fuel cells
PET	poly(ethylene terephthalate)
PF	phenol formaldehyde
PH	polysaccharide
PHB	poly(3-hydroxybutyrate)
PIB	polyisobutylene
PLA	poly(lactic acid)
PLGA	poly lactic-co-glycolic acid
PMMA	poly(methyl methacrylate)
PP	polypropylene
PPO	poly(phenylene oxide)
PS	polystyrene
PTFE	polytetrafluoroethylene
PVAC	poly(vinyl acetate)
PVC	poly(vinyl chloride)
PVDF	poly(vinylidene fluoride)
PVE	poly(vinyl ether)
PVK	poly(N-vinylcarbazol)
PVOH	poly(vinyl alcohol)
PVP	poly(N-vinyl-pyrrolidone)
PVPP	polyvinyl pyrrolidone

SCTZ	sodium salt of chitosan succinylamide
SDBS	sodium dodecylbenzenesulfonate
SLWIP	second layer of the water-insoluble polymer
SOFC	solid oxide fuel cells
STM	scanning tunneling microscope
SWNT	single-walled nanotubes
TCNE	tetracyanoethylene
TEA	triethanolamine
TEMPO	2,2,6,6-tetramethylpiperi-din-1-oxyl
TESPT	triethoxysilylpropyl-tetrasulfide
TNF	trinitrofluorene

PREFACE

This book focuses on important aspects of materials chemistry by providing an overview of the theoretical aspects of materials chemistry, by describing the characterization and analysis methods for materials, and by explaining physical transport mechanisms in various materials. Not only does this book summarize the classical theories of materials chemistry, but also it exhibits their engineering applications in response to the current key issues. This book contains 14 chapters and provides practical equations, figures, and references, providing suitable complement to the text. On the other hand, this book struggles between its breadth and the depth of each aspect. It tackles most key issues in materials science and physical chemistry; while the readers may have to consult more specific references if they need to delve very deeply into a special topic.

This book is designed to provide important information for scientists and engineers on experimental research in materials chemistry using modern methods. The methods and instrumentation described represent modern analytical techniques useful to researchers, product development specialists, and quality control experts in polymer synthesis and manufacturing.

ABOUT THE EDITORS

Klaus Friedrich, PhD, is currently research director at the Institute for Composite Materials (IVW), University of Kaiserslautern, Germany. He acts also as a scientific board member of various international journals in the fields of materials science, composites, and tribology, and he has published more than 800 journal/conference papers, and several books related to fracture mechanics and tribology of polymer (nano)composites. He was a visiting assistant professor with the Center for Composite Materials, University of Delaware, USA, and has worked as a professor for polymers and composites at the Technical University Hamburg-Harburg, Germany. He graduated with a degree in mechanical engineering and in 1978 he earned his PhD in materials science at the Ruhr University Bochum in Germany.

Gennady E. Zaikov, DSc, is head of the Polymer Division at the N.M. Emanuel Institute of Biochemical Physics, Russian Academy of Sciences, Moscow, Russia, and Professor at Moscow State Academy of Fine Chemical Technology, Russia, as well as professor at Kazan National Research Technological University, Kazan, Russia. He is also a prolific author, researcher, and lecturer. He has received several awards for his work, including the Russian Federation Scholarship for Outstanding Scientists. He has been a member of many professional organizations and on the editorial boards of many international science journals.

A. K. Haghi, PhD, holds a BSc in urban and environmental engineering from University of North Carolina (USA); a MSc in mechanical engineering from North Carolina A&T State University (USA); a DEA in applied mechanics, acoustics and materials from Université de Technologie de Compiègne (France); and a PhD in engineering sciences from Université de Franche-Comté (France). He is the author and editor of 165 books as well as 1000 published papers in various journals and conference proceedings. Dr. Haghi has received several grants, consulted for a number of major corporations, and is a frequent speaker to national and international audiences. Since 1983, he served as a professor at several universities. He is currently editor-in-chief of the *International Journal of Chemoinformatics and Chemical Engineering* and *Polymers Research Journal* and on the editorial boards of many international journals. He is a member of the Canadian Research and Development Center of Sciences and Cultures (CRDCSC), Montreal, Quebec, Canada.

CHAPTER 1

PRODUCTION OF MICROTUBES FROM A CHITOSAN SOLUTION IN GLYCOLIC ACID

TATIANA S. BABICHEVA, NATALIA O. GEGEL, and ANNA B. SHIPOVSKAYA

Educational and Research Institute of Nanostructures and Biosystems, Saratov State University, 83 Astrakhanskaya St., Saratov 410012, Russian Federation;
E-mail: Tatyana.babicheva.1993@mail.ru, GegelNO@yandex.ru, ShipovskayaAB@rambler.ru

CONTENTS

ABSTRACT

Microtubes were prepared from chitosan solutions in glycolic acids by dry molding methods with salting-out agents of different nature: NaOH (an organic base), $N(C_2H_4OH)_3$ (an inorganic base), and $C_{12}H_{25}C_6H_4SO_3Na$ (an anionic surfactant). Their morphology, biocompatibility, elastic-deformation and physico-mechanical properties were examined. The influence of the salting-out agent nature on the mechanism of the chemical reaction proceeding during the microtube wall formation was estimated. The usage of sodium dodecylbenzenesulfonate (SDBS) as a salting-out agent allows obtaining microtubes with high strength properties. High adhesion and high proliferative activity of the epithelial-like MA-104 cellular culture on the surface of our microtubular substrates in model in vitro experiments were revealed. Certain physicochemical and biochemical parameters are comparable with the similar characteristics of human blood vessels.

1.1 INTRODUCTION

Nowadays, implants made of biocompatible biodegradable polymers are widely used as pins, screws, plates, membranes, and meshes in reconstructive surgery (orthopedics, maxillofacial surgery, dentistry) to stimulate osteosynthesis and to improve the healing of bones and soft tissues. In recent years, materials made of biodegradable polymers have begun to be used as prostheses, stents, and grafts in cardiovascular, urological and ophthalmic surgery. The design and usage of such medical materials are based on the idea of replacing the damaged sections of hollow tubular organs (blood vessels, ureter, tear ducts, etc.) by their bioartificial analogs. Over time, such a biodegradable tube-frame becomes covered with the human connective and endothelial tissues and is excreted from the body. After the final bioresorption of the polymer framework, a new living organ (or part thereof) is formed and the functional quality of the old tissue completely restores.

In this connection, the design of preparation methods of hollow cylindrical structures with the width of their walls in the micron range (microtubes) from biodegradable polymers for the use as prostheses of biomedical purposes, particularly, as analogs of blood vessels, is urgent. It is known from the literature of the preparation of such materials from biodegradable synthetic and natural polymers. For example, approaches are known to the preparation of tissue-engineering hollow cylindrical structures by electrospinning of poly(L-lactide-co-glycolide) and poly(ε-caprolactone) [1, 2], a blend of polydioxanone and soluble elastin [3], poly(diol citrate) as a biodegradable polyester elastomer [4], and self-assembled hydrogel plates based on poly(ethylene glycol) diacrylate [5]. Blood vessel analogs are also made of heteropolysaccharides, such as alginates, gellan gum, Kappa carrageenan, hyaluronan, and chitosan [6–9]. However, the methods and techniques of obtaining microtubular structures described in Refs. [1–9] are multistage and often involve aggressive solvents. The latter circumstance, in turn, would adversely affect the biocompatibility and other biochemical properties of the finished material, and would limit its use in clinical practice. In addition, the functional transformations of a bioprosthesis related to its resorption in the human body and, consequently, to cell division on it and tissue repairing, determine specific requirements to the physico-mechanical properties of the material. For example, the strength and elasticity indicators of the original structures must exceed those of vascular prostheses and patches made of non-biodegradable polymers. In this connection, the search for new approaches to designing hollow cylindrical frame structures (blood vessel analogs) with desired parameters is topical.

The aminopolysaccharide chitosan is a promising polymer for making biodegradable materials of biomedical purposes. Due to such its properties as biocompatibility with human tissues, non-toxicity, non-allergic nature, the ability to regenerate tissues, bioresorbability in the metabolic environment, chitosan can be used for making biodegradable prostheses for reconstructive surgery [10–12]. In Refs. [13, 14] we describe the production of 3D microtubes from chitosan solutions by the wet and dry molding processes. These methods are based on phase separation processes and the polymer-analogous transformation polysalt → polybase. Aqueous solutions of NaOH and triethanolamine (TEA) were used as transfer agents (salting-out agents) to convert chitosan in microtubes from its polysalt form into the water-insoluble polybasic form. Citric and lactic acids were used as solvents in the forming chitosan solutions. The microtubes obtained from chitosan solutions in citric acid are found to have a fragile porous inner layer. For the microtubes obtained from chitosan solutions in lactic acid, their morphology, biocompatibility, elastic-deformation and physico-mechanical properties were evaluated. The strength of the microtubes obtained by the dry method is much higher than in the case of the wet one. It is suggested that chitosan microtubes are promising as vascular prostheses. To continue our research in this direction, glycolic acid was used to prepare forming chitosan solutions in this work. Its choice is due to the fact that it falls into the class of pharmacopeial acids [15], and the polyesters and copolymers based thereon are not only actively studied to produce materials for surgery, tissue engineering, and bioartificial organs [16–18], but are already widely used in regenerative medicine [19, 20].

The aim of this work was to obtain microtubes from a chitosan solution in glycolic acid by the dry spinning method with the usage of salting-out agents of different chemical nature, to study their morphology, mechanical and biochemical properties.

1.2 EXPERIMENTAL PART

1.2.1 MATERIALS AND REAGENTS

Chitosan with a molecular weight of 700 KDa and a deacetylation degree of 80 mol. % was received from Bioprogress Ltd. (Russian Federation). Glycolic acid (GA) was used as an organic solvent, $C_{GA} = 1.5\%$. The physicochemical characteristics of glycolic acid are: mp 79°C, the dissociation constant in water $K = 1.48\times10^{-4}$ (25°C). Chitosan solutions with C_{CH} = 3.5–4.0 wt. % were prepared by dissolving a sample of air-dried powder of the polymer in the presence of the acid on a magnetic stirrer at room temperature during 5–7 h. The resulting solution was left to remove air bubbles for 24 h and used to form microtubes by both wet and dry methods. To convert the chitosan in microtubes from its polysalt form to the water-insoluble polybasic one, several salting-out agents were used (in various combinations): an organic base – 50% TEA ($K = 4.5\times10^{-7}$, 20°C), an inorganic base – 5% NaOH ($K = 5.9$, 25°C), and an anionic micelle-forming surfactant – 0.1 M SDBS (CMC = 1.3 mM), capable of forming ionic bonds with protonated amino groups [21]. Bidistilled water was taken to prepare solutions of the acid, alkalis, triethanolamine, and SDBS.

Xenopericardial biomaterial was gotten from bovine pericardium, the manufacturer Cardioplant Ltd. (Russian Federation).

1.2.2 MICROTUBE PREPARATION TECHNIQUE

Figure 1.1 presents a diagram of microtube preparation from chitosan by the dry molding method. The microtube formation process was performed as follows. A glass rod with a circular cross-section (the diameter of 5 mm) was vertically immersed into a chitosan solution and kept during 1 min for its surface to be coated with a uniform layer of the solution. The rod coated with a chitosan solution layer was then immersed in an aqueous solution of the salting-out agent for 1 min to form a first layer of the water-insoluble polymer (FLWIP), removed and dried in an oven at 45–50°C until complete evaporation of the solvent (4–5 h). A condensed phase of chitosan glycolate was formed. Then, the rod was again immersed in the aqueous solution of the salting-out agent to form a second layer of the water-insoluble polymer (SLWIP) and, correspondingly, to form the wall of the microtube as a hollow cylinder. Samples of the microtubes obtained by both dry methods were removed from the rod, washed with distilled water until neutral reaction, and stored in a swollen state in distilled water. The volume fraction of the solid phase in swollen samples was ~80 wt. %.

1.2.3 METHODS OF EXAMINATION

The morphology of sample chitosan microtubes was evaluated visually for color, transparency, and the uniformity of the inner and outer surfaces.

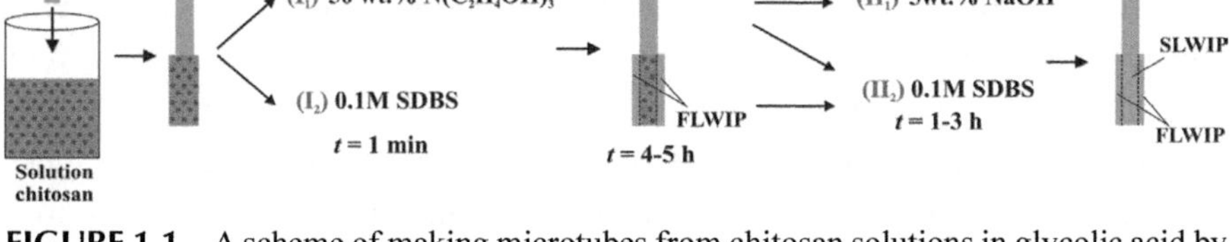

FIGURE 1.1 A scheme of making microtubes from chitosan solutions in glycolic acid by the dry molding method, the salting-out agent being 5% NaOH, 50% TEA or 0.1 M SDBS.

The samples were photographed with a digital camera Canon EOS 650D Kit (China).

SEM photos were obtained on a MIRA/LMU scanning electron microscope (Tescan, Czech Republic), a 15 kV accelerating voltage, a 2 vA current. Cuts of the samples were prepared for microscopic examination as follows: a microtube was taken out from distilled water and put into a special holder; a ring of a 3 mm length was cut with a sharp blade and dried in air for 24 h. The prepared sample was then removed from the holder and used in experiments. A 5 nm thick layer of gold was deposited on the obtained sample with a K450X Carbon Coater device (Deutschland).

The wall thickness (*d*, mm) of our microtubes was measured with an electronic digital outside micrometer CT 200–521 (China) with a scale division of 10 μm. Measurements were carried out several times at different sites of every microtube, and the thickness was then averaged.

The humidity of the samples was determined on a moisture analyzer and MX-50 (Japan), the weighing accuracy being ±0.0001.

Elastoplastic properties were examined on a tensile testing machine of uniaxial tension Tinius Olsen H1KS (Deutschland) with a 50 N load cell. Microtubes were cut along, unfolded and fixed in the terminals of the machine as plates. Every sample was tested at least 5 times. The breaking force and elongation at break were calculated according to standard procedures. The data obtained were plotted as stress–strain curves $\sigma = f(\varepsilon)$. The tensile stress (σ, MPa) was calculated on the basis of the cross-sectional area of original samples. The elongation at break (ε, %) was calculated using the original length of sample microtubes.

To assess the biocompatibility of our microtubes, the MA-104 culture of epithelial-like cells (rhesus monkey embryonic kidney) was used. Cells were cultivated under sterile conditions: in a laboratory specially equipped with a complex of “clean” laminates (second class protection, Nuaire, USA). A microtube sample was sterilized in a 70% ethanol solution for 20 min, placed into sterile Petri dishes (FalconBD), filled with the DMEM culture medium (Biolot, RF) supplemented with 10% fetal bovine serum (FetalBovineSerum, HyClone, UK) and an antibiotic-antimycotic mixture, and a suspension of the cell culture was introduced at a concentration of 1×10^4 cells/cm^3. Culturing was performed in a

SanyoMCO-18 M CO_2 incubator (Sanyo, Japan) in a 5% CO_2 atmosphere at 37°C. The culture medium was replaced on changing the color of the medium indicator. Cells were cultivated during 7 days. The adhesion and proliferation of the cell culture on our microtubes were evaluated daily on an inverted Biolam P microscope (Russian Federation) with a digital CCD camera DMS 300 Scopotek (China): a 10X lens, 3 Mpx resolution.

1.3 RESULTS AND DISCUSSION

To form the FLWIP–SLWIP of the chitosan microtube walls by the dry spinning process, several combinations of the salting-out reagents were used: TEA–NaOH ($I_1 - II_1$), TEA–0.1 M DDBSN ($I_1 - II_2$), and DDBSN–DDBSN ($I_2 - II_2$) (Figure 1.1). The interaction mechanism of chitosan macromolecules with the molecules of the salting-out agent is as follows.

In the initial aqueous acidic solution, chitosan exists in its polysalt form ($\sim NH_3^+$) due to protonation of its side NH_2 groups:

$$\sim NH_2 + HOCH_2COOH \rightarrow \sim NH_3^+ \, ^-OOCCH_2OH.$$

When organic or inorganic bases are used to form microtube walls, in the thin layer of the chitosan solution in glycolic acid deposited on the glass rod, the polymer-analogous polysalt → polybase conversion proceeds accompanied by phase separation. When FLWIP of the microtube wall is being formed, the chemical reaction proceeds at the interface of the two liquid phases: chitosan solution–salting-out agent solution. When SLWIP of the microtube wall is being formed, the reaction occurs at the solid–liquid interphase: condensed chitosan glycolate layer–salting-out agent solution. Note that the solid (condensed) polysalt phase is formed at the drying stage (Figure 1.1) as a result of evaporation of water molecules, and probably some of the molecules of glycolic acid, unbound ionically with chitosan amino groups.

In an alkaline medium (NaOH) the salt groups are converted to amine ones to form a polybase (~ -NH2), and the polymer loses its solubility (↓). The conversion scheme is shown below:

$$\sim NH_3^+ \, ^-OOCCH_2OH + Na^+ \, ^-OH \rightarrow \sim NH_2(\downarrow) + HOCH_2COONa + H_2O.$$

The polymer-analogous reaction of polysalt → polybase conversion proceeding in the solid polymer phase is due to the directed diffusion of low-molecular-weight ions (OH^-, Na^+) to the condensed polymer layer due to a concentration gradient at the $\sim NH_3^+ \mid NaOH$ interface, resulting in the equalization of the chemical potential of the low-molecular-weight component over the system.

Similar processes also proceed when an aqueous TEA solution is used as the salting-out agent. Glycolic-triethanolamine ester is one of the products of this chemical interaction:

$$(HOCH_2CH_2)_3\text{–}N + H_2O \rightarrow [(HOCH_2CH_2)_3\text{–}NH]^{+}\,{}^{-}OH$$
$$\sim NH_3^{+}\,{}^{-}OOCCH_2OH + [(HOCH_2CH_2)_3\text{–}NH]^{+} + {}^{-}OH \rightarrow$$
$$\sim NH_2(\downarrow) + (HOCH_2CH_2)_2\text{–}N\text{–}CH_2CH_2OOCCH_2OH + 2H_2O.$$

As TEA is a weak base ($K = 4.5\times10^{-7}$) the formation rate of the polybase is significantly lower than for a stronger base, NaOH. The possibility of such a reaction is evidenced by Ref. [22], which shows the possibility of the formation of a similar ester by reacting TEA with acetic acid.

When SDBS is taken, the microtube wall is formed by the ionotropic gelation mechanism as a result of frontal diffusion of the surfactant through the interphase. When FLWIP or SLWIP is being formed, the chitosan glycolate macromolecules interact with the surfactant molecules in the liquid phase or the solid (condensed) polysalt phase, respectively. In both cases, a polyelectrolyte-surfactant complex is formed according to the following reaction:

$$\sim NH_3^{+}\,{}^{-}OOCCH_2OH + C_{12}H_{25}C_6H_4SO_2O^{-}\,{}^{+}Na$$
$$\rightarrow \sim NH_3^{+}\,{}^{-}OSO_2C_6H_4C_{12}H_{25} + HOCH_2COONa.$$

Figure 1.2 (a–c) shows photos of our samples of chitosan microtubes obtained using several salting-out agents. The microtubes have no seams and are characterized by smooth external and internal wall surfaces without visual defects.

All samples are characterized by the almost uniform distribution of wall thicknesses along the length of the sample. The wall thickness of our microtubes depends on the nature of the salting-out agent and reaches its

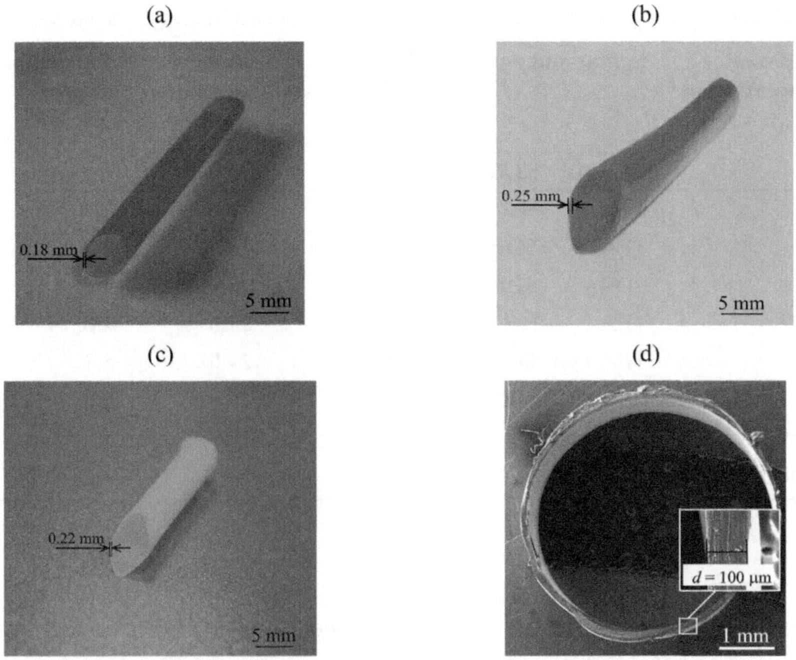

FIGURE 1.2 Photos of our microtubes obtained from a 4.0 wt. % chitosan solution in glycolic acid using the following combinations of salting-out agents: (a) 50% TEA–5% NaOH; (b) 50% TEA–0.1 M SDBS; (c) SDBS 0.1 M–0.1 M SDBS (c); (d) a SEM photo of a cut of the dried tube shown in figure (a).

maximum value for SDBS (Figure 1.2a–c; Table 1.1). This is probably due to some peculiarities of the chemical and physicochemical properties of the salting-out agents used.

Let us consider possible supramolecular structures formed by the interaction of chitosan glycolate with salting-out agents of different chemical nature. In all cases, for example, whether a base or a surfactant is used as a salting-out agent, the structure of the surface layers of the microtube wall is a semipermeable membrane, for example, that permeable to the molecules of low-molecular-weight substances (water). The application of an organic or inorganic base to form SLWIR and FLWIR of the microtube wall leads to the formation of a relatively dense supramolecular structure of the polybasic condensed phase (Figure 1.2d) stabilized with a network of intramolecular and intermolecular contacts (hydrogen bonding).

TABLE 1.1 Physico-Mechanical Properties of Our Chitosan Microtube Samples

Chitosan concentration (C_{CH}), wt. %	Salting-out agent		Microtube wall thickness (*d*), mm	Relative elongation (ε), %	Elongation at break (σ), MPa
	FLWIP	SLWIP			
3.5	TEA	NaOH	0.09	22.0	3.5
	TEA	SDBS	0.17	10.0	9.7
	SDBS	SDBS	0.18	28.0	4.0
4.0	TEA	NaOH	0.16	21.0	2.0
	TEA	SDBS	0.25	29.0	5.7
	SDBS	SDBS	0.31	29.0	6.8
Xenopericardial biomaterial			0.28	20.0	5.7
Human blood vessels (vienna, artery)			0.25–0.71	5–10	1.4–11.1*

* Data from Ref. [23].

When a micellar surfactant solution is used (with a SDBS concentration well above its CMC), the supramolecular structure of the microtube wall can be represented as a loose 3D network formed by micelle-like clusters of the surfactant-polyelectrolyte complexes. The size of the internal cavities in such a structure is known to be determined by the characteristic size of the polyelectrolyte-surfactant complex measured as the average distance between the micelle-like aggregates in the gel [24, 25]. The size of the latter, during ionotropic chitosan gelation in the presence of anionic surfactants such as sodium dodecyl sulfate, can reach 3.5–7.5 nm [26]. Therefore, the differences in the wall thickness of the microtubes obtained with TEA, NaOH or SDBS are logically explained by the different character of the supramolecular organization of the solid phase of polybase and polycomplex.

Figure 1.3 shows the deformation curves $\sigma = f(\varepsilon)$ of chitosan microtube samples obtained by using various reagents to convert the protonated amino groups of the polysalt into the basic amino groups of the polybase.

Figure 1.3 (curve *1*) shows the deformation curve of the sample obtained from a chitosan solution with $C = 3.5$ wt. % with the TEA–NaOH pair of salting-out agents to form FLWIP and SLWIP of the microtube wall (Figure 1.1, $I_1 - II_1$). It is evident that the stress-strain dependence has

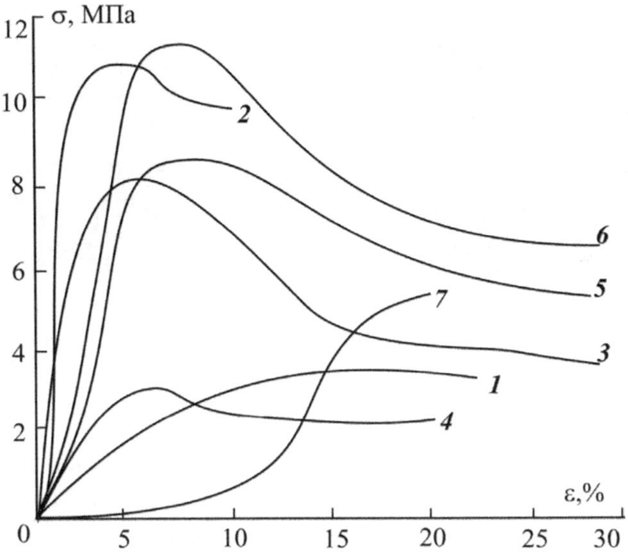

FIGURE 1.3 Stress–strain curves: for microtube samples obtained from chitosan solution in glycolic acid at polymer concentrations C = 3.5 (curves *1–3*) and 4.0 wt. % (curves *4–6*) and using the following combinations of salting-out agents: 50% TEA–5% NaOH (1:4) 50% TEA–0.1 M SDBS (2, 5), and SDBS 0.1 M–0.1 M SDBS (3, 6); and for a xenopericardial sample, a conventional material for bioprostheses and patches of blood vessels (7).

an elastic deformation fragment, whose value is about 4–5%, and a plastic deformation one, this deformation occurring uniformly without yield. Such $\sigma = f(\varepsilon)$ curves are typical of viscoplastic materials. The stress–strain curves of the samples obtained from the same solution but using TEA–SDBS (Figure 1.1, $I_1 - II_2$; Figure 1.3, curve 2), SDBS–SDBS (Figure 1.1, $I_2 - II_2$; Figure 1.3, curve 3) to form FLWIP–SLWIP, and the solutions with C = 4.0 wt. % (Figure 1.3, curves *4–6*) show forced-elastic deformation characteristic of viscoelastic materials. These samples are deformed at tension to form a neck due to the flow effect.

It should be noted that the use of SDBS to form the microtube wall leads to significantly stronger samples (Figure 1.3, curves *2*, *3*, *5*, and *6*; Table 1.1) as compared to using TEA and NaOH for the chemical polysalt → polybase reaction (curves *1* and *4*). The elasticity of these samples does not change and remains sufficiently high, ~28–30% (curves *3*, *5*, and *6*). The maximum values of σ = 11.5 MPa and ε_p = 29.5% were

FIGURE 1.4 Fetal rhesus kidney epitheliocytes (MA-104) after 4 (a) and 7 days (b) of culturing on the surface of a microtube obtained from a 4.0 wt. % chitosan solution in glycolic acid using the TEA-NaOH pair of salting-out agents. Light microscopy. Magnification ×100.

observed for the microtube obtained from the chitosan solution with $C = 4.0$ wt. % with the use of SDBS–SDBS (curve *6*). Such high values of tensile strength are perhaps due to peculiarities of the supramolecular structure of the polymer system formed by the surfactant-polyelectrolyte complexes [21, 24, 25]. Moreover, hardening of the structure is facilitated by heat treatment at one of the stages of forming microtubes (Figure 1.1), it enhances the interaction between surfactant molecules and chitosan macromolecules [27].

It should be noted that the elastic-deformation characteristics and strength values of microtubes are significantly higher than σ_p and ε_p of xenopericardial biomaterial samples traditionally used in reconstructive surgery of the heart and blood vessels (Figure 1.3, curve *7*) and are comparable with the same characteristics of human blood vessels [23].

In model in vitro experiments, the biocompatibility of our chitosan microtubes was assessed with the epithelial cell culture MA-104 as an example. The morphology of the cells cultured for one week on the microtube surface is illustrated in Figure 1.4. The experimental results show good adhesion and high proliferative activity of epithelial cells on the surface of the microtubular substrates.

1.4 CONCLUSION

The possibility of producing hollow cylindrical structures with a wall thickness in the micron range (microtubes) from chitosan solution in glycolic acid by the dry molding method is shown. The influence of the salting-out agent nature (a base or a surfactant) on the supramolecular ordering, elastic-plastic and physico-mechanical properties of the samples was evaluated. The usage of an organic or inorganic base as a salting-out agent leads to the formation of a densely-packed supramolecular structure of the microtube wall formed like a physical network of entanglements and stabilized by hydrogen bonding. When SDBS is used as a salting-out agent, a network structure of the microtube wall is formed, consisting of water-insoluble electrostatic complexes between the oppositely charged macromolecules of the polyelectrolyte (chitosan) and the surfactant molecules. The use of SDBS, in comparison with other salting-out agents, significantly raises the strength of the samples. The mechanical properties of our microtubes are superior to those of xenopericard, a biomaterial successfully used in cardiac surgery for several decades, and are comparable to native blood vessels (veins and arteries). In vitro experiments have shown high biocompatibility of our chitosan microtubes with epithelial-like cells. The results obtained allow us to consider chitosan microtubes as promising bioartificial materials to repair damaged human blood vessels.

ACKNOWLEDGMENTS

The work was supported by the Russian Ministry of Education and Science, State Task No 4.1212.2014/K.

KEYWORDS

- **biodegradable vascular prostheses**
- **chitosan**
- **microtube**
- **molding**
- **polymer-analogous transformations**
- **salting-out agent**

REFERENCES

1. Panseri, S., Cunha, C., Lowery, J., Carro, U. D., Taraballi, F., Amadio, S., Vescovi, A., Gelain, F. Electrospun micro- and nanofiber tubes for functional nervous regeneration in sciatic nerve transections. BMC Biotechnol. 2008, Vol. 8, 39.
2. Vaz, C. M., Tuijl, S., Bouten, C. V. C., Baaijens, F. P. T. Design of scaffolds for blood vessel tissue engineering using a multi-layering electrospinning technique. Acta Biomater. 2005, Vol. 1, No. 5, 575–582.
3. Smith, M. J., McClure, M. J., Sell, S. A., Barnes, C. P., Walpoth, B. H., Simpson, D. G., Bowlin, G. L. Suture-reinforced electrospun polydioxanone–elastin small-diameter tubes for use in vascular tissue engineering: A feasibility study. Acta Biomaterialia. 2008, Vol. 4, 58–66.
4. Yang, J., Motlagh, D., Webb, A. R., Ameer, G. A. Novel biphasic elastomeric scaffold for small-diameter blood vessel tissue engineering. Tissue engineering. 2005, Vol. 11, No. 11–12, 1876–1886.
5. Baek, K., Jeong, J. H., Shkumatov, A., Bashir, R., Kong, H. In situ self-folding assembly of a multi-walled hydrogel tube for uniaxial sustained molecular release. Adv. Mater. 2013, DOI: 10.1002/adma.201300951.
6. Barros, A. A., Duarte, A. R. C., Pires, R. A., Lima, A., Mano, J. F., Reis, R. L. Tailor made degradable ureteral stents from natural origin polysaccharides. Materials Tenth Conference on Supercritical Fluids and Their Applications. 2013, 1–6.
7. Lepidi, S., Grego, F., Vindigni, V., Zavan, B., Tonello, C., Deriu, G. P., Abatangelo, G., Cortivo, R. Hyaluronan biodegradable scaffold for small-caliber artery grafting: preliminary results in an animal model. Eur. J. Vasc. Endovasc. Surg. 2006, Vol. 32, 411–417.
8. Kong, X., Han, B., Wang, H., Li, H., Xu, W., Liu, W. Mechanical properties of biodegradable small-diameter chitosan artificial vascular prosthesis. J. Biomed. Mater. Res. Part, A. 2012, DOI: 10.1002/jbm.a.34136.
9. Zhu, C., Fan, D., Duan, Z., Xue, W., Shang, L., Chen, F., Luo, Y. Initial investigation of novel human-like collagen/chitosan scaffold for vascular tissue engineering. J. Biomed. Mater. Res. 2009, Vol. 89A, 829–840.
10. Rinaudo, M. Chitin and chitosan: properties and applications. Prog. Polym. Sci. 2006, Vol. 31, No. 7, 603–632.
11. Kim, I.-Y., Seo, S.-J., Moon, H.-S., Yoo, M.-K., Park, I.-Y., Kim, B.-C., Cho, C.-S. Chitosan and its derivatives for tissue engineering applications. Biotech. Adv. 2008, Vol. 26, No. 1, 1–21.
12. Huang, C., Chen, R., Ke, Q., Morsi, Y., Zhang, K., Mo, X. Electrospun collagen–chitosan–TPU nanofibrous scaffolds for tissue engineered tubular grafts. Colloids and Surfaces B: Biointerfaces. 2011, Vol. 82, No. 2, 307–315.
13. Gegel, N. O., Shipovskaya, A. B., Vdovykh, L. S., Babicheva, T. S. Preparation and properties of 3D chitosan microtubes. J. Soft Matter. Volume 2014, Article ID 863096. 9 p.
14. Kuchanskaya, L. S., Gegel, N. O., Shipovskaya, A. B. Preparation of chitosan-based microtubes. Modern Perspectives in the study of chitin and chitosan: Materials XI-th International conf. Murmansk. 2012, 59–63.

15. General Pharmacopeia, Article № 42–0070–07.
16. Makadia, H. K., Siegel, S. J. Poly lactic-*co*-glycolic acid (PLGA) as biodegradable controlled drug delivery carrier. Polymers. 2011, Vol. 3, No. 3, 1377–1397.
17. Carletti, E., Motta, A., Migliaresi, C. Scaffolds for tissue engineering and 3D cell culture. Methods in Molec. Bio. 2011, Vol. 695, 17–39.
18. Pan, Z., Ding, J. Poly(lactide-*co*-glycolide) porous scaffolds for tissue engineering and regenerative medicine. Interface Focus. 2012, Vol. 2, No. 3, 366–377.
19. Biodegradable systems in tissue engineering and regenerative medicine/Reis, R. L., San Roman, J. CRC Press. Boca Raton. Florida. 2004, 592 p.
20. Nair, L. S., Laurenein, C. T. Biodegradable polymers as biomaterials. Prog. Polym. Sci. 2007, Vol. 32, No. 8–9, 762–798.
21. Kildeeva, N. R., Babak, V. G., Vichoreva, G. A. Ageev, E. P. et al. A new approach to the development of materials with controlled release of the drug substance. Herald of Moscow University. Series 2. Chemistry (in. Rus.). 2000, Vol. 41, No. 6, 423–425.
22. Stepanova, T. J. Modification of surface properties of warp yarns from natural and synthetic fibers. Apriori. Series: natural and technical sciences. 2014, No. 3. 17 p.
23. Kumar, V. A., Caves, J. M., Haller, C. A., Dai, E., Liu, L., Grainger, S., Chaikof, E. L. A cellular vascular grafts generated from collagen and elastin analogs. Acta Biomater. 2013, Vol. 9, No. 9, 8067–8074.
24. Chitin and chitosan. Ed. K. G. Skriabin, G. A. Vikhoreva, V. P. Varlamov. 2002, 211–216 [in Rus.].
25. Rinaudo, M., Kildeyeva, N. R., Babak, V. G. Surfactant-polyelectrolyte complexes based on chitin derivatives. Russian Chemical Journal. 2008, Vol. LII, No. 1, 84–90.
26. Babak, V. G., Merkovich, E. A., Galbraikh, L. S., Shtykova, E. V., Rinaudo, M. Kinetics of diffusionally induced gelation and ordered nanostructure formation in surfactant–polyelectrolyte complexes formed at water/water emulsion type interfaces. Mendeleev Communs. 2000, Vol. 10, No. 3, 94–95.
27. Masubon, T., McClements, D. J. Influence of pH, ionic strength, and temperature on self-association and interactions of sodium dodecyl sulfate in the absence and presence of chitosan. Langmuir. 2005, Vol. 21, No. 1, 79–86.

CHAPTER 2

PHYSICO-CHEMICAL LAWS OF SURFACE OXIDATION OF POLYPROPYLENE AND POLYETHYLENE

M. V. BAZUNOVA,[1] S. V. KOLESOV,[1] R. F. TUKHVATULLIN,[1] E. I. KULISH,[1] and G. E. ZAIKOV[2]

[1]*Bashkir State University, 32 Zaki Validi Street, 450076 Ufa, Republic of Bashkortostan, Russia*

[2]*Institute of Biochemical Physics Named N.M. Emanuel of Russian Academy of Sciences, 4 Kosygina Street, 119334, Moscow, Russia*

CONTENTS

ABSTRACT

Physico-chemical laws of surface oxidation of materials of polypropylene, polyethylene and their waste are studied. Found that the background polyolefin material has virtually no effect on the course of its oxidative modification.

2.1 INTRODUCTION

The purpose of chemical modification of polymers is to change chemical structure by introducing the functional groups with different chemical nature into the macromolecules. In some cases, it is necessary to improve the characteristics of a polymer surface by chemical modification while retaining the properties of materials in the volume and shape of the polymer material (fiber, film, and bulk product). It is necessary, for example, when changing the wettability, sorption, adhesion and electrical characteristics of materials in the desired direction [1].

An effective method of modifying of the surface of polymeric material is oxidation. It is known that at temperatures below 90°C the reaction takes place mainly on the surface and is not accompanied by oxidative degradation of the polymer in the bulk. Obvious, that it is impractical to create complex chemical surface modification technology simultaneously saving money on technical costs of creating the necessary forms of materials. Oxidation is also a convenient method of pre-activation of the polymer surface, which leads to the appearance of oxygen-containing functional groups capable of being active centers during the further chemical modification. For example, it is known that the polyolefins oxidation is accompanied by the formation of hydroperoxide (HP) groups (Figure 2.1). Further thermal decomposition of the HP-groups leads to the appearance of free radicals on the surface and initiate growth of the grafted chains.

Therefore, it seems reasonable to study the simple and technologically advanced methods of surface oxidation of the waste fibrous and film

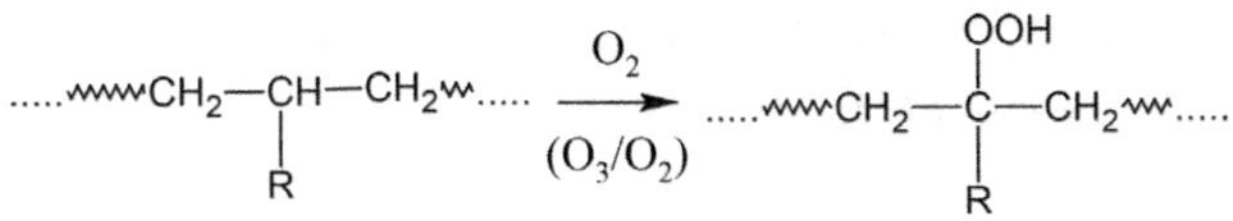

FIGURE 2.1 Oxidation of carbon-chain polymers (R = $-CH_3$, -H).

materials of polypropylene (PP) and polyethylene (PE) as a method of creating of the secondary polymeric materials.

2.2 EXPERIMENTAL PART

Previously cleaned and dried waste PP-fibers (original PP-fiber GOST 26574–85), waste PE-film (original PE-film GOST 17811–78) have been used as an object of research in this chapter.

Concentration of HP-groups on the surface of oxidized samples is defined by modified iodometric method [2] with the photo-electric calorimeter КФК-2МП.

The limited wetting angle is defined by the standard procedure [3].

The adsorption capacity (A) of the samples under static conditions for condensed water vapor, determined by method of complete saturation of the sorbent by adsorbate vapor in standard conditions at 20°C [4] and calculated by the formula: A=m/(M•d), wherein m – mass of the adsorbed water, g; M – mass of the dried sample, g; d – density of the adsorbate, g/cm^3.

2.2.1 PP-FIBERS, PE-FILMS, AND THEIR WASTES OXIDATION PROCEDURE (IN AN AQUEOUS MEDIUM)

About 1.5 g of the particulate polymer material charged into a round bottom flask equipped with reflux condenser, thermometer, mechanical stirrer and a bubbler for air supplying. 120 mL of H_2O_2 solution and the necessary amount of $FeSO_4 \cdot 7\ H_2O$ was added, then heated under stirring to 70–90°C and supplied by air from the compressor during 3–20 hrs. After the process gone, hydrogen peroxide solution was drained, sample thoroughly washed by water (first tap, then distilled) and dried in air at room temperature.

2.2.2 PP-FIBERS, PE-FILMS, AND THEIR WASTES SOLID-PHASE OXIDATION PROCEDURE

A weighted sample of the particulate polymer material placed into the thermostated reactor (a glass tube with diameter of 15 mm and a length of 150 mm). The reactor was fed by heated air from the compressor, at a rate of 7.2 L/h. Oxidation was conducted at 85°C for 4–10 hrs.

2.2.3 PP-FIBER PE-FILMS AND THEIR WASTES OZONATION PROCEDURE

a) 1.5 g of the particulate polymer material charged into the thermostated reactor (a glass tube with diameter of 15 mm and a length of 150 mm). For 60 min ozone-oxygen mixture, produced by an ozonizer (design is described in Ref. [5]), with flow rate of 30 L/h at ambient temperature was passed through the sample. Ozonator's performance by ozone is 12.5 mmol/h. After the ozonation sample is purged with an inert gas to remove the residual ozone and analyzed for HP-groups content.
b) 1.5 g of the particulate polymer material charged into a round bottom flask equipped with thermometer, mechanical stirrer and a bubbler for air supplying. 120 mL of distilled CCl_4 is added, the ozone-oxygen mixture is passed through the reaction mixture for 60 min at a flow rate of 30 L/h at room temperature. After the ozonation sample is purged with an inert gas to remove the residual ozone, solvent is merged. The sample was washed with chloroform and distilled water, dried in air at room temperature and analyzed for HP-groups content.

2.3 RESULTS AND DISCUSSION

The literature describes the oxidation of the PP-fibers in the medium of toluene by air at a temperature of 70–120°C in the presence of radical initiators and without them. In practice, these processes lead to the pollution of atmosphere by organic solvents vapor and require the inclusion of the solvent regeneration step in the production scheme. Previously, [6], we describe the method of oxidation of the waste PP-fibers by air oxygen in an aqueous medium in the presence of H_2O_2/Fe^{2+} initiating system. This method is also applied to the surface oxidation of the PE-film, and found that when carrying out the oxidation process at 85°C for 4 h in the presence of 2.7 mol/L H_2O_2 the content of HP-groups in the oxidized material is 0.38×10^{-5} mol/sm^2 (Table 2.1).

Uninitiated solid-phase surface oxidation of PP-fiber, PP- and PE-film with atmospheric oxygen at a temperature of 85°C was studied for the first time. Found that the content of HP-groups on the surface of oxidized samples, achieved during the process for 4 h, is 3.42×10^{-5}mol/sm^2 for PP-fibers and 1.86×10^{-5} mol/sm^2 for PE-film (Table 2.1).

The most effective method of the oxidation isozonation that even at room temperature during the process for 1 hour, gives the same results as in the previous cases: accumulation of HP-groups on the surface of the studied samples (Table 2.1).

The oxidation conditions and properties of the obtained samples are shown in Table 2.1.

Kinetic curves of HP-groups accumulation during solid-surface oxidation of waste PP-fiber are shown in Figures 2.2–2.3.

As follows from the data presented in Figure 2.1, the storage character of HP groups on the surface of the original PP-fiber having no exploitation, and accordingly, the period of aging in vivo and on the surface of the waste PP-fibers, subjected to aging in vivo, generally, identical. Only the initial parts of the kinetic curve of recently produced PP-fibers have the minimum. It can be caused by an intensive breakage of HP-groups emerged in the material during its processing, which dominates over the accumulation of HP-groups during thermal oxidation. It must be concluded that the background of polyolefin material has virtually no effect on the course of its oxidative modification.

Dyeability by azo-dyes and wettability of the modified samples are studied to confirm the changing of their surface properties. Dyeability of all oxidized samples obtained from waste PP-fiber, PP- and PE- films by azo-dyes is significantly better than dyeability of unmodified materials. Data on reducing of the limited wetting angle of the modified samples indicates increasing of hydrophilic surface. For example, the limited wetting angle of the oxidized waste PE-film surface by water is 10–12° less than the original film's.

TABLE 2.1 Terms of Oxidation of the Waste Polyolefins and Their Properties

Object	Oxidation terms					HP-groups concentration in the material, 10^{-5} mol/sm^2	Limited wetting angle	A, cm^3/g
	Medium	Oxidant	Initiator	Time, hour	T, °C			
PP-fiber (GOST 6574–85)	—	—	—	—	—	0.45	84°	1.10
Waste PP-fiber	—	—	—	—	—	1.5	81°	1.30
Waste PP-fiber	—	air	—	4	85	3.4	69°	1.38
Waste PP-fiber	H_2O	air	2,7 MH_2O_2, 0.37 mg/mL $FeSO_4$*7H_2O	4	85	3.2	71°	1.39
Waste PP-fiber	—	O_3/O_2	—	1	20	3.3	71°	1.39
PE-film (GOST 17811–78)	—	—	—	—	—	0.31	87°	1.08
Waste PE-film	H_2O	air	2.7 MH_2O_2,0,37 mg/mL $FeSO_4$*7H_2O	4	85	0.38	69°	1.37
Waste PE-film	—	air	—	4	85	1.8	72°	1.49
Waste PE-film	—	O_3/O_2	—	1	20	2.8	72°	1.49

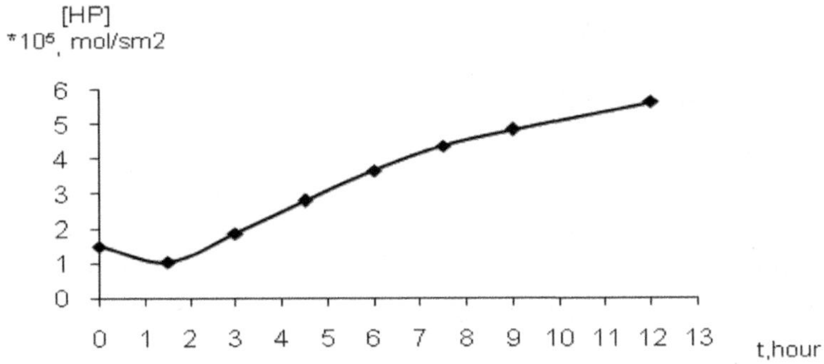

FIGURE 2.2 The dependence of the HP-groups concentration (mol/sm^2) on the surface of the waste PP-fiber from the oxidation time (temperature 85°C).

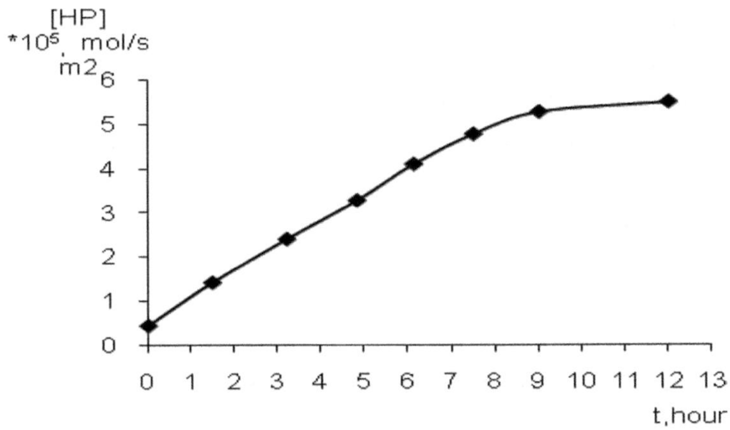

FIGURE 2.3 Dependence of the HP-groups concentration (mol/sm^2) on the surface of PP-fibers (GOST 26574–85) from the oxidation time (temperature 85°C).

Also, sorption activity of the modified and unmodified samples measured to evaluate the properties of the polymeric materials obtained by the oxidative modification (Table 2.1). As can be seen from Table 2.1, the materials subjected to the oxidative modification (aging or oxidation by atmospheric oxygen in an aqueous medium) have greater sorption capacity for water vapor than an unmodified one. This is probably due, first, to the accumulation of oxygen-containing groups and, respectively, with increasing of hydrophilic properties of the surface, and secondly, with the change in the form of supermolecular features on the surface of the polymer macromolecules, leading to the softening of the structure.

2.4 CONCLUSIONS

Thus, comparing the results of oxidative surface modification (oxidation by air oxygen in an aqueous medium or in gas phase, ozonation) of waste fibrous and film materials of PP and PE, we can see that the obtained modified samples has better adhesion and absorption properties than unmodified ones. So we can suggest that there is possibility of their usage as multifunctional additives in various composite materials, polymer-bitumen compositions, etc.

KEYWORDS

- **adhesion**
- **chemical modification**
- **chemical modification of polymers**
- **oxidative modification**
- **polyethylene**
- **polypropylene**
- **polyolefin material**
- **sorption**
- **surface oxidation**
- **waste of polypropylene and polyethylene**

REFERENCES

1. Povstugar, V. I., Kondolopov, V. I., Mikhailov, S. S. The Structure and Surface Properties of Polymeric Materials. Moscow: Khimiya, 1988, 190 p.
2 Antonovskii, V. L., Buzlanova, M. M. Analytical Chemistry Organic Peroxide Compounds. Moscow: Khimiya, 1978.
3. Putilova, I. N. Guide to Practical Work on Colloidal Chemistry. Moscow: Visshaya Shkola, 1981, 292 p.
4. Keltsev, N. V. Fundamentals of Adsorption Technology. Moscow: Khimiya, 1984, 595 p.
5. Vendillo, V. G., Emelyanov, Y. M., Filippov, Y. The laboratory setup for ozone, Zavodskaya Laboratoriya, 1950, 11, p. 1401.
6. Bazunova, M. V., Kolesov, S. V., Korsakov, A. V. Journal of Applied Chemistry. 2006, 79(5), 865–867.

CHAPTER 3

LIQUID FILLING WITH RUBBER SOLUTION POLYMERIZATION

ILIA VALERIEVICH PANKOV,[1] VIKTOR PETROVICH YUDIN,[2] and VIKTOR NIKOLAEVICH VEREZHNIKOV[1]

[1]*Voronezh State University, Russian Federation, 394006, Voronezh, Universitetskaya pl, No.1, Russia*

[2]*Voronezh Affiliated Societies of Science Researching Synthetic Rubber, Russian Federation, 394014, Voronezh, Mendeleeva st, No.3, Building B, Russia, E-mail: skilful25@mail.ru*

CONTENTS

ABSTRACT

Development of the master batch production process is bond to necessity of simplification of tread rubber manufacturing process used for "green" tires as well as elimination of toxic carbon white "dusting." An insertion of nitrous fragments to the polymer chain, which interact with carbon white surface, was an evolution of research work over Li-PBSR rubber. Carbon white, modified by various nitrous and silicon additives, can be produced on an industrial scale as master batch with Li-PBSR. In this case, carbon white "dusting" is prevented while vulcanizate remain its properties. Thus, it can be assumed that the developed compositions can be applied to production of tire rubbers using "green" tire technology.

3.1 INTRODUCTION

The use of silica extenders (white carbon) in recipe of rubber mixture used for tire treads, is one of the greatest developments in the range of tire production technology. The use of silica allows to improve tire traction with a road surface (with a wet, as well with icing), to decrease rolling losses, to increase wear resistance, etc. In domestic rubber industry the use of high-dispersed silica extenders is not wide spreaded. But it is possible to say today, that in the near future an essential increase of the high-dispersed grades of carbon white in Russian Federation is predicted. The single, wide used in the industry, method of silica extended rubbers is their preparation in the internal mixer or in mills [1].

The above-mentioned method has some drawbacks, particularly – a low mixing quality and high energy and working expenses. These drawbacks are specified by a low compatibility of the hydrophobic besieged silica filler (BSF) and as a result – a number and time of mixing stages is increased. The alternative version to the "dry" mixing is rubber extending with silica extenders (master batch) at the liquid phase. The world and domestic experience generalization in the range of mixtures production, using liquid phase method demonstrated the advantage of the liquid phase process in comparison to the traditional:

1. the uniform ingredients distribution in extended rubber;
2. the energy consumption decrease till 50% in case of rubber mixtures production on the base of rubber, extended at liquid phase;
3. the improvement of quality characteristics of the rubber mixtures and vulcanized rubber.

In comparison with the classical conditions of the rubber extending with BSF in the industry, which is carried out in mixers at high temperatures (at 160°C), the liquid phase technology doesn't require such external conditions [2].

During the masterbatches preparation the silane coupling agent Si-69 (bis-(triethoxysilylpropyl)-tetrasulfide – TESPT) is used, which has some disadvantages, connected with its structure:

1. evolution during its reaction with carbon white a byproduct as – ethanol, which is not preferred at production;
2. Si-69 is high-priced at its use in a quite great quantities in standard recipes of rubber mixtures preparation (10 mass parts/100 mass parts of carbon white).

The above-mentioned disadvantages are eliminated via azeotrope during in the organic solvents to remove ethyl alcohol molecules and to prevent side reactions. For the particularly, replacement of Si-69 the use of the low molecular nitrogenized PB-N in amount of 10 mass parts/100 mass parts of extender was proposed and as a result of this the silane coupling agent content was decreased till 3 mass parts [3]. The raw materials expenses were decreased till 600 USD of finished product.

For the silica extender use it is necessary to solve following tasks:

- to provide stability of the prepared suspension;
- to obtain the uniform suspension distribution in the rubber solution;
- to provide effective isolation of the prepared system "rubber-silica" without rubber or silica losses.

The development of the masterbatches production process is also connected with a necessity of the tread rubber process, used for "green" tires manufacturing, simplification, as well as elimination of the toxic carbon white "dusting." This makes production of the extended compositions ecology friendly [4].

3.2 EXPERIMENTAL PART

Two butadiene-styrene rubbers were used:

1. Li-PBSR – non-branched.
2. Li-PBSR-SSBR – block copolymer.

These rubbers were obtained via anionic polymerization, with a narrow and corrected molecular mass distribution (M_w/M_n=1.2–1.6), in this case the vulcanization net has a more regulated structure during vulcanization process.

The base technology of such rubber production process is developed at the scientific-research institute of synthetic rubber Voronezh branch (Voronezh branch of the Federal state unitary enterprise NIISK).

Li-PBSR rubber has following advantages:

1. The presence in macromolecules of the short end blocks of polystyrene gives the raw rubber mixture an increased cohesive resistance and in tread rubber-increased tire traction with a wet road surface.
2. The functional compounds, contained N-, OH-groups may be introduced in the Li-PBR molecule, which increase rubber interaction with extender.

The method of polydiene synthesis via anionic polymerization, using Li-organic compounds was used. The n-BuLi, as well as a product of the two-stage styrene and butadiene copolymerization, namely, polybutadienyl-polystyryl lithium (PB-PS-Li) – 35% solution in Nefras $M_w = 20,000$, active Li-content Li^+ – 0.0021N were used. Butadiene was charged batchwise, polymerization was carried out at the temperature of 60–90°C. Nefras was used as a solvent.

During Li-PBSR-SSBR-N block copolymers synthesis polymerization product of butadiene with n-BuLi as initiator, as well as proton-donor additive M-11 – tetraoxypropylene-diamine (modified lapramol) were used. In this case active polymer was deactivated via introduction in the reaction mixture of α-pyrrolidone as a solution toluene or ε-caprolactame solution in toluene. The rubbers contain also antioxidant Agidol-2 (2,2′ methylene-bis(4-methyl-6-tretbuthylphenol). The obtained polymerizate and their characteristics are presented in the Table 3.1.

While the best results were obtained during Li-PBSR (non-branched structure), namely, № 4 sample use, which has the best microstructure and Mooney viscosity values, further the rubber compounds preparation on the base of a one of recipes was carried out, using sample № 4.

The SSBR-N samples preparation was carried out via anionic polymerization, using 35% PB-PS-Li (active Li^+ content 0.0021N) solution in Nefras, as a modifying agent the proton-donor additive M-11 was used.

TABLE 3.1 Composition and Properties of the Synthesized Rubbers

Parameter	Sample Number			
	1	2	3	4
Butadiene, mL	1400	2500	1350	1200
Styrene, mL	250	250	80	20
Dry residue, %	17	23	15	15
M_w	100,000	100,000	185,000	320,000
Content polystyrene, %	26	19.5	8	8
1,2-links, %	54	64.7	11.2	10
Mooney viscosity, arb. Units	76	87	47	94

The process was deactivated via introduction in polymerization mixture of the compound on the base of carbonylic nitrogenized compound of α-pyrrolidone or ε-caprolactame. The obtained polymerizate and their characteristics are presented in the table.

BSF of "Rhodia" company – Zeosil-1165, as well as silane coupling agent Si-69 were used in this work. The silica extender preparation was carried out via azeotropic drying in organic solvents in the mixture with Si-69 with acetic acid addition as catalyst for interaction of hydrophobizators and silica surface.

The "carbon white" dispersion was treated with a low molecular polybutadiene with a low molecular polybutadiene (M_w = 2000–2500), modified with nitrogenized end groups of the "carbon white" surface. The low molecular polybutadiene was synthesized via anionic polymerization by deactivation with carbonylic nitrogenized compounds as α-pyrrolidone, ε-caprolactame, N-methylpyrrolidone.

As this preparation stage a dry modified silica as extender is used with the trade grades of rubbers, used in standard recipes of rubber mixture preparation. The carbon white content was changed from 50 mass parts/100 mass parts of Li-PBSR. Then silica filler was mixed with Li-PBSR-N solution and modified DSSK-N with following isolation. The extended rubber isolation was carried out via two methods:

1) using water-steam degassing; and
2) using waterless degassing.

The degassing and isolation methods of the prepared compositions are very important in this case, and it is necessary to pay great attention to these methods. Because of hydrophility of carbon white the use of traditional methods of degassing is very difficult. As a result, it is necessary to modify surface of carbon white and/or use agglomerating agents. But using generally accepted degassing and isolation flow diagrams a new technology will be low-price and quickly realized.

The use of water-steam degassing allows to minimize capital expenditures, which are necessary for introduction of developed technology in industry. Data, obtained for PB-N, used as modifying agent for carbon white demonstrate, that extender losses at such isolation method are low – in the range of 0.5–1.5%, and this is agreed with all norms.

One of the advantages of waterless degassing is quite simple equipment design and as a result of this – production of the finished products, which is not require additional drying, as well as decrease of energy losses.

Masterbatch, isolated, use in one of the indicated methods, was mixed on mills with another ingredients of rubber mixture. Optimal composition of masterbatch are presented in Table 3.2.

Three versions of rubber mixture were prepared on mills at 70–80°C during 30 minutes. Vulcanization was carried out at 150°C during 25–50 minutes. The obtained results show, that the use of a more active vulcanization group is resulted in improvement of all indices of tread rubber (in sample B – amount of diphenylguanidine and sulfenamide-T is increased 1.5 times).

3.3 RESULTS AND DISCUSSION

Composition of rubber compounds and results of testing of rubber compounds are presented in Tables 3.3 and 3.4.

From experimental data it is clear that the use of the developed compositions allows, in whole, to improve physical-mechanical and hysteresis properties of vulcanizates in comparison to the well-known production technology of rubber compounds, using trade grades of rubbers. During the obtained compositions use, the process energy was decreased more then 1/3 (third).

TABLE 3.2 Optimal Composition of Masterbatch

Rubber	Amount, Mass Parts
SSBR-N[a]	80
Li-PBSR-N[b]	20
SiO_2[c]	50

Note:
[a]styrene content – 30%; 1,2-butadiene units in diene part – 70%;
[b]with end NH-CH_3 groups;
[c]treatment by 3% Si-69 and 10% polybutadiene with NH-R end groups (M_w=2500).

TABLE 3.3 Composition of Rubber Compounds

Composition of mixture, g	Standard	Sample Number	
		A	B
Li-PBSR + SSBR-25–60	100	—	—
Zeosil-1165	50	—	—
Masterbatch		150	150
TU N-330	15	15	15
Sulfur	1.3	1.3	1.3
Diphenylguanidine	1.5	1.5	2.25
ZnO	3	3	3
Stearic acid	1	1	1
Oil PN-6	6	6	6
OP deafen	1	1	1
Si-69	8	—	—
Sulfenamide-T	1.7	1.7	2.55

TABLE 3.4 Results of Testing of Rubber Compounds

Results	Standard	Sample Number					
		A			B		
Mooney viscosity, arb. units	82	90			75		
Optimum cure time, min	15.5	25	35	50	25	35	50
M 300%, MPa	9.6	9.5	9.7	10.7	11.7	12.2	11.4

TABLE 3.4 Continued

Results	Standard	Sample Number					
			A			B	
σ, MPa	16.9	17.1	18.1	18.9	20.1	21.9	20.1
L, %	440	540	550	530	510	500	520
l, %	19.6	35	34	32	24	22	22
Resilience, %(20/100°C)	18=32–14	—	20=34–14	—	—	21=34–13	—
Tear resistance, kN m^{-1}	62		35			107	

On the base of the obtained results it is possible to advise the before mentioned compositions for manufacturing of various tire types: lorry and car tires, as well as farm service and industrial tires.

3.4 CONCLUSION

In accordance with a forecast the world consumption in the nearest time will be near one hundred thousand ton/year and will increase, because of ecology problems.

As it was estimated before the economic benefit during extended compositions production is more then 600 USD per ton of the finished product, taking in consideration only expenses for raw materials. The energy expenses during this production process carrying out are decreased double.

The economic process indices may be evaluated after organization of trial production, which will be realized on the base of Voronezh Affiliated Societies of Science Researching Synthetic Rubber (NIISK) in 2015. The development shops of enterprise have possibility to produce more then 2000 ton/year of extended compositions for "green tires."

KEYWORDS

- **green tires**
- **master batch**

- **rubbers**
- **silica filler**
- **solution polymerization**

REFERENCES

1. Rakhmatullin, A. I. Liquid filling with rubber solution polymerization of the silica filler: Thesis PhD, Candidate of Technology Sciences. Kazan, 2010, p. 152.
2. Pankov, I. V., Verezhnikov, V. N., Yudin, V. P. A development of masterbatch mixture based on lithium polybutadiene Li-PBR for "green" tire production. All Russian Youth Conference. Synthesis, properties study, modification and recycling of macromolecular compounds. 11–14 September 2012, Ufa.
3. Pankov, I. V., Yudin, V. P., Verezhnikov, V. N. Liquid phase filling of butadiene-styrene rubbers, obtained by solution polymerization, with carbon white, modified with silicon and nitrous compounds. 6th All Russian Kargin Conference "Polymers – 2014." Vol. II. Compilation of Reports Theses in Two Sections. Section One. 27–31 January 2014, Moscow.
4. Grishin, B. S. Materials Rubber Industry. Vol. 1. Kazan, KSTU, 2010, p. 506.

CHAPTER 4

INTENSIFICATION OF THE PROCESSES OF OIL AND GAS MIXTURES HEATING IN HORIZONTAL PIPE

Z. SALIMOV,[1] O. ISMAILOV,[1] SH. SAYDAHMEDOV,[2] and G. E. ZAIKOV[3]

[1]*Institute of General and Inorganic Chemistry, Uzbek Academy of Sciences, 77-a M. Ulugbek Street, 100 170, Tashkent, Uzbekistan; E-mail: ismoilovnmpi@mail.ru*

[2]*Fergansky Refinery, Russian Academy of Sciences, 4 Kosygin Street, 119 334 Moscow, Russia*

[3]*N.M. Emanuel Institute of Biochemical Physics, Russian Academy of Sciences, 4 Kosygin Street, 119 334 Moscow, Russia; E-mail: chembio@sky.chph.ras.ru*

CONTENTS

ABSTRACT

The results of experimental studies to identify the degree of intensification of the processes of oil and gas mixtures heating in horizontal pipe due to changes in hydrodynamic regimes.

4.1 INTRODUCTION

The problem of rational and efficient use of energy resources is one of the most important. With increasing of energy capacity and output more and more weight and dimensions of heat exchangers used are increasing. For their operation a huge amount of electricity and heat is in use. Improving the efficiency of heat transfer and reducing the weight and dimensions of these devices depends on the rational organization in their hydrodynamic regimes of heat exchanging traffic flows. Horizontal pipe and tubular devices are characterized with a fact that they can organize any hydrodynamic regimes of flow: laminar ($Re < 2300$), transitional ($2300 < Re < 10{,}000$) and turbulent ($Re > 10{,}000$), which differ with varying intensity of hydro, thermal and mass transfer processes [1].

At refineries of the Uzbekistan as a hydrocarbon feedstock oil-gas mixture is used in which the share of gas condensate in the mixture varies in the redistribution of 20–80%.

4.2 OBJECTS AND METHODS

To study the heating of hydrocarbons (oil, gas condensate, and mixtures thereof) with vapors of gas condensate and to establish the influence of operating parameters of the efficiency of heat transfer in tubular heat exchangers, we collected the experimental setup the concept of which is shown at Figure 4.1. The experimental setup consists of steam generator 15, the heat exchanger of the "tube in tube" with pipes 8 and 9 are mounted on the supporting stands 16, meters for metering gas 13 and hydrocarbon feed 4, the centrifugal pump 2 for pumping of raw material, expenditure feedstock tank 1 and the dipstick 11 collecting the heated feed. The unit is equipped with valves 3 and 12 to control the flow of natural gas and

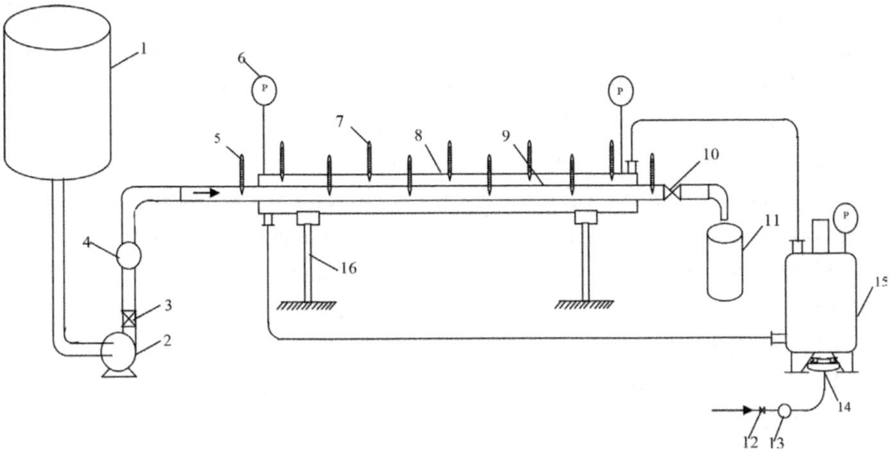

FIGURE 4.1 Experimental setup: 1 – The storage container for hydrocarbon raw materials; 2 centrifugal pump; 3, 10 and 12 – valve; 4 – counter flow of raw materials; 5 – thermometers for raw materials; 6 – pressure gauges; 7 – thermometers for heating agent; 8 – outer pipe (casing); 9 – inner tube; 11 – dimensional container heated raw materials; 13 – gas flow meter; 14 – burner gas; 15 – steam generator; 16 – support column.

hydrocarbon mercury thermometers 5 and 7 for measuring the temperature of the raw material in the inner tube and the heating medium in the annulus apparatus and gauges 6 for measuring the differential pressure of heating medium in the end portions of the heat exchanger. The heat transfer tube 9 has an inner diameter of 20 mm and effective length 2000 mm. The housing unit 8 is made of pipes with an inner diameter of 50 mm [2].

To the surface of the inner tube with pitch size 500 mm five pockets with outer sleeve are welded, into which in an oil bath thermometers are laid for temperature measurement control of hydrocarbon feedstock flowing through it. A cylindrical gap formed between the pocket layer and the liner is filled with insulating material. Such an arrangement allows the pocket to minimize the influence of temperature heating medium that is streamlined sleeve for temperature measurement accuracy of the hydrocarbon feedstock in the inner tube in the experiments. To the outer tube with 400 mm step six pockets are welded with oil for bookmark thermometers that measure the temperature of the heating agent in the annulus of the machine. To prevent heat losses to the environment the entire outer surface of the pilot plant and the steam generator is coated with glass wool and sealed with aluminum foil.

In experiments gas condensate vapors were used as a coolant, basic parameters of which are given at Table 4.1. The experiments were performed at a rate of hydrocarbon feedstock in the inner pipe 0.1061; 0.2123; 0.3184; 0.4246; 0.5307; 0.6369; 0.7431 and 0.8492 m/s. The process was organized in countercurrent movement of heat exchanging flows. The limits of speed variation of hydrocarbons provide establishing of various hydrodynamic modes of their motion in the experimental apparatus.

Using the results of the experiments, the basic physical, chemical and thermal properties of the investigated materials and coolant were studied and then the heat transfer coefficients of the heating medium to the outer wall of the inner tube and α_1 from the inner wall of the pipe to the heated feedstock α_2 along the length L of the device were calculated.

Criterion equation to calculate the heat transfer coefficient α_1 is selected based on the value of the Reynolds number *Re* [3, 4]:

$$Re = \upsilon\,(D_{in} - d_{ex})\,\rho/\mu \qquad (1)$$

where $\upsilon = 4V/\pi(D^2_{in-}d^2_{ex})$ – average speed of heating medium (steam condensate) in the annulus system, m/s; V – volumetric flow rate, m^3/s; D_{in} – inner diameter of the housing unit, m; d_{ex} – the outer diameter of the inner tube, m; ρ and μ, respectively, the density (kg/m^3) and dynamic viscosity (Pa. s) of coolant.

Re meaningfully turbulent motion of vapor condensate in the annulus experienced exchanger was set, as the Reynolds number is in the redistribution of 147,683 ÷ 162,985 (see Tables 4.2–4.5). In this case, the criterion equation to determine the Nusselt number is:

$$Nu = 0023 \cdot \mathrm{Re}^{0.8} \cdot \mathrm{Pr}^{0.4} \cdot \left(\frac{D_{in}}{d_{ex}}\right)^{0,45} \qquad (2)$$

TABLE 4.1 The Main Parameters of Gas Condensate

Heating agent	P Pressure, kPa	Temperature, t, °C	Velocity ω, m/s	Density at 20°C temperature, ρ, kg/m³	Kinematic viscosity at 20°C temperature, ν, 10^{-6} m²/sec
Vapors of gas condensate	250	120	6.8	759	1.03

where Pr = (Sμ)/λ – Prandtl number; C and λ, respectively – specific heat (kJ/kg) and the coefficient of thermal conductivity (W/m.K) of the coolant.

According to the calculated values of the criteria Nu heat transfer coefficient α1 from the heating medium to the outer wall of the heat exchange tube was determined [1]:

$$\alpha_1 = \frac{Nu \cdot \lambda}{d_{eq}} \tag{3}$$

where d_{eq} – equivalent diameter of the annulus section of the apparatus, $d_{eq} = d_{in} - d_{ex} = 0.025$ m.

Values of the coefficient of heat transfer from the inner wall of the pipe unit to a heated environment, α_2 are determined depending on the mode of flow in a horizontal pipe.

The value of the Reynolds number, which determines the modes of movement of hydrocarbons in a horizontal pipe system, was determined by the well-known formula $\text{Re}_{li} = \frac{\upsilon_{in} \cdot d_{ex}}{\nu_{in}}$, where υ_{in} – speed test liquid, m/s; ν_{in} – the kinematic viscosity, mm²/s.

To calculate the Nusselt number in laminar flow in a horizontal pipe the following criterion equation was used [3]:

$$Nu_{li} = 0{,}17\,\text{Re}_{li}^{0,33}\,\text{Pr}_{li}^{0,43}\,Gr_{li}^{0,1}\left(\frac{\text{Pr}_{li}}{\text{Pr}_{wa}}\right)^{0,25} \cdot \varepsilon_t \tag{4}$$

An approximate calculation of the criterion Nu_{li} when forced movement of the fluid flow in the pipe in the transition mode is performed by the following equation:

$$Nu_{li} = 0{,}008\,\text{Re}_{li}^{0,9}\,\text{Pr}_{li}^{0,43} \tag{5}$$

Criterion equation of heat transfer for turbulent movement of oil in double-tube apparatus is as follows:

$$Nu_{li} = 0{,}021\,\text{Re}_{li}^{0,8}\,\text{Pr}_{li}^{0,43}\left(\frac{\text{Pr}_{li}}{\text{Pr}_{wa}}\right)^{0,25} \cdot \varepsilon_t \tag{6}$$

where $Nu_{li} = d_{in} \cdot \alpha_2 / \lambda$ – is a Nusselt number characterizing the similarity of heat transfer processes at the interface between the wall and the flow of the sample liquid; $Pr_{li} = C \cdot v \cdot \rho / \lambda$ – is a Prandtl number at an average temperature of fluid flow; $Gr_{li} = g \cdot d^3 / v$ – is a Grashof criterion at an average temperature of fluid flow; $Pr_{wa} = C \cdot v \cdot \rho / \lambda$ – is a Prandtl number at the average temperature of the pipe wall; $\varepsilon_t = f(L/d)$ – is a coefficient taking into account the effect of the size of the heat exchange tube to increase the heat transfer coefficient α_2.

Coefficient of heat transfer from the wall to the liquid heat exchange tube α_2 is calculated from the following expression:

$$\alpha_2 = \frac{Nu_{ij} \cdot \lambda}{d_{in}} \tag{7}$$

The value of heat transfer coefficient K for 1 m length of experienced heat exchanger was defined by the equation set out in the source [4]:

$$K = \frac{1}{\frac{1}{\alpha_1 r_{ex}} + \frac{1}{\alpha_2 r_{in}} + \frac{1}{\lambda} 2,31\lg\frac{r_{ex}}{r_{in}} + \frac{1}{r_{polex}} + \frac{1}{r_{polin}}} \tag{8}$$

where r_{ex} and r_{in} – are outer and inner radius of inner tube, m; $r_{pol\ ex}$ and $r_{pol\ in}$ – are external and internal thermal conductivity of contamination of the inner tube, $W/(m^2K)$.

To determine the amount of heat transferred from the coolant to heated liquid the basic heat transfer equation was used:

$$Q = KL\Delta t_{av} \tag{9}$$

where K is the heat transfer coefficient, $W/(m^2K)$; Δt_{av} – the average temperature difference, °C; L – length of horizontal pipe, m.

The degree of intensification (or increasing of the efficiency of heat exchange) with heating of oil and gas mixtures by changing the hydrodynamic regime is determined from the following relations:

$$i_\alpha = \frac{\alpha_2^{''}}{\alpha_2^{'}};\ i_K = \frac{K^{''}}{K^{'}};\ i_q = \frac{Q^{''}}{Q^{'}} \tag{10}$$

where i, i_K and i_Q – are the degree of intensification of heat transfer and heat transfer coefficients K, increasing of the amount of heat transferred Q due to changes in hydrodynamic regimes; $\alpha_2^{'}$, K' and Q' – accordingly, are heat transfer, heat transfer and the quantity of heat transferred in a laminar flow regime coefficients; $\alpha_2^{''}$, K'' and Q'' – values of factors transfer and heat transfers, and also quantity of transferable heat – with transitional and turbulent motion of oil and gas mixtures in a horizontal pipe.

4.3 RESULTS AND DISCUSSION

The results of studies on determination of the degree of influence of hydrodynamic conditions on the efficiency of heat irradiation and heat transfer coefficients as well as the amount of heat transferred at oil and gas flow moving in a horizontal pipe were summarized and presented at Tables 4.2–4.5.

The data in Table 4.2 indicate that the regimes of the oil in the inner tube of a double-tube heat exchanger do smoothly transit from laminar (at υ_{ex} = 0.1061÷0.4246 m/s) to the transition mode (υ_{ex} > 0.5307 m/s). However, under the conditions of the experiment the turbulent regime was not formed. In laminar flow heat transfer coefficient of the wall to the heated oil α_2 increases from 420 to 681 W/(m^2 K), the coefficient of heat transfer K increases from 7.3 to 10.9 W/(m^2K), and the amount of transferred heat Q varies in the redistribution of 8.7 to 219 watts. In the transient mode the coefficient of heat transfer from a wall to the heated oil α_2 increases from 681 to 1011 W/(m^2K), the coefficient of heat transfer – from 10.9 to 14.9 W/(m^2K), and the amount of transferred heat varies at the redistribution of 219.5 to 1162.9 watts.

The data in Table 4.3 shows that the regimes of gas condensate movement in the inner tube of a double-tube apparatus smoothly transit from the transitional regime (at υ_{ex} = 0,1061÷0,3184 m/s) to turbulent (at υ_{ex} > 0,4246 m/s). In the transient mode the coefficient of heat transfer from a wall to the heated oil α_2 increases from 475 to 1349 W/(m^2K), the coefficient of heat transfer K – from 8.1 to 18.1 W/(m^2K), and the amount of transferred heat Q changes in the redistribution of 28.4 to 297.2 watts. At a turbulent flow the coefficient of heat irradiation from the wall to the heated oil α_2 increases from 1958 to 3087 W/(m^2K), the coefficient of heat transfer – from 13.5 to 42.8 W/(m^2K), and the amount of transferred heat varies within the redistribution of 607.5 to 2439.6 watts.

At heating of oil and gas condensate mixture (20% oil + 80% gas) in a horizontal pipe, the driving modes seamlessly move from transitional

TABLE 4.2 Improving the Efficiency of Heat Exchange with Heating Oil Due to Changes in Hydrodynamic Regimes

Heat transfer agent (vapors of gas condensate)					Heated medium (oil)						Performance of heat transfer		
Change in the temperature of heat carrier, °C		Re	Nu	α_1, W/m²K	Rate of hydrocarbons raw materials, w, m/s	Change in the temperature of heat carrier, °C		Re	Nu	α_1, W/m²K	K, W/m²K	Δt_{av}, °C	Q, W
$t_{н}$	$t_{к}$					$t_{н}$	$t_{к}$						
120	76	165,138	4122	3495	0.1061	20	66	615	66.23	420	7.3	0.6	8.69
120	72	163,872	4073	3457	0.2123	20	63	1136	84.1	516	8.7	1.2	20.96
120	71	163,241	4029	3447	0.3184	20	61	1826	94.2	587	9.7	4.7	91.45
120	67	162,581	4012	3428	0.4246	20	56	2021	102.4	681	10.9	10	219.5
120	63	196,257	3975	3386	0.5307	20	50	2326	125.4	762	11.9	15.8	378.7
120	58	158,364	3962	3378	0.6369	20	42	2456	142.7	976	14.4	23.4	677.5
120	54	156,435	3921	3354	0.7431	20	34	2531	153.8	1012	14.8	30.5	905.9
120	50	156,421	3892	3298	0.8492	20	23	2623	146.2	1011	14.9	39.3	1162.9

TABLE 4.3 Improving the Efficiency of Heat Transfer by Heating Gas Condensate Due to Changes in Hydrodynamic Regimes

Heat transfer agent (vapors of gas condensate)					Heated medium (oil)						Performance of heat transfer		
Change in the temperature of heat carrier, °C		Re	Nu	α_1, W/m²K	Rate of hydrocarbons raw materials, w, m/s	Change in the temperature of heat carrier, °C		Re	Nu	α_1, W/m²K	K, W/m²K	Δt_{av}, °C	Q, W
$t_н$	$t_к$					$t_н$	$t_к$						
120	57	157,354	3921	3335	0.1061	20	82	3412	69.52	475	8.1	1.75	28.4
120	53	155,231	3901	3321	0.2123	20	80	6913	136.8	889	13.5	2.93	79.1
120	49	152,123	3756	3301	0.3184	20	77	9152	161.2	1349	18.1	8.21	297.2
120	45	152,872	3834	3127	0.4246	20	72	12894	279.3	1958	22.5	13.5	607.5
120	41	151,293	3885	3246	0.5307	20	66	14589	326.8	2258	24.7	19.3	953.4
120	36	150,251	3812	3241	0.6369	20	58	16492	365.2	2654	26.9	26.9	1447.2
120	32	143,284	3735	3184	0.7431	20	50	17294	397.5	2816	27.5	34.0	1870
120	28	147,683	3725.3	3112	0.8492	20	39	18657	439.6	3087	28.5	42.8	2439.6

TABLE 4.4 Improving of the Efficiency of Heat Transfer by Heating a Mixture of Oil and Gas Condensate Consisting of 20% Oil and 80% Gas Condensate, Due to Changes in Hydrodynamic Regimes

Heat transfer agent (vapors of gas condensate)						Heated medium (oil)					Performance of heat transfer		
Change in the temperature of heat carrier, °C		Re	Nu	α_1, W/m²K	Rate of hydrocarbons raw materials, w, m/s	Change in the temperature of heat carrier, °C		Re	Nu	α_1, W/m²K	K, W/m²K	Δt_{av}, °C	Q, W
$t_н$	$t_к$					$t_н$	$t_к$						
120	63	159,267	3980	3383	0.1061	20	77	2750	63.05	431	7.5	1.75	26.2
120	59	157,814	3950	3357	0.2123	20	75	5600	118.7	811	12.5	2.34	58.5
120	55	156,382	3919	3331	0.3184	20	72	8032	167.1	1144	16.2	7.62	246.8
120	51	154,968	3890	3306	0.4246	20	67	10240	256.4	1757	21.5	12.9	554.7
120	47	153,575	3860	3281	0.5307	20	61	12129	301.0	2066	23.6	18.7	882.6
120	44	152,542	3838	3262	0.6369	20	51	13306	340.3	2343	25.2	26.4	1330.6
120	38	150,510	3794	3226	0.7431	20	45	14710	376.5	2596	26.5	33.4	1770.2
120	34	149,180	3766	3201	0.8492	20	34	15231	407.8	2821	27.6	42.3	2334.9

TABLE 4.5 Improving of the Efficiency of Heat Transfer by Heating of a Mixture of Oil and Gas Condensate Consisting of 40% Oil and 60% Gas Condensate, Due to Changes in Hydrodynamic Regimes

Heat transfer agent (vapors of gas condensate)					Heated medium (oil)						Performance of heat transfer		
Change in the temperature of heat carrier, °C		Re	Nu	α_1, W/m²K	Rate of hydrocarbons raw materials, w, m/s	Change in the temperature of heat carrier, °C		Re	Nu	α_1, W/m²K	K, W/m²K	Δt_{av}, °C	Q, W
$t_н$	$t_к$					$t_н$	$t_к$						
120	67	160,739	4011	3409	0.1061	20	73	2059	72	481	8.2	1.17	19.2
120	63	159,267	3980	3383	0.2123	20	70	4231	105.3	704	11.2	2.34	52.4
120	59	157,814	3950	3357	0.3184	20	68	6069	148.3	992	14.6	7.62	222.5
120	55	156,382	3919	3331	0.4246	20	63	7737	187.7	1258	17.3	12.9	446.3
120	51	154,968	3890	3306	0.5307	20	57	9164	223.2	1498	19.4	18.7	725.56
120	48	153,921	3867	3287	0.6369	20	48	10144	315.8	2124	23.9	25.8	1233.2
120	42	151,860	3823	3250	0.7431	20	41	11114	348.7	2352	25.3	33.4	1690
120	38	150,510	3794	3225	0.8492	20	30	11509	377.6	2553	26.3	42.3	2224.9

(with υ_{ex} = 0.1061÷0.3184 m/s) to turbulent (at υ_{ex} > 0.4246 m/s). In the transient mode the coefficient of heat irradiation from the heated wall to the sample liquid α_2 increases from 431 up to 1.144 W/(m^{2}K), the coefficient of heat transfer K – from 7.5 to 16.2 W/(m^{2}K), and the quantity of heat transferred Q changes in the redistribution of 26.2 to 246.8 watts. At a turbulent flow the coefficient of heat irradiation from the heated wall to fluid α_2 does increase from 1757 to 2821 W/(m^{2}K), the coefficient of heat transfer – from 21.5 to 27.6 W/(m^{2}K), and the amount of transferred heat varies within the range of 554.7–2334.9 watts.

The data in Table 4.5 shows that the regimes of hydrocarbon feedstock (a mixture of oil and gas condensate 40% oil + 60% gas) in the inner tube of a double-tube unit smoothly transit from moving (at υ_{ex} = 0.1061 m/s) to the turbulent regime (when υ_{ex} > 0.2123 m/s). In the transient mode the coefficient of heat irradiation from the heated wall to the sample liquid α_2 is 481 W/(m^{2}K), the coefficient of heat transfer K – 8.2 W/(m^{2}K), and the amount of transferred heat Q has reached 19.2 W. In the turbulent flow the heat irradiation coefficient from the heated wall to the feedstock α_2 increases from 704 to 2553 W/(m^{2}K), the coefficient of heat transfer – from 11.2 to 26.3 W/(m^{2}K), and the amount of transferred heat varies in a range of 52.4–2224.9 watts.

From analysis of the experimental data presented in Tables 4.2–4.5, it follows that the efficiency of heat transfer by heating the oil-gas condensate mixtures depends on their flow regimes. For example, improved

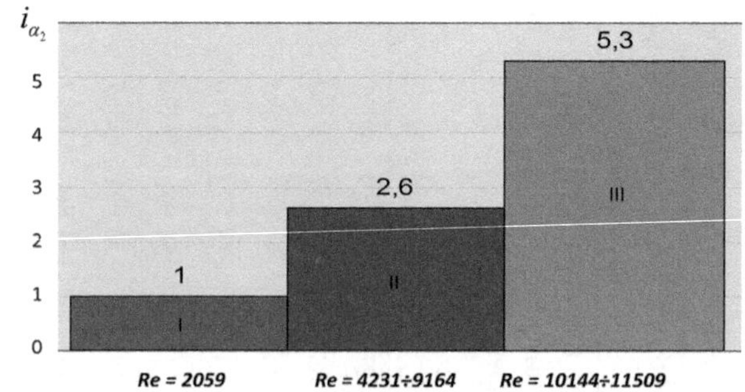

FIGURE 4.2 Increase of the heat irradiation coefficient α_2 at heating of oil and gas condensate mixture with a change in hydrodynamic conditions: I – laminar; II – transitional regime; III – the turbulent regime.

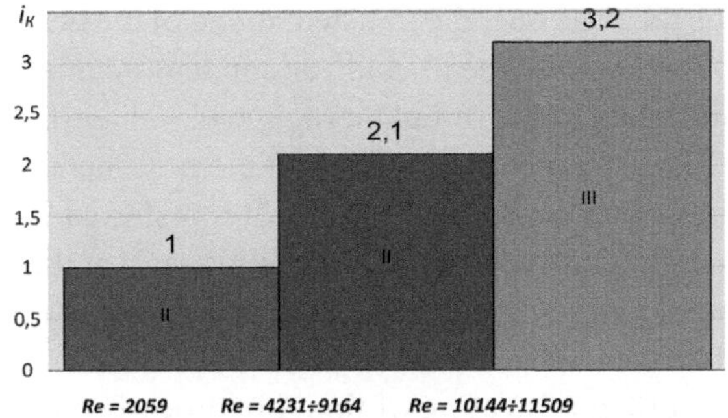

FIGURE 4.3 Increase of the value of heat transfer coefficient K by heating of a mixture of oil and gas condensate at the hydrodynamic regimes change: I – laminar regime; II – transitional regime; III – the turbulent regime.

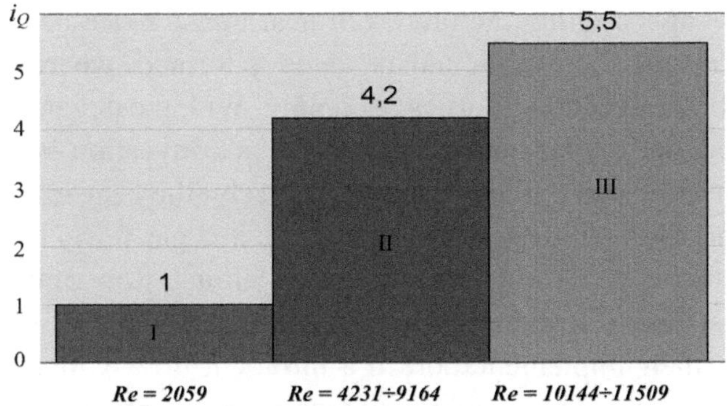

FIGURE 4.4 Increase of the number of values to overeat heat Q at heating of oil and gas condensate mixture (40% oil and 60% gas) with a change in hydrodynamic regimes.

efficiency of heat transfer of oil and gas condensate by heating a mixture consisting of 40% oil and 60% of gas condensate (i.e., increasing of α_2, K, Q values) due to changes in hydrodynamic conditions are illustrated in the diagrams (see Figures 4.2–4.4).

4.4 CONCLUSION

Based on these studies it can be noted that the hydrodynamic regimes of oil and gas mixtures in the horizontal pipe strongly influence the efficiency of heat exchange during heating. If the efficiency of heat transfer in laminar regime could be taken as one, the transient mode of the heat transfer coefficient from the inner wall to the oil and gas condensate mixture α_2 would be increased twice, while at the turbulent regime α_2 value would be increased by 5.3 times. Thus, the heat transfer coefficient K increases accordingly 2.1 and 3.2 times, and the amount Q of heat transferred increases significantly with changing of hydrodynamic conditions. If at the transition from laminar to transitional regime, the Q value increases 4.2 times, then at the turbulent regime it increases 5.5 times.

The positive effect of heating of liquid hydrocarbons in the horizontal pipe is explained as follows. At the core of the flow the heat transfer is carried out simultaneously with conduction and convection effects. The mechanism of heat transfer in the core of the flow in the turbulent motion of the medium is characterized by intensive mixing due to turbulent fluctuations. As approaching the tube wall heat transfer rate decreases, since near the wall the thermal boundary layer is formed, where increasingly important thermal conductivity is becoming. With the development of turbulence boundary layer becomes so thin that convection begins to exert a dominant influence on heat transfer. Whereby the heat irradiation coefficient is increased at turbulent regime of oil and gas flows in a horizontal pipe more then 5 times in comparison with laminar flow of the liquid.

So at oil and gas mixtures heating processes' projects it is desirable to consider their implementation in a turbulent flow, which will help to develop energy-saving technology of thermal preparation of hydrocarbons to the primary distillation in refineries by optimizing the hydrodynamic conditions in tubular heat exchangers.

KEYWORDS

- **amount of transferred heat**
- **calorific efficiency**
- **degree of intensification of the process**
- **heat exchange efficiency**
- **heat transfer coefficient**

- **laminar**
- **oil and gas mixture**
- **physico-chemical properties**
- **similarity criteria**
- **transitional**
- **turbulent mode**

REFERENCES

1. Zakharov, A. A., Bakhshiyeva, L. T., Kondaurov, B. P., et al. Processes and Devices of Chemical Technology. Edited by prof. AA Zakharova. M.: Publishing Center "Academy", 2006. P. 30–53.
2. Salimov, Z. S., Ismailov, O. Yu., Radzhibaev, D. P. Effect of movement of oil and gas condensate in the heat transfer coefficient in a double-tube unit. Uzbek Oil and Gas Journal, 2014, № 1, 39–42.
3. Kasatkin, A. G. Basic Processes and Devices of Chemical Technology: Textbook for Universities. 8th ed., Rev. M.: Chemistry, 1971, 42, 297–301.
4. Pavlov, K. F., Romankiv, P. G., Noskov, A. A. Examples and Tasks at the Rate of Processes and Devices of Chemical Technology. Textbook for High Schools. Ed. by Romankova P.G. 10th ed., Rev. and ed. Leningrad: Chemistry, 1987, 149–169, 531.

CHAPTER 5

CADMIUM AND LEAD IN ACID SOD-PODZOLIC SOILS

LYUDMILA N. SHIKHOVA[1] and EUGENE M. LISITSYN[1,2]

[1]*Vyatka State Agricultural Academy, 133 Oktyabrsky Prospect, Kirov, 610017, Russia*

[2]*North-East Agricultural Research Institute of Rosselkhoz Academy, 166-a Lenin St., Kirov, 610007, Russia, E-mail: shikhova-l@mail.ru*

CONTENTS

ABSTRACT

Estimation of total content of heavy metal in different horizons of acid sod-podzolic soils of Kirov region (Russia) covered with natural vegetation (forest, meadow) and used as arable land is spent by a method of atomic-absorption spectroscopy. The content of total cadmium in different horizons of sod-podzolic and podzolic loamy soils fluctuated from 0.66 to 1.11 mg/kg. The content of exchangeable cadmium in soils fluctuated from 0.01 to 0.30 mg/kg in different horizons. The greatest content of exchangeable Cd is noted in accumulating-eluvial parts of a profile and in mother rock, the least – in illuvial horizons. In humus horizons of meadow and fallow land soils its content is statistically higher than in arable land soils (0.17±0.02 and 0.08±0.02 mg/kg, accordingly). Ratio of exchangeable cadmium in its total content varied largely. In the top part of a profile it fluctuated from 10 to 40%, in illuvial parts drop down to 1–11% and increased again at approach to mother rock. During growth season content of exchangeable Cd in the top horizons of forest podzolic soil can vary by 4–5 times sometimes reaching the critical values close to maximum allowable concentration (MAC) (0.2 mg/kg). In fallow land sod-podzolic soil it maximum is revealed in the middle of a season (1.0 mg/kg) that is almost 10 times higher than in autumn. On an arable land under clover crops total cadmium content is minimum in the beginning and the middle of a season and sharply rises in the autumn to the values close to MAC (0.8 mg/kg). It is not revealed any significant correlations between content of cadmium with other agrochemical properties of arable land soils. In covering loams of different districts of Kirov region the amount of total Pb fluctuated from 5.0 to 43.0 mg/kg. The least content of total lead is characteristic to granulometricaly light mother rocks. Significantly higher content of exchangeable forms of lead is in loamy soils (0.45–3.16 mg/kg) than in sandy one (0.45–2.22 mg/kg). The content of exchangeable lead in alluvial sandy deposits does not exceed 0.95 mg/kg. Arable horizons of loamy agricultural soils contain 26–30 mg/kg Pb. Sod-podzolic soils of region accumulate insignificant amounts of exchangeable Pb. Forest podzolic and sod-podzolic soils contain about 1 mg/kg of exchangeable lead in litter horizon. Sod-podzolic arable and meadow soils contained 1.86±0.13 mg/kg of exchangeable lead in humus horizons. High significant coefficients of correlation between contents of exchangeable lead and exchangeable acidity (from 0.81 to 0.99) are received. Influence of acidity on input of heavy metals (HM) in plants, interactions of aluminum stress and action of HM has been investigated with use of methods of sand and soil culture. Treatment with cadmium and lead leads to accumulation of heavy metals by plants. This accumulation reaches 30–35 mg/kg of dry mass. The most amount of cadmium is collected in roots and stems of cereals, and slightly less – in leaves. The basic part of lead was absorbed by roots. In variants with HM treatments the delay of germination for 2–3 days was marked; leaf blades were narrower than in control variants. By 5th week of growth the bottom leaves especially in a variant with lead treatment began to dry up and die off. Decrease in plant height is noted for all oats (by 41–61%) and barley varieties (by 40–48%). The greatest depression under the influence of HM is noted at such parameter as plant dry mass. At cadmium stress depression of plant dry mass makes 25.0–64.2%. Joint influence of cadmium and aluminum has appeared more considerably than toxic action of sole cadmium. Deviation from control values in a pigment complex of oats lies within limits 12.2% to 50.4% and of barley – within limits 18.2% to 66.0%. Changes had the general character and near identical intensity for a chlorophyll of both fraction (*a* and *b*). Changes in carotenoids content have similar character. As a whole level of Al-resistance and resistance to toxic action of high concentration of HM coincide on the majority of considered parameters.

5.1 INTRODUCTION

Numerous facts of environmental contamination by heavy metals as a result of dispersion of industrial emissions through atmosphere or in the form of a waste (slag, slime) and the polluted industrial waters are described in the literature [1, 2]. The Kirov region is included into first ten subjects of the Russian Federation on a ratio of farmland with excess of maximum allowable concentration on heavy metals [3]. The volume of anthropogenous input of the most toxic metals – lead and cadmium – into environment reaches considerable values (451.64 and 4.05 t, accordingly) [4]. The number of the publications devoted to problems of technogenic pollution of soils of Kirov region with HM for the last years has considerably increased. But it is not always possible to be guided confidently in this data as the territory of region is poorly studied for background content of HM in soils and mother rock.

Long application of mineral fertilizers leads to accumulation of metals in soil [5–8]. Cadmium input is 2 times more than its removal by agricultural plants. It leads to excess of MAC of metals in soils and plants. Under the available data, concentration of cadmium, nickel, lead and chrome in agricultural production of some regions is already above their MAC [9]. Background content of Cd in soil is appreciably defined by soil type. In podzolic and sod-podzolic soils it makes 0.70–2.31 mg/kg, in gray wood soil – 0.65 mg/kg [10]. Concentration of metal in humus horizon of sod-podzolic loamy soils of the European Russia is about 0.14 mg/kg, of chernozems – 0.24 mg/kg [9]. The content of the element in humus horizons of the basic types of soils of Western Siberia is 0.074 mg/kg [11]. Loamy chernozems of southwest Altai contain 0.135 mg/kg of cadmium, alluvial sandy and sandy soils – 0.121 mg/kg, sod-podzolic sandy soils – 0.116 mg/kg [12]. Alluvial soils of Middle Ob' contain 0.1–1.8 mg/kg of cadmium in the top horizons [12].

The highest mobility and availability to plants cadmium possesses in acid soils with low pH level, low capacity of absorption, fulvatic structure of humus, and washing water mode [13–14]. Lowering of mobility of cadmium in arable soils is promoted by periodic liming [15–17].

Many researchers studying toxicity of heavy metals for agricultural crops notice that cadmium is 2–20 times more toxic then other metals. At comparison of metals on toxicity in their equal doses they have following order: Cd > Zn > Cu > Pb or Cd > Ni > Cu > Zn > Cr > Pb [17, 18]. Investigations of cadmium treatment in a wide range of doses have allowed to establish that in each botanical family there are plants with various level of resistance to this element [17].

In soils of Russian Plain the background content of lead fluctuates from 2.6 to 43.0 mg/kg. On the average, top horizons of sod-podzolic and gray wood soils contain 14.6 and 17.2 mg/kg Pb according to [19] and up to 26–37 mg/kg in arable loamy soils [20]. Sod-podzolic soils of Udmurtiya contain total Pb: 13.9 mg/kg – in sandy and sandy loam soils; 25.6 mg/kg – in loamy soils. In a litter and humus horizon of sod-podzolic soil of Moscow Region there are 36 and 41 mg/kg of total lead according to [21, 22]. There is data that in acid soils essential part of Pb (II), up to 70%, is in an exchange condition, apparently, as a part of organic-mineral complexes [23]. In acid and strong acid media ions of Pb actively migrate within a soil profile and can be taken out for its limits. Movement of Pb in depth of a profile occurs in a chelate form.

The greatest concentration of lead is found out in the top layer of non-processed soils enriched with organic substance [21, 24]. The second maximum of lead content meets in illuvial horizons of sod-podzolic soils [10].

At last time lead involves a great attention as one of the main components of chemical pollution of environment and as an element, toxic for plants. Though there is no data testifying that lead is vital for growth of any kinds of plants, it is a lot of publications shown stimulating action of low concentration of some lead salts (mainly $Pb(NO_3)_2$) on plant growth. Moreover, effects are described of braking of plants metabolism because of low levels of lead in medium.

Values of MAC of lead for plants lie in wide enough intervals – from 0.5 – 1.2 mg/kg [25] to 10.0–20.0 mg/kg [26]. The natural content of lead in above ground parts of grasses makes from 1.5 to 40.0 mg/kg of dry matter [24].

Use of possibilities of plants to resist against stress action in a zone of roots (lowered pH of soil solution, presence of toxic ions of metals) is a theoretical basis of the ecological breeding directed on increase of resistance of agricultural crops to action of edaphic stressors [27].

Therefore the purpose of the given work was (1) estimation of background contents of exchangeable ions of HM (cadmium and lead) in acid sod-podzolic soils of the Kirov region; (2) studying of relations between HM content and some properties of soils; and (3) interrelations of resistance of oats and barley plants to aluminum stress and action of heavy metals.

5.2 EXPERIMENTAL PART

Objects of research were as well as various varieties of oats and barley, used in agricultural production on these soils. For the characteristic of the profile content of heavy metals soil samples were taken by soil horizons to 70–150 cm depth from soil cuts in different areas of surveyed territory. Soil samples were processed according to the standard methods [28]. Estimation of total content of HM is spent by a method of atomic-absorption spectroscopy. Exchangeable fractions of HM were determined in acetate-ammonium buffer solution, pH 4.8 [28, 29].

During growth season (May–September) sample of soils from three top horizons of arable and fallow land podzolic loamy soil were collected monthly.

For studying of influence of acidity on input of HM in plants, interactions of aluminum stress and action of HM on parameters of resistance of oats and barley plants a series of experiments has been made with use of methods of sand and soil culture. Previously level of potential aluminum resistance of some oats and barley varieties was established [30] by method described earlier [31]. The oats and barley varieties different by level of aluminum resistance (in decreasing mode) – oats Krechet > Fakir > Ulov and barley Dina > Abava > Elf were objects of research.

The scheme of experiment in sand culture:

1. Control – without HM, a background – a full dose of Knop's solution;
2. Background + cadmium sulfate (100 μM for oats, 50 μM for barley);
3. Background + lead chloride (500 μM for an oats and barley).

Eight seeds of each variety were sown in pots with 0.5 kg of sand; after 7 days five most developed plants were left in each pot. Duration of experiment was 5 weeks (up to tillering stage). Analyzed traits: height of plants, volume of root system, cation exchange capacity of roots, dry weight of leaves, stems and roots, specific leaf area, content of pigments in leaves [32], the cadmium and lead content in leaves, stems and roots. Each variant of experiment had 6 replications for each variety.

The scheme of experiment in soil culture:

1. Control soil (without entering HM);
2. Control soil + 1 mg/kg of cadmium;
3. Control soil + 5 mg/kg of cadmium;

4. Control soil +10 mg/kg of cadmium;
5. Control soil + 60 mg/kg of lead;
6. Control soil + 300 mg/kg of lead;
7. Control soil + 600 mg/kg of lead.

Each pot contains 3.5 kg of soil (agrozem with the high content of organic substance (3.5%) and pH 7.25). Initially soil was carefully sifted, acidified with solutions of acids (sulfuric and hydrochloric depending on a pollution variant), treated with corresponding quantity of salt of metal (cadmium sulfate or lead chloride) and mixed.

Ten seeds of each variety of oats or barley were sown in each pot; seven most developed plants were left 7 days later. Duration of experiment is 5 weeks (up to tillering stage). Analyzed traits: height of plants, dry weight of leaves, stems and roots, a ratio of leaves, stems and roots in a plant, the content of pigments, the cadmium and lead content in leaves, stem and roots. Each variant of experiment had 6 replications for each variety.

For studying of joint action of high HM content and aluminum on plants a series of experiments in sand culture has been made. The scheme of experiment included:

1. Control – without HM, a background – a full dose of Knop's solution;
2. Background + aluminum sulfate (1 mM);
3. Background + aluminum sulfate (1 mM) + cadmium sulfate (100 μM for oats, 50 μM for barley).

Each variant of experiment had 6 replications for each variety.

At tillering, ear tube formation, and earing stages samples of leaves were collected for estimation content of pigments in leaves in triple replications was spent. The content and a ratio of a chlorophyll a and b, carotenoids in acetone extract [32] were defined.

Statistical processing of the data is spent by methods of dispersion and correlation analyzes with application of software packages STATGRAPHICS Plus for Windows 5.1.

5.3 RESULTS AND DISCUSSION

5.3.1 SEASONAL AND PROFILE DYNAMICS OF CADMIUM

The territory of the Kirov region is characterized with various enough soil cover. The basic part of territory of area is occupied with soils of podzolic type. The area of podzolic and sod-podzolic soils makes 77.9% of total region area [33]. Sod-podzolic and podzolic soils of heavy mechanical structure are generated basically on covering deposits. In natural state they possess low enough level of fertility, have high acidity, the low contents of organic substance of fulvatic type, a low saturation with bases of a soil-absorbing complex.

The content of total cadmium in soils of the Kirov region is in the limits known for other regions, though some above than across Russia as a whole (Table 5.1).

TABLE 5.1 Content Cd in Profiles of Some Podzolic and Sod-Podzolic Loamy Soils of the Kirov Region

Location	Soil horizon	Depth, cm	pH	Content of Cd (mg/kg)	
				Total	Exchangeable
Murashi district, arable land	A_p	0–24	5.70	0.88	0.13
	A_{he}	24–33	3.96	0.78	0.10
	AB	33–53	3.80	0.97	0.04
	B	53–86	3.82	0.74	0.04
	BC	86–118	3.85	0.93	0.01
	C	118–130	3.94	0.76	0.16
Murashi district, meadow	O_e	0–4	4.59	0.26	0.11
	A_h	4–15	4.05	0.70	0.29
	A_{he}	15–20	3.87	0.86	0.10
	AB	20–36	3.80	0.86	0.05
	B	36–70	3.84	0.77	0.02
Murashi district, forest	O_i	3–6	3.89	0.73	0.15
	A_{he}	6–35	3.85	0.81	0.23
	AB	35–70	3.76	1.08	0.04
	B	70–80	3.88	0.72	0.08
	BC	80–115	3.69	0.66	0.12

TABLE 5.1 Continued

Location	Soil horizon	Depth, cm	pH	Content of Cd (mg/kg)	
				Total	Exchangeable
Afanas'evo district, arable land	A_p	0–28	5.87	0.94	0.04
	A_{he}	28–48	4.22	0.91	0.21
	AB	48–70	4.03	0.79	0.11
	B	70–96	4.10	0.96	0.11
	BC	96–120	4.21	1.20	0.13
	C	120–150	4.34	1.01	0.13
Afanas'evo district, forest	O_i	0–7	4.33	1.03	0.13
	A_{he}	7–30	3.91	1.03	0.19
	AB	30–51	4.09	1.02	0.01
	B	51–86	5.18	1.08	0.05
	BC	86–120	7.07	1.02	0.07
Falenki district, forest	O_i	0–7	3.95	0.84	0.07
	A_{he}	7–20	3.87	0.79	0.15
	AB	20–30	4.02	0.69	0.01
	B	30–70	4.34	0.74	0.12
Zuevka district, fallow land	A_h	0–27	4.54	0.82	0.07
	A_{he}	27–45	4.04	0.74	0.19
	AB	45–70	4.05	0.84	0.05
	B	70–103	4.09	1.09	0.10
	BC	103–120	4.38	0.88	0.13
	C	120–130	4.51	0.99	0.12

For different horizons of sod-podzolic and podzolic loamy soils its content fluctuated from 0.66 to 1.11 mg/kg. Biogenic accumulation of total Cd in the studied profiles is expressed slightly that some authors explain with considerable migration of cadmium downwards on a profile in soils of humid landscapes [24]. Eluvial-illuvial redistribution of total Cd in profiles was often observed.

The content of exchangeable cadmium in soils fluctuated from 0.01 to 0.30 mg/kg in different horizons. The greatest content of exchangeable Cd is noted in accumulating-eluvial parts of a profile and in mother rock, the least – in illuvial horizons. In some humus and eluvial horizons the content of exchangeable cadmium exceeds known level of MAC equal to 0.2 mg/kg. Redistribution of Cd on a profile is considerable influenced with exchangeable organic substance; cadmium content is higher in those horizons where content of labile organic matter is higher too.

Ratio of exchangeable cadmium in its total content varied largely. In the top part of a profile it fluctuated from 10 to 40%, in illuvial parts drop down to 1–11% and increased again at approach to mother rock. Statistically significant correlation between the content of total and exchangeable cadmium was absent. Content of exchangeable cadmium in organogenic horizons of the studied podzolic and sod-podzolic soils is insignificant as a whole. However, during growth season content of exchangeable Cd in the top horizons of forest podzolic soil can vary by 4–5 times sometimes reaching the critical values close to MAC (Figure 5.1).

Sod-podzolic soils of Kirov region are characterized by the minimum content of exchangeable cadmium (Table 5.1). Thus in humus horizons of meadow and fallow land soils its content is statistically higher than in arable land soils (0.17±0.02 and 0.08±0.02 mg/kg accordingly). The similar content of exchangeable cadmium in arable layer of sod-podzolic loamy soil was noted by [34] – 0.11 mg/kg. Content of exchangeable Cd in arable horizons of sod-podzolic soils of Yarano-Vyatka interfluve is 0.09–0.28 mg/kg,

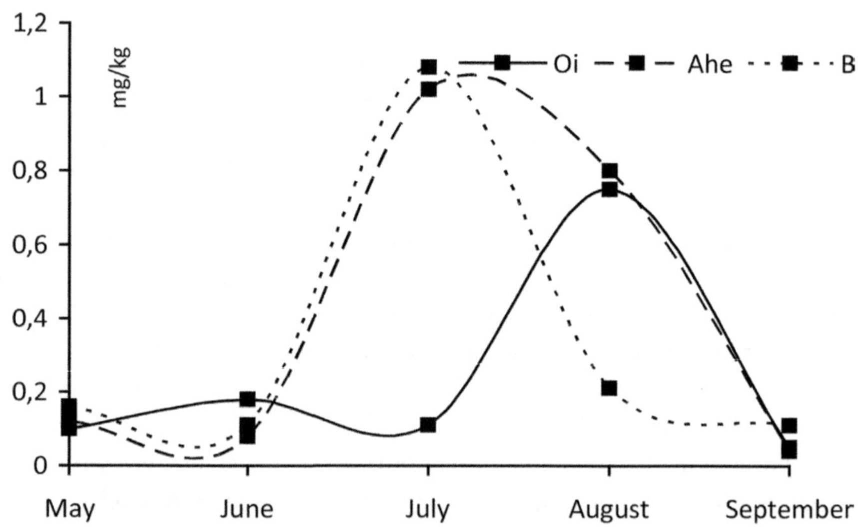

FIGURE 5.1 Seasonal dynamics of content of exchangeable Cd in forest podzolic soil of the middle taiga.

of sod-gleyey soils – 0.12–0.19 mg/kg [35]. Unfortunately authors do not specify in what extractant exchangeable cadmium was defined.

Distribution of exchangeable cadmium on soil profiles of arable sod-podzolic soils in different districts of Kirov region has the same features as on forest sod-podzolic and podzolic soils (Figure 5.2).

Arable horizons as a rule contain a little amount of exchangeable Cd. The maximum content is in eluvial horizons and the minimum one is almost always in illuvial parts of profiles.

It is not revealed any significant correlations between content of cadmium with other agrochemical properties of arable land soils probably because of constant anthropogenous influences. In meadow soils content of exchangeable Cd positively correlates with pH (r = 0.60) that obviously

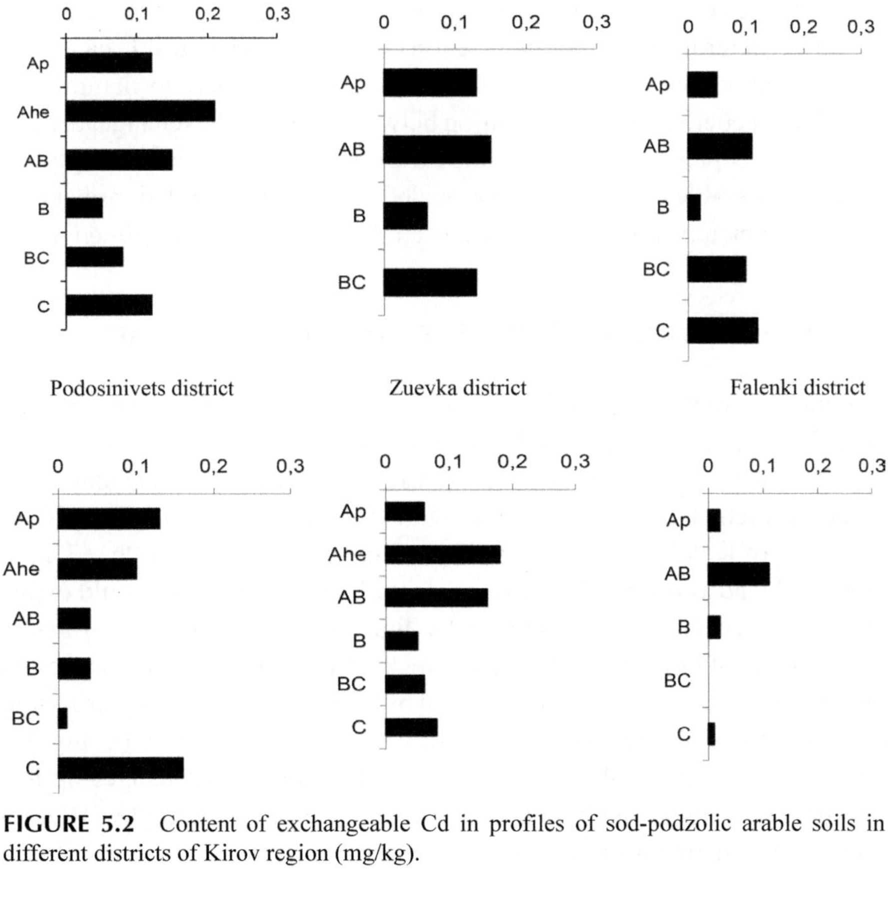

FIGURE 5.2 Content of exchangeable Cd in profiles of sod-podzolic arable soils in different districts of Kirov region (mg/kg).

is connected with conditions of mineralization of organic matter. Besides it is known that exchangeable cadmium is presented in loamy soils basically as exchangeable forms therefore there is a competition between cadmium and calcium, magnesium for exchange sites in soil-absorbing complex. Coefficient of pair correlation between content of Cd on one hand and calcium and magnesium on the other hand is as much as $r = -0.57$.

High coefficient of variation of exchangeable Cd content in soils may be partly explained with fluctuation of the content of the element during a growth season. For example in fallow land sod-podzolic soil it maximum is revealed in the middle of a season (1.0 mg/kg) that is almost 10 times higher than its content in autumn (Figure 5.3).

On an arable land under clover crops cadmium content is minimum in the beginning and the middle of a season and sharply rises in the autumn to the values close to MAC (0.8 mg/kg). Therefore, soil samples collected at different terms can strongly differ on content of exchangeable cadmium.

In fallow land soil where soil processes come nearer to natural high significant coefficients of correlation between content of exchangeable cadmium and content of total and labile carbon (0.64 and 0.73 accordingly) are noted. For arable soil there is not revealed any significant relations between these parameters; apparently it is caused by anthropogenous influence.

5.3.2 SEASONAL AND PROFILE DYNAMICS OF LEAD

In mother rocks and soils of the Kirov region the content of lead (Table 5.2) did not fall outside the limits known for other regions.

In covering loams of different districts of Kirov region the amount of total Pb fluctuated from 5.0 to 43.0 mg/kg. Eluvial-illuvial horizons of Perm deposits of Kotelnich and Orlov districts contain a little total Pb – 7.8 and 9.0 mg/kg accordingly. The least content of total lead as one would expect is characteristic to granulometrically light mother rocks. Fluvial-glacial and ancient alluvial deposits of Kotelnich, Orlov and Orichi districts of the Kirov region contain 1.9 to 3.75 mg/kg of lead. Significantly higher content of exchangeable forms of lead is also in loamy soils (0.45–3.16 mg/ kg) than in sandy one (0.45–2.22 mg/kg). The content of exchangeable lead in modern alluvial sandy deposits does not exceed 0.95 mg/kg. It is not revealed statistically significant correlations between the contents of total

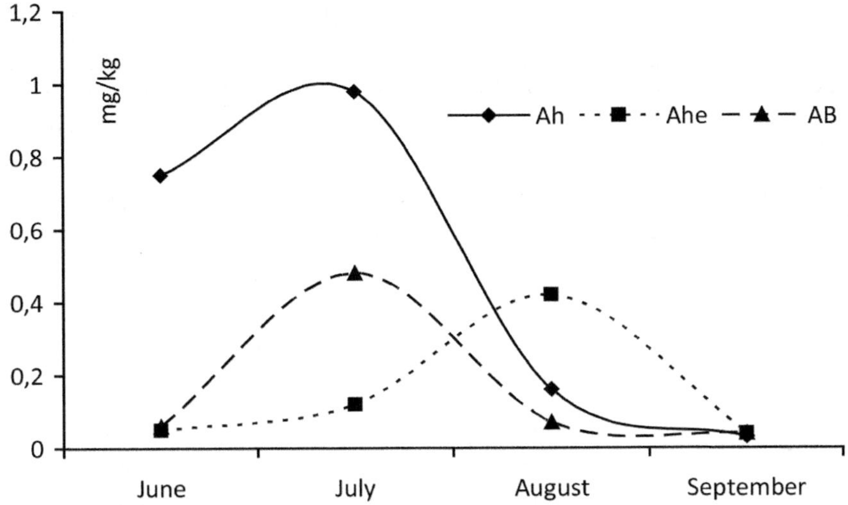

FIGURE 5.3A Seasonal dynamics of content of exchangeable Cd in fallow land sod-podzolic soil of the Middle taiga.

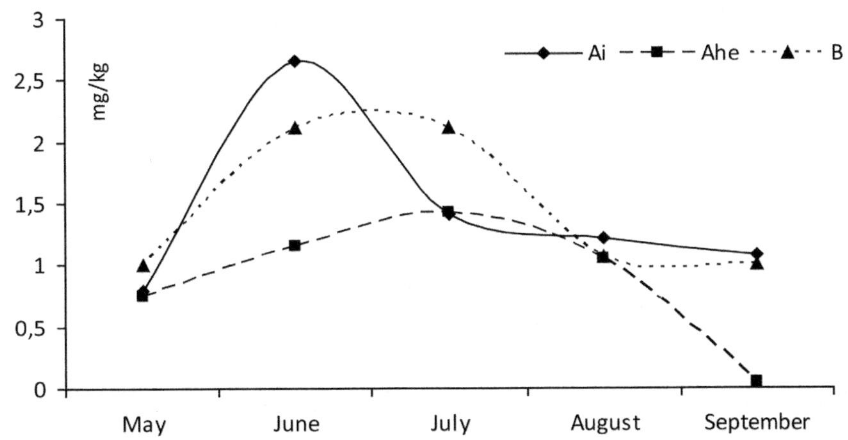

FIGURE 5.3B Seasonal dynamics of exchangeable Pb content in podzolic soils of the middle taiga.

and exchangeable Pb in mother rocks. The amount of exchangeable Pb both in loamy and in light soils is not connected significantly with their content of organic substance, pH value, and level of exchange acidity. As a whole the content of total and exchangeable Pb in basic mother rocks of region can be characterized as an average, which is not beyond literary known concentration for sedimentary rocks.

TABLE 5.2 Content of Pb in Profiles of Some Podzolic and Sod-Podzolic Loamy Soils of the Kirov Region

Location	Soil horizon	Depth, cm	pH	Content of Pb (mg/kg)	
				Total	Exchangeable
Murashi district, arable land	A_p	0–24	5.70	30	1.38
	A_{he}	24–33	3.96	17	1.25
	AB	33–53	3.80	15	1.40
	B	53–86	3.82	26	3.20
	BC	86–118	3.85	23	3.11
	C	118–130	3.94	22	4.25
Murashi district, meadow	O_e	0–4	4.59	24	2.15
	A_h	4–15	4.05	20	2.00
	A_{he}	15–20	3.87	16	1.44
	AB	20–36	3.80	29	1.68
	B	36–70	3.84	15	3.26
Murashi district, forest	O_i	3–6	3.89	27	6.75
	A_{he}	6–35	3.85	24	5.50
	AB	35–70	3.76	20	1.42
	B	70–80	3.88	20	1.42
	BC	80–115	3.69	20	3.20
Afanas'evo district, arable land	A_p	0–28	5.87	26	1.28
	A_{he}	28–48	4.22	34	1.45
	AB	48–70	4.03	31	0.40
	B	70–96	4.10	32	0.40
	BC	96–120	4.21	25	1.11
	C	120–150	4.34	38	1.00
Afanas'evo district, forest	O_i	0–7	4.33	32	2.42
	A_{he}	7–30	3.91	37	2.42
	AB	30–51	4.09	27	2.62
	B	51–86	5.18	33	2.62
	BC	86–120	7.07	35	2.41
Falenki district, forest	O_i	0–7	3.95	22	2.51
	A_{he}	7–20	3.87	20	1.45
	AB	20–30	4.02	14	0.84
	B	30–70	4.34	5	1.12

TABLE 5.2 Continued

Location	Soil horizon	Depth, cm	pH	Content of Pb (mg/kg)	
				Total	Exchangeable
Zuevka district, fallow land	A_h	0–27	4.54	28	2.50
	A_{he}	27–45	4.04	19	0.75
	AB	45–70	4.05	25	0.50
	B	70–103	4.09	27	1.20
	BC	103–120	4.38	30	1.56
	C	120–130	4.51	20	2.41

Arable horizons of loamy agricultural soils contain 26–30 mg/kg Pb. Sod-podzolic soils of region accumulate insignificant amounts of exchangeable Pb. Forest podzolic and sod-podzolic soils contain about 1 mg/kg of exchangeable lead in litter horizon (Table 5.2). Thus litter horizons of loamy soils contain statistically more lead than litter horizons of sandy soils. Element distribution on profiles of forest soils has chaotic enough character. Samples of loamy soils contain as a whole more exchangeable fractions of an element than of light soils.

The maximum content of an element is not always marked in the top horizons. It is obvious that absence of laws of distribution of an element on a profile is caused by seasonal dynamics of its content. The amount of lead during a season can differ by 2–3 times (Figure 5.4). Loamy forest soils contain significantly more exchangeable lead than sandy soils.

Sod-podzolic arable and meadow soils contained 1.86±0.13 mg/kg of exchangeable lead in humus horizons on the average (Table 5.2). Element distribution on profiles of arable soils has not strongly pronounced biogenic-accumulative character. Sometimes the top part of a profile is poor in exchangeable lead in comparison with underlying horizons. The increase in content of exchangeable fractions of an element in mother rocks is more expressed. The minimum of content of an element is almost always noticed in eluvial horizons.

In arable soils it is not revealed statistically significant correlations between content of exchangeable Pb and pH, total carbon, exchangeable acidity, and humidity also.

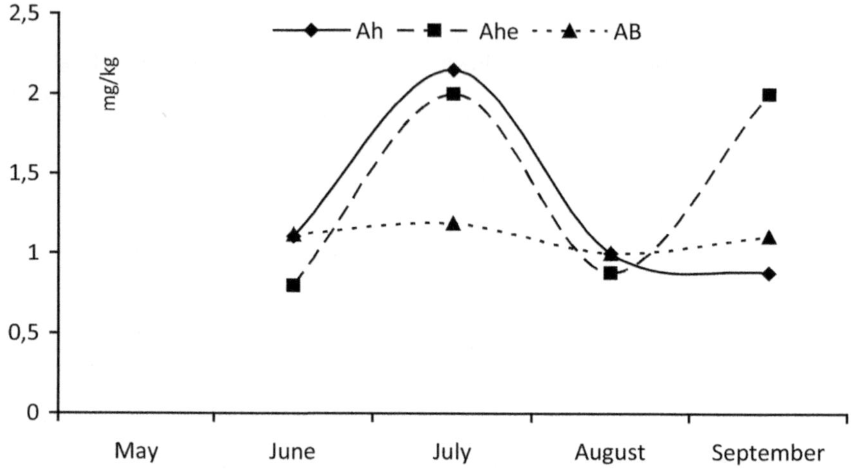

FIGURE 5.4 Seasonal dynamics of exchangeable Pb in fallow land podzolic soil of the middle taiga.

Seasonal dynamics of exchangeable lead in arable and fallow land sod-podzolic soil are various (Figures 5.4 and 5.5). Dynamics of the content of exchangeable Pb significantly correlates with ratio of labile organic matter in humus horizons of both locations but with different signs. The correlation coefficient is equal to $r = -0.90$ on arable land and $r = 0.95$ on a fallow land. Such different values of correlation coefficients possibly testify various dynamics of organic matter in annually processed and fallow soils and with different structure of organic matter.

In both sites high significant coefficients of correlation between contents of exchangeable lead and exchangeable acidity (from 0.81 to 0.99) are received. It is caused by a competition between ions Al^{3+}, H^+ and Pb^{2+} for exchange sites in soil-absorbing complexes when the acidity increase conducts to replacement of Pb^{2+} into solution.

5.3.3 INFLUENCE OF IONS OF HEAVY METALS OF OATS AND BARLEY PLANTS

Data of Table 5.3 indicate that input of cadmium and lead (in a dose of 4.5 and 41.0 mg/kg of sand accordingly) in the conditions of sandy culture leads to accumulation of heavy metals by plants. This accumulation

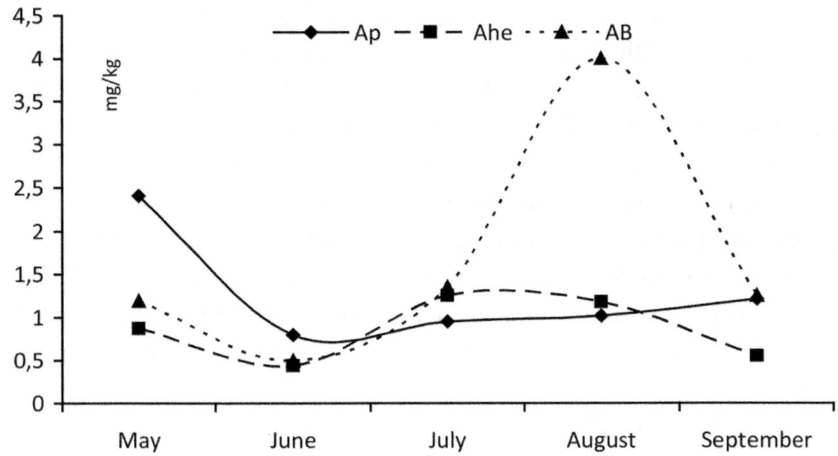

FIGURE 5.5 Seasonal dynamics of exchangeable Pb in arable podzolic soil of the middle taiga.

TABLE 5.3 Accumulation of Heavy Metals in Plants of Oats and Barley

Variety	Treatment	Cd, mg/kg of dry mass			Pb, mg/kg of dry mass		
		Roots	Stems	Leaves	Roots	Stems	Leaves
Oats							
Ulov	Control	1.08	0.60	0.42	27.61	4.13	3.44
	Heavy metal	13.45	27.08	24.22	65.46	11.11	9.49
Fakir	Control	0.27	0.52	0.49	21.24	4.10	4.29
	Heavy metal	11.70	26.11	21.63	61.41	29.99	8.63
Krechet	Control	0.43	0.40	0.47	40.12	2.84	3.42
	Heavy metal	17.73	38.92	21.44	56.65	13.53	9.49
Barley							
Elf	Control	0.75	0.44	0.17	15.05	3.61	6.87
	Heavy metal	28.34	49.15	17.17	63.72	21.25	8.71
Abava	Control	0.57	0.66	0.28	30.95	7.14	6.55
	Heavy metal	24.16	13.08	12.90	48.12	20.21	12.47
Dina	Control	0.47	0.23	0.41	35.81	8.22	3.45
	Heavy metal	33.30	36.41	8.14	59.34	24.3	11.35

exceeds control level by some ten times and reaches 30–35 mg/kg of dry mass. Distribution of heavy metals on parts of oats and barley plants is different. The most amount of cadmium is collected in roots and stems of cereals, and slightly less – in leaves. The basic part of lead was absorbed by roots. Varieties of oats and barley could not be differentiated on degree of heavy metal accumulation in their vegetative organs.

During growth season plants of oats and barley in variants of heavy metal treatment did not differ significantly from control variants on development of morphometric traits. Unique difference is noted at oats plants. It consisted in various terms of approach of a full phase of panicle formation. In variants with lead treatment approach of this phase has been noted for 3 days before control variants, and at cadmium treatment – for 3 days later.

Experiments in sand culture have shown that mutual action of aluminum and cadmium on plants of oats is shown in decrease in plant height by 28.3–32.3%. Significant decrease in plant height among barley was observed at variety Dina (in cadmium + aluminum treatment – by 6.8%, in aluminum treatment – by 16.2%) and Elf (in cadmium + aluminum treatment – by 7.6%).

Significant decrease in dry mass of plants was observed for oats Fakir and Krechet (by 25.2 and 34.8% accordingly) at mutual treatment with cadmium and aluminum (Figure 5.6). Thus the mass shortage is basically characteristic for such parts of plant as stems and roots; but the mass ratio of leaves is increased in comparison with a control treatment. At barley decrease in dry mass of plants is noted also at mutual action of cadmium and aluminum for varieties Elf (by 11.8% in comparison with the control treatment) and Dina (by 19.3%) (Figure 5.6). Unlike oats, in barley depression is characteristic for roots and leaves – change in stems has not

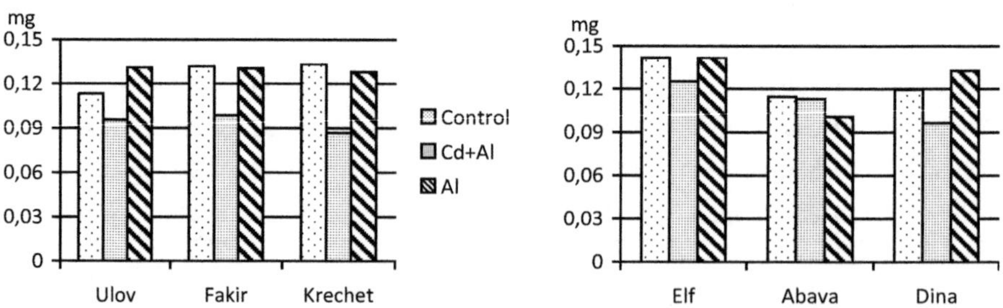

FIGURE 5.6 Influence of Cd and Al on dry mass of oats and barley plants.

concerned; in a background of decrease in total mass of plant the mass ratio of stems has even grown. At oats Ulov and barley Abava the mass of plants did not change significantly.

In a pigment complex of leaves significant disturbances have occurred only in oats Fakir in Cd+Al treatment: content of chlorophyll b increase by 28.3%, but content of chlorophyll a has decreased a little; therefore change of the content of total chlorophyll (a + b) has appeared insignificant.

Among barley disturbances in pigment content of leaves are revealed for varieties Abava and Dina but changes are characteristic only for variants of aluminum treatment and have diverse character: variety Abava has increased content of chlorophyll a and at the expense of it the sums of chlorophyll; variety Dina on the contrary has lowered it. At mutual influence of cadmium and aluminum there were not any statistically significant changes in a pigment complex of barley.

Both in oats and in barley carotenoids of leaves have appeared more stable – their changes are insignificant.

Possibly used concentration of cadmium (4.5 mg/kg of sand) and aluminum (11 mg/kg of sand) are insufficient for depression of development of cereal plants under conditions of sand culture. Considering that MAC of total cadmium in sandy and sandy loam soils makes 0.5 mg/kg, it is possible to notice that nine fold excess of MAC does not lead to considerable depression of oats and barley plants' growth and development even in sand culture when there is no practically absorbing ability (buffer action of soil in relation to pollutant) and all metal is in the exchangeable form accessible to plants. Possibly given cereal crops possess effective mechanisms of cadmium detoxication.

To overcome protective mechanisms of plants concerning heavy metals concentration of cadmium and lead has been increased up to 10 and 640 mg/kg of sand accordingly that exceed their MAC levels by 20 times.

With increase in heavy metal and Al concentration depression of almost all estimated parameters began to have significant character. In variants with HM treatments the delay of germination for 2–3 days was marked; leaf blades were narrower than in control variants. By 5th week of growth the bottom leaves especially in a variant with lead treatment began to dry up and die off. Decrease in plant height in comparison with the control treatment is noted for all oats varieties (Ulov has lowered plant height on the average by 61%, Fakir – by 49%, and Krechet – by 41%) and barley varieties (Elf – by 48%, Abava – by 40%. and Dina – by 42%) (Figure 5.7).

Joint action of HM and aluminum was observed only in barley (basically in cadmium + aluminum treatment). In variants with lead treatment for barley and in all treatments for oats it is noted significant difference between action of single metal and joint action of metal and aluminum. Thus varieties of oats and barley are classified by degree of growth depression under condition of sand culture in the same order as by their resistance to high acidity and aluminum content in natural soils.

The greatest depression under the influence of HM is noted at such parameter as plant dry mass (Figure 5.8). Oats plants demonstrate decrease in plant dry mass in lead and lead + aluminum treatments by 76.2–85.8%; variety Ulov has most suffered and Krechet – the least. At cadmium stress

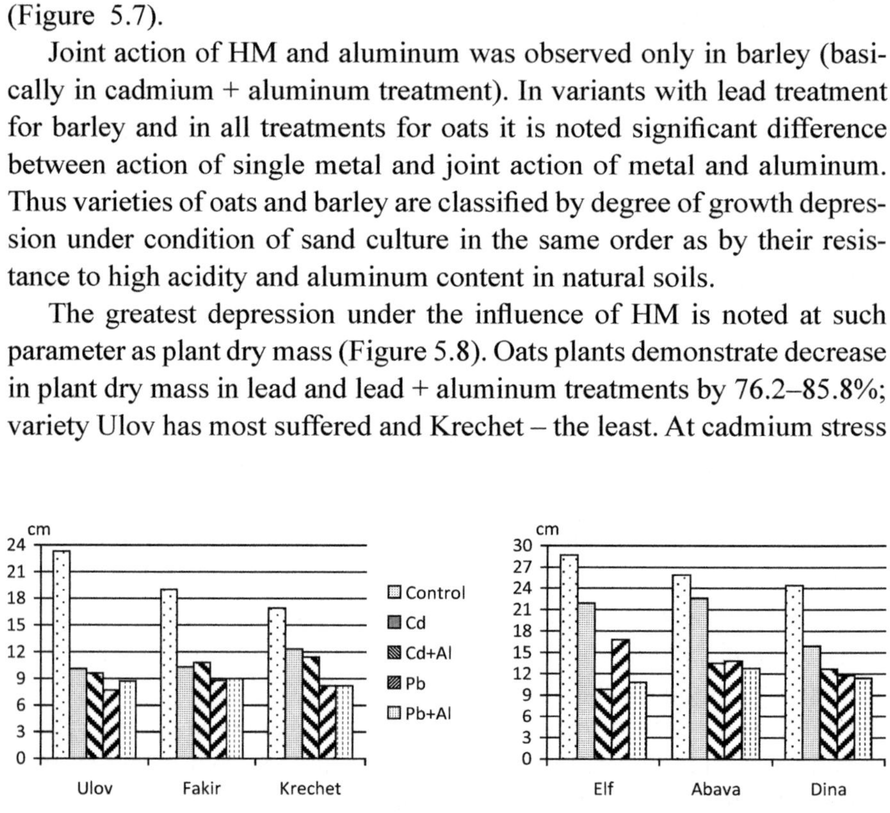

FIGURE 5.7 Influence of Al and high doses of Cd and Pb on height of plants of barley and oats.

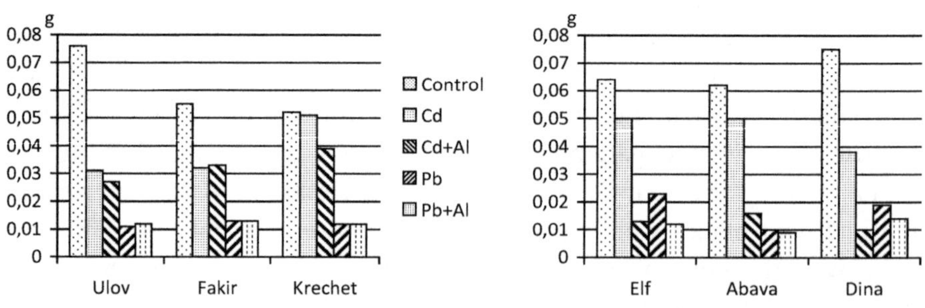

FIGURE 5.8 Influence of Al and high doses of Cd and Pb on dry mass of plants of oats and barley.

depression of plant dry mass is less and makes from 64.2% in variety Ulov (cadmium + aluminum treatment) to 25% in variety Krechet in the same variant. Thus resistance of oats varieties estimated by dry mass of plants coincides with their field aluminum resistance. The difference between treatments with sole HM and HM + Al is insignificant except for a case of variety Krechet considered above.

A same different pattern is observed in barley (Figure 5.8). Resistance of varieties to HM action by given trait does not coincide with their Al-resistance: barley Elf has less suffered. The greatest decrease in plant dry mass is noted in cadmium + aluminum, lead, and lead + aluminum treatments – by 63.3–87.1%. In sole cadmium treatment depression of plant mass makes no more than 50%. Joint influence of cadmium and aluminum has appeared more considerably than toxic action of sole cadmium.

Under the influence of high HM concentration changes have occurred in a pigment complex of leaves as well. Basically decrease in content of chlorophyll takes place but stimulation effect however was marked too.

Deviation from control values in a pigment complex of oats lays within limits 12.2% to 50.4% and is observed in different variants of treatment. Joint action of HM and aluminum was showed both in decrease, and in increase of chlorophyll content; these changes had the general character for chlorophyll of both fractions (*a* and *b*) (Figure 5.9); however content of chlorophyll a varied in a greater degree to what significant decrease in the ratio a/b in comparison with control values (by 7.1–18.0%) testifies.

In a pigment complex of barley a deviation from control values lays within limits 18.2% to 66.0%. Under cadmium treatment stimulation of

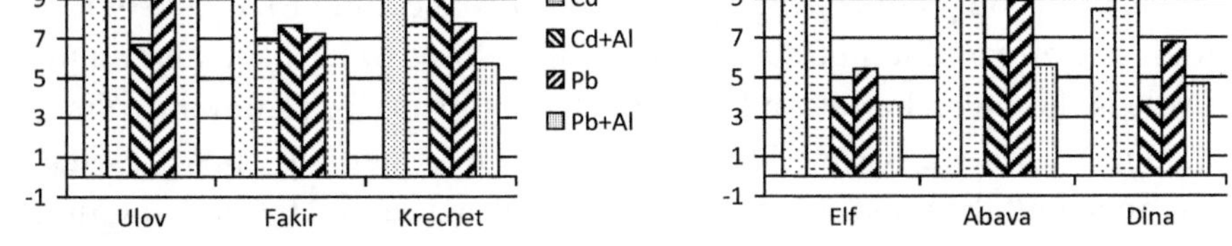

FIGURE 5.9 Influence of Al and high doses of Cd and Pb on content of total chlorophyll in oats and barley leaves.

synthetic of pigments was observed but in all other treatments – its depression. Changes had the general character and near identical intensity for a chlorophyll of both fraction (*a* and *b*). Changes in carotenoids content have similar character. Pigment complex of barley Abava and Dina was more resistant to toxic action of HM that as a whole coincides with their characteristic as edaphic resistant varieties.

5.4 CONCLUSION

Thus soils of Kirov region are characterized in comparison with other part of the European Russia by raised content of total Cd and insignificant amount of exchangeable cadmium. High content of exchangeable and total Cd in some soil samples are caused possibly with anthropogenic influence. Exchangeable cadmium accumulates in top accumulating-eluvial parts of soil profile as a part of labile organic matter.

The studied region does not differ considerably from other regions of Russia on content of total and exchangeable lead. Loamy mother rocks and soils contain more lead than sandy and sandy loam soils. Biogenic-accumulative character of distribution on a profile and elluvial-iluvial differentiation is more characteristic for total lead than for its exchangeable forms. Profile distribution of exchangeable lead is chaotic that is caused by considerable seasonal dynamics of its content. It was not possible to reveal accurate dependences between the content of lead and the content of organic matter. High correlation coefficients between content of organic matter and the content and mobility of lead in arable and fallow land sod-podzolic soils serve as indirect evidence of considerable influence of organic matter on the content of lead at studying of their seasonal dynamics.

Such heavy metals as cadmium and lead have obvious toxic effect on cereal crops only at very high concentration exceeding MAC in some tenfold. The effect of joint treatment with high doses of HM and aluminum as a rule is shown on barley and is more characteristic for cadmium + the aluminum treatment; barley Abava which are intermediate on Al-resistance has appeared the most resistant to joint toxic action of HM and aluminum.

KEYWORDS

- **acid soil**
- **aluminum**
- **barley**
- **cadmium**
- **carotenoids**
- **chlorophyll**
- **lead**
- **oats**
- **seasonal and profile dynamics**
- **soil horizons**

REFERENCES

1. Abii, T. A., Okorie, D. O. Assessment of the level of heavy metals [Cu, Pb, Cd and Cr] contamination in four popular vegetables sold in urban and rural markets of Abia State Nigeria. Continental Journal of Water Air and Soil pollution (Nigeria). 2011, Vol. 2(1), 42–47.
2. Machado, S., Rabelo, T. S., Portella, R. B., Carvalho, M. F., Magna, G. A. M. A study of the routes of contamination by lead and cadmium in Santo Amaro, Brazil. Environmental Technology. 2013, Vol. 34, 559–571.
3. State (national) report on condition and use of land in Russian Federation for 1995, Moscow: Russlit Publ., 1996, 120 p. [in Russian].
4. Burkov, N. A. Applied ecology. Kirov: Publishing House Vyatka, 2005, 271 p. [in Russian].
5. Oyedele, D. J., Asonugho, C., Awotoye, O. O. Heavy metals in soil and accumulation by edible vegetables after phosphate fertilizer application. Electronic Journal of Environmental Agricultural Food Chemistry 2006, Vol. 5(4), 1446–1453.
6. Guzman, E. T. R., Regil, E. O., Gutierrez, L. R. R., Albericl, M. V. E., Hernandez, A. R., Regil, E. D. Contamination of corn growing areas due to intense fertilization in the high plane of Mexico. Water, Air and Soil Pollution. 2006, Vol. 175, 77–98.
7. Mendes, A. M. S., Duda, G. P., Nascimento, G. W. A., Silva, M. O. Bioavailability of cadmium and lead in a soil amended with phosphorus fertilizers. Scientia Agricola. 2006, Vol. 63, 328–332.

8. Dissanayake, C. B., Chandrajith, R. Phosphate mineral fertilizers, trace metals and human health. J. Natn. Sci. Foundation Sri Lanka. 2009, Vol. 37(3), 153–165.
9. Zink and cadmium in the environment. Ed. V. V. Dobrovolsky. Moscow: Nauka ("Science" in Russian), 1992, 200 p. [in Russian].
10. Zolotareva, B. N., Scripnichenko, I. I., Geletjuk, N. I., Sigaeva, E. V., Piunova, V. V. Content and distribution of heavy metals (lead, cadmium, and mercury) in soils of European USSR. Genesis, fertility and amelioration of soils. Pushchino: Scientific Center of Academy of Sciences, 1980, 77–90 [in Russian].
11. Il'in, V. B. Estimation of mass-flow of heavy metals in system soil-agricultural crop. Agrochemistry. 2006, N 3, 52–59. [in Russian].
12. Gorjunova, T. A. Heavy metals (Cd, Pb, Cu, Zn) in soils and plants of south-west part of Altai Territory. Siberian ecological journal. 2001, N 2, 181–190. [in Russian].
13. Titova, V. I. Optimization of plant nutrition and ecological-agrochemical estimation of fertilizers use on soils with high contents of exchangeable phosphorus. DSc Thesis. Saint-Petersburg, 1998, 340 p. [in Russian].
14. Gavrilova, I. P., Bogdanova, M. V., Samonova, O. A. Experience of square estimation of degree of Russian soil pollution by heavy metals. Herald of Moscow University. Soil Science. 1995, N1. 48–53. [in Russian].
15. Mineev, V. G., Gomonova, N. F. Estimation of agrochemistry ecological functions on behavior of Cd in agrocenosis on sod-podzolic soil. Herald of Moscow University. Soil Science. 1999, N1, 46–50. [in Russian].
16. Tsiganjuk, S. I. Influence of long-term application of phosphorus and lime fertilizers on accumulation of heavy metals in soil and plant products. PhD Thesis. Moscow: All Russian Institute of Fertilizers and Agrochemistry, 1994, 126 p. [in Russian].
17. Yagodin, B. A., Govorina, V. V., Vinogradova, S. B., Zamarajev, A. G., Chapovskaya, G. V. Accumulation of Cd and Pb by some agricultural crops on sod-podzolic soils of different degree of cultivation. Transactions of Timiryazev Agricultural Academy. 1995, V. 2, 85–100. [in Russian].
18. Chemistry of heavy metals, arsenic, and molybdenum in soils. Ed. Zyrin, N. G., Sadovnikova, L. K. Moscow: Publishing House of Moscow University, 1985, 208 p. [in Russian].
19. Zyrin, N. G., Chebotaryova, N. A. About forms of copper, zinc, and lead complexes in soils and their availability to plants. Content and form of microelements complexes in soils. Moscow: Publishing House of Moscow University, 1979, P. 350–386 [in Russian].
20. Egoshina, T. L., Shikhova, L. N. Lead in soils and plants of North-East of European Russia. Herald of Orenburg State University. 2008, N10, 135–141. [in Russian].
21. Travnikova, L. S., Kakhnovich, Z. N., Bolshakov, V. A., Kogut, B. M., Sorokin, S. E., Ismagilova, N. H., Titova, N. A. Importance of analysis of organic-mineral fractions for estimation of pollution of sod-podzolic soils with heavy metals. Pochvovedenie ("Soil science" in Russian). 2000, N 1, 92–101. [in Russian].
22. Volgin, D. A. Background level and the content of heavy metals in a soil cover of the Moscow area. Herald of Moscow State Region University. Geography. 2011, N 1, 26–33. [in Russian].
23. Ponizovsky, A. A., Mironenko, E. V. Mechanisms of lead (II) absorption by soils. Pochvovedenie ("Soil science" in Russian). 2001, N 4, 418–429. [in Russian].

24. Kabata-Pendias, A., Trace Elements in Soils and Plants, Fourth Edition. CRC Press. Boca Raton FL. 2010, 548 p.
25. Lukina, N. V., Nikonov, V. V. State of spruce biogeocenoses of North under conditions of technogenic pollution. Apatity: Academy of Science of USSR, 1993, 132 p. [in Russian].
26. Sauerbeck, D. Welche Schwerrmetallgehalte in Pflanzen durfen nicht uberschritten werden, um Wachstumsbeeintrachtigungen zu vermeiden?. Landwirtschaftliche Forschung: Kongressband. 1982, Heft 16. S, 59–72. [in German] [Content of some heavy metals in plants does not reach critical values: can plants avoid growth damages? Agricultural Researches: Materials of Congress].
27. Lisitsyn, E. M., Shchennikova, I. N., Shupletsova, O. N. Cultivation of barley on acid sod-podzolic soils of north-east of Europe. in: Barley: Production, Cultivation and Uses. Ed. Steven, B. Elfson. New York: Nova Publ. 2011, P. 49–92
28. Arinushkina, E. V. Handbook on chemical analysis of soils. Moscow: Publishing House of Moscow University, 1970, 488 p. [in Russian].
29. Methodical indications for estimation of microelements in soils, fodders, and plants by atomic-absorption spectroscopy. Moscow: CINAO Publ., 1995, 95 p. [in Russian].
30. Lisitsyn, E. M., Batalova, G., Shchennikova, I. N. Creation of oats and barley varieties for acid soils. Theory and practice. Palmarium Academic Publishing, Saarbrucken, Germany, 2012, 228 p. [in Russian].
31. Lisitsyn, E. M. Intravarietal Level of Aluminum Resistance in Cereal Crops. Journal of Plant Nutrition. 2000, V. 23(6), 793–804.
32. Lichtenthaler, H. K., Bushmann, C. Chlorophylls and carotenoids: measurement and characterization by UV-VIS spectroscopy. Current Protocols in Food Analytical Chemistry. 2001, F4.3.1–F4.3.8.
33. Tjulin, V. V., Gushchina, A. M. Features of soils of Kirov region and their use at intense agriculture. Kirov: Publishing house Vyatka, 1991, 94 p. [in Russian].
34. Yulushev, I. G. System of fertilizers use in crop rotation. Kirov: Publishing House Vyatka, 1999, 154 p. [in Russian].
35. Prokashev, A. M. Soils with Complex Organic Profiles of South of Kirov Region. Kirov: Publishing House Vyatka, 1999, 174 p. [in Russian].

CHAPTER 6

STRUCTURAL FRAGMENTS OF ASPHALTENES FROM CRUDE OIL

R. N. SHIRYAEVA, F. H. KUDASHEVA, and D. I. SHAFIGULLINA

Bashkir State University, 9, Ufa, Republic of Bashkortostan, 450076, Russia

CONTENTS

ABSTRACT

The structural fragments of asphaltenes of Gerasimov oil have been studied by method of IR-spectroscopy. It was found that asphaltenes, which were separated from the oil by the method of the Institute of Oil and SARA-Analysis, differ in aromaticity and branching aliphatic chains.

6.1 INTRODUCTION

Asphaltenes are high molecular weight components of the oil. Molecules of asphaltenes are characterized by the presence of condensed aromatic nucleus, small aliphatic side chains, and the presence of heteroatoms and metals [1–3]. In quantitative terms, asphaltenes do not predominate in the oil: their content can range from trace amounts in light oils and up to 16–20% in highly viscosity. Interest in studding of asphaltenes is due to their negative impact on the properties of the oil disperse systems. Oil with a high content of asphaltenes are characterized by high viscosity, which requires the use of expensive technologies for enhanced oil recovery and facilitate its transportation. Less viscous oil with a low content of asphaltenes are unstable due to the precipitation of asphaltenes, which leads to the plugging of wells and pipelines.

Asphaltenes play a major role in structuring of the oil disperse systems and determine the stability of the colloidal structure of crude oil. Therefore, the studying of the composition and molecular structure of asphaltenes, identifying of promising directions of their processing is relevant.

In this chapter, we present the results of studying the structural fragments of asphaltenes from crude oil of Gerasimov field. Oil characteristics are presented in Table 6.1.

The oil is characterized by a high content of paraffin-naphthenic hydrocarbons, resins and asphaltenes.

For separation of asphaltenes from crude oil to study we use their properties to fall out of the oil system when adding low molecular weight alkanes. The amount and composition of asphaltenes depends on the

TABLE 6.1 Physico-Chemical Characteristics of Oil

Indices	
Density at 20°C, kg/m^3	908.8
Hydrocarbon composition, % mass	
Paraffin-naphthenic	36.9
Aromatic	48.4
Monocyclic	17.1
Bicyclic	8.0
Polycyclic	23.3
Resin benzene	4.3
Resin alcohol-benzene	7.8
Asphaltenes	2.6

volume ratio of oil: the precipitator. The selection of asphaltenes from the original oil was carried out by the method of SARA and research Institute of oil. According to the Institute of petroleum asphaltenes were separated by sedimentation of 40-fold excess of petroleum ether (fraction 40–70). In SARA-method heptanes was used as a precipitator of asphaltenes.

The IR spectra of crude oil and asphaltenes were removed with spectrometer of "Shimadzu" firm in the form of films, made between the plates of NaCl in the range of 4000–650 cm^{-1}.

The IR spectra of crude oil and asphaltenes are presented in Figures 6.1–6.3, and the calculated spectral coefficients are in Table 6.2.

*$C_1=A_{1600}/A_{720}$ (aromaticity); $C_2=A_{1710}/A_{1460}$ (oxidation); $C_3=A_{1380}/A_{1460}$ (branching); $C_4=A_{720}+A_{1380}/A_{1600}$ (aliphaticity); $C_5=A_{1030}/A_{1460}$ (osamenost).

Gerasimov oil contains significant amounts of CH_3, CH_2 – groups (2920, 2850, 1460, 1380 cm^{-1}). Asphaltene contains more terminal methyl groups (1380 cm^{-1}). The intensity of the absorption band at 1600 cm^{-1} is more for asphaltenes, selected by the method of SARA. Asphaltenes precipitated by the method of the research Institute of oil are more aromatic and oxygenated. But the yield of asphaltenes is less in comparison with the method of SARA. The structural features of asphaltenes isolated

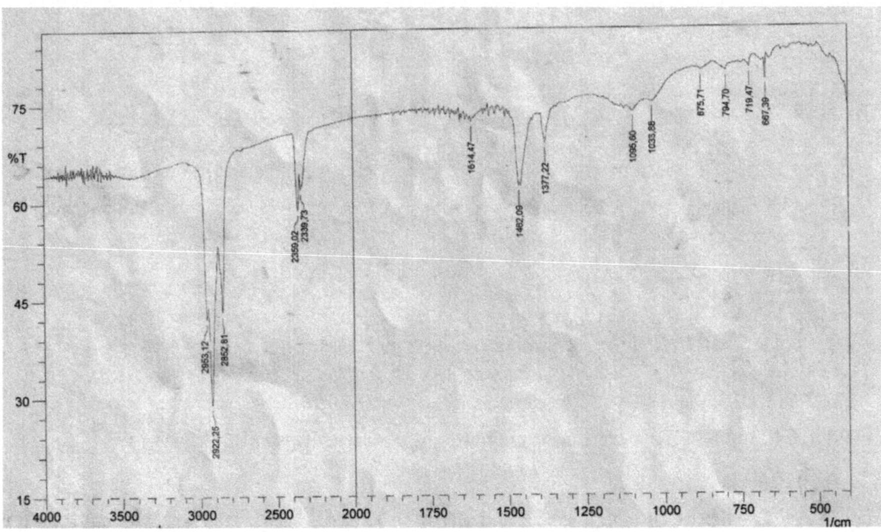

FIGURE 6.1 The IR spectrum of the oil of Gerasimov field.

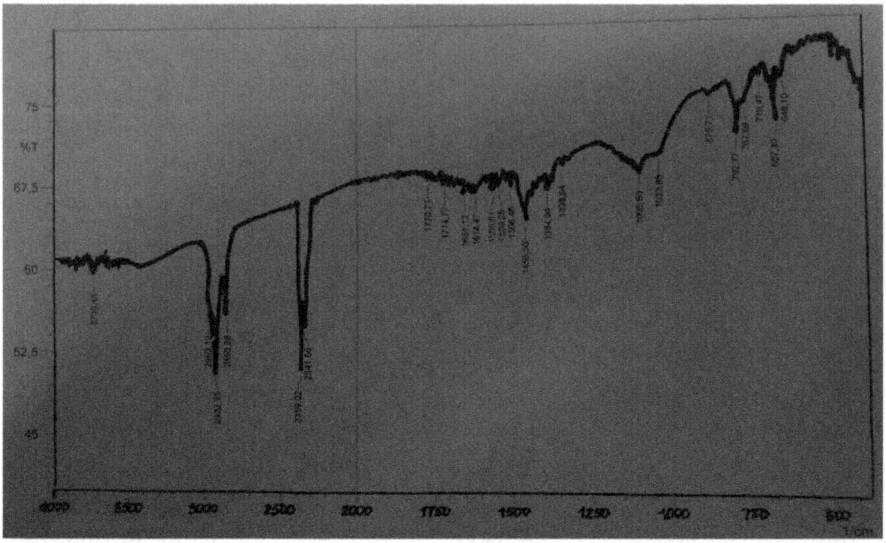

FIGURE 6.2 The IR spectrum of asphaltenes of Gerasimov oil (research Institute of petroleum).

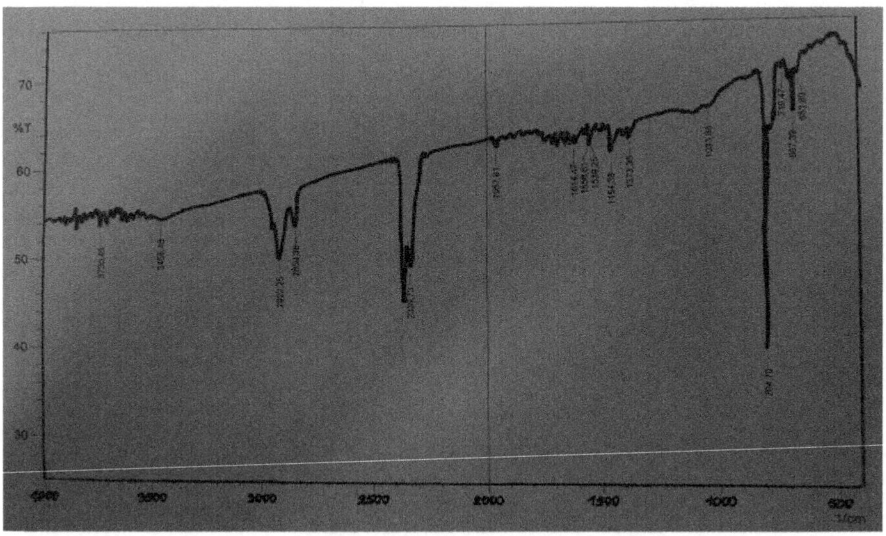

FIGURE 6.3 The IR spectrum of asphaltenes of Gerasimov oil (SARA).

by SARA-method are low aromaticity kernel and its high substitution aliphatic chains.

TABLE 6.2 The Spectral Coefficients for Oil and Asphaltenes

Sample	Optical density in the maximum of absorption bands ν, cm^{-1}						Spectral coefficients				
	1710	1600	1460	1380	1030	720	C_1	C_2	C_3	C_4	C_5
Original oil	—	0.14	0.21	0.16	0.12	0.10	1.41	—	0.76	1.23	0.57
Asphaltene (SARA)	—	0.21	0.21	0.20	0.18	0.15	1.37	—	0.95	1.10	0.86
Asphaltene (Research Institute oil)	0.17	0.18	0.19	0.17	0.16	0.11	1.60	0.89	0.89	1.05	0.84

Evaluation of the colloidal stability of the oil (S) to be precipitated asphaltenes were carried out according to the formula: GAIN = (aromatics + resins)/(paraffins + asphaltenes). The colloidal stability of the oil is 1.53, the ratio of content of resins and asphaltenes is 4,65. Thus, the average molecules of asphaltenes, highlighted by two methods differ in aromaticity and branching aliphatic chains.

KEYWORDS

- **asphaltenes**
- **infrared spectroscopy**
- **oil**
- **stability of colloidal system**

REFERENCES

1. Ganeeva, U. M., Yusupova, T. N., Romanov, G. V. Asphaltene nanoaggRegates: Structure, Phase Transformations, the Impact on Petroleum Systems. USP. 2011, 80(10), 1034–1050.
2. Dmitriev, D. E., Golovko A. K. Modeling of molecular structures petroleum resins and asphaltenes and calculation of their thermodynamic stability. Chemistry for Sustainable Development. 2010, No. 18, 177–187.
3. Akhmetov, B. R., Evdokimov, I. N., Eliseev, N. K. Features of the optical absorption spectra of crude oils and petroleum asphaltenes. Science and Technology of Hydrocarbons. 2002, No. 3, 25–30.

CHAPTER 7

OXIDATION OF APPLE PECTIN UNDER THE ACTION OF HYDROGEN PEROXIDE

YU. S. ZIMIN,[1] N. S. BORISOVA,[1] and G. R. TIMERBAEVA[2]

[1]*Bashkir State University, 32 Z, Validi Street, Ufa, 450076, Republic of Bashkortostan. Russia,*
E-mail: ZiminYuS@mail.ru, NSGuskova@rambler.ru

[2]*Branch of Ufa State Aviation Technical University, 5 Youth Md. Tuimazy, 452750, Republic of Bashkortostan, Russia,*
E-mail: guzel.timerbaeva@mail.ru

CONTENTS

ABSTRACT

It is shown that the oxidation of apple pectin under the action of hydrogen peroxide and ozone-oxygen mixture in water solutions goes by a radical mechanism and is accompanied by a destruction of polysaccharide macromolecules. We study the kinetics of oxidative transformations (functionalization and destruction) of apple pectin under a wide range of process conditions.

7.1 INTRODUCTION

Nowadays natural polysaccharides, pectin are widely used in various industries (food, pharmaceutical, cosmetic, etc.) and their need is rising each year. Despite this fact there are no large industries producing pectin in Russia. The polysaccharide is usually produced from industrial waste of plant materials (citrus peel, apple pomace, sugar beet pulp, sunflower baskets) [1]. Thus, unused apple wastes (apple meal) with 80% initial pectin substances are of great value for launching industrial pectin production in Russia. Data on apple pectin oxidative modification may also serve an additional advantage for its production.

Due to its high sorption and complexing abilities the oxidized pectin may be applied as a matrix carrier of drug and biologically active substances. One of the ways of pectin modification is the oxidation of the polysaccharide in water medium using non-toxic and easily removable oxidants. So being quite topical, the study of kinetic regularities of oxidative destruction and functionalization of apple pectin is aimed at receiving further polymer products of the given molecular weight.

The work under consideration presents the data on studying the kinetics of apple pectin (AP) oxidation under the action of hydrogen peroxide (reaction system "AP + H_2O_2 + O_2 + H_2O") and ozone-oxygen mixture (reaction system "AP + O_3 + O_2 + H_2O") and also determining the nature of these processes. The use of the above-mentioned oxidants in the experiment is caused by the necessity of obtaining modified low-molecular apple pectin products with no additional cleansing thus reducing their cost.

7.2 EXPERIMENTAL PART

For experiments apple pectin of Herbstreith & Fox KG Pektin-Fabrik Neuenburg Ltd. (Germany) with the molecular weight 213 kDa ([η] = 3.5 dl/g, 25 ± 1°C) was used. Fresh redistilled water served as a solvent. Ozone was received by the ozonizer of the known construction [2]. Ozone concentration in the gas mixture at the outlet of the ozonizer amounted to 1–2 vol. %. The feed rate of the ozone-oxygen mixture ($VO_{3+}O_2$) was changed from 5.8 to 6.8 L/h. Hydrogen peroxide was used of an "analytical reagent grade" type.

Dissolved apple pectin was oxidized by hydrogen peroxide (or ozone-oxygen mixture) in a glass thermostated reactor of a bubbling type with periodical samples taking. The concentration of carboxyl groups was determined by potentiometric titration. The intrinsic viscosities of pectin solutions were measured in a Ubbelohde viscosimeter with the hung level. Molecular weights of the initial and oxidized biopolymer were calculated by intrinsic viscosity values using Mark-Kuhn-Hauvink equation [3].

7.3 RESULTS AND DISCUSSION

7.3.1 OXIDATIVE FUNCTIONALIZATION OF APPLE PECTIN

Dissolved apple pectin is not subjected to oxidation and destruction while heating and air bubbling through a polysaccharide solution up to 60–90°C. However, if hydrogen peroxide or ozone-oxygen mixture is added to water pectin solution it leads to oxidative transformations of the biopolymer demonstrated by considerate increase of carboxyl group (-COOH) concentration in the solution (Figure 7.1).

From Figure 7.1 it is seen that in oxidizing apple pectin by ozone-oxygen mixture the process of forming functional groups is more intensive as compared with hydrogen peroxide. It is explained by strong ozone oxidative ability (standard oxidative ozone potential is +2.07 B and hydrogen peroxide equals +1.77 B [4]). Kinetic accumulation curves of COOH-groups point to the decreasing speed of pectin oxidation. The initial speed

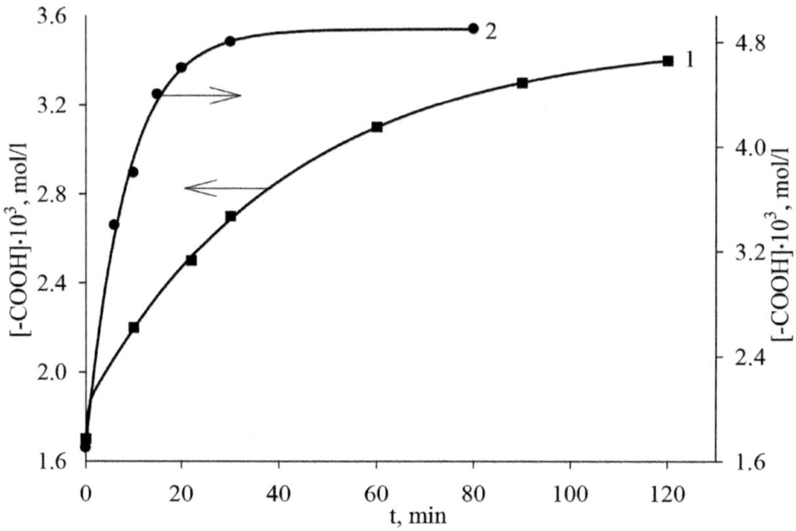

FIGURE 7.1 Kinetic curves of COOH-groups accumulation in the systems "AP + H_2O_2 + O_2 + H_2O" (1) ($[H_2O_2]_0$ = 1 mol/L) and "AP + O_3 + O_2 + H_2O" (2) ($Vo_{3+}o_2$ = 6.8 L/h); 70°C, $[AP]_0$ = 0.5% mass.

of carboxyl group accumulation (V_{COOH}) determined by the polynomial analysis was used as a quantitative characteristic of the process.

In the present work the influence of the process conditions (initial concentration of reagents and temperature) on the kinetics of carboxyl group accumulation is studied.

The research of V_{COOH} dependence from the initial concentration of apple pectin was carried out on the example of the reaction system "AP + H_2O_2 + O_2 + H_2O." It is shown that the speed of carboxyl groups accumulation increases linearly and is accompanied by $[AP]_0$ growing (Figure 7.2). It demonstrates the first order of the pectin reaction.

While the initial concentration of hydrogen peroxide ("AP + H_2O_2 + O_2 + H_2O") and the speed of ozone-oxygen mixture bubbling ("AP + O_3 + O_2 + H_2O") is growing, the accumulation speed of COOH-groups is raising either (Figure 7.3).

The initial speed of carboxyl group accumulating linearly depends on $[H_2O_2]_0$ that points to the first order of the reaction on hydrogen peroxide. The mode of oxidative transformations plays an important role in studying the dependence of V_{COOH} from $Vo_{3+}o_2$ (Figure 7.3). At low speed of

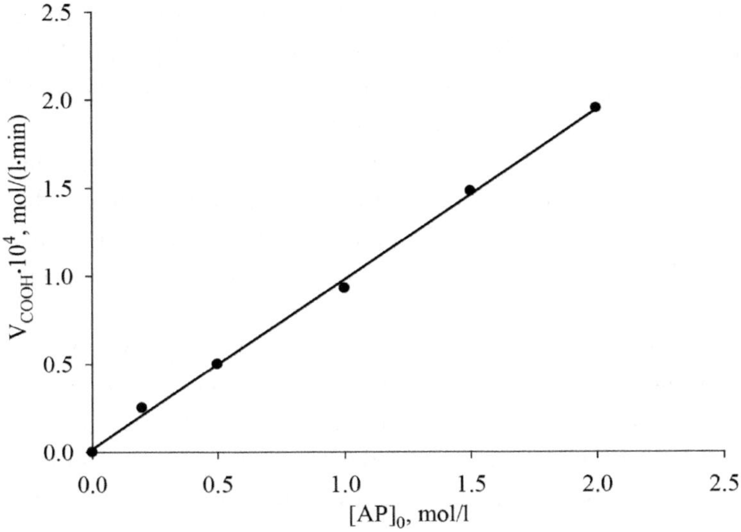

FIGURE 7.2 Dependence of initial speed of accumulating carboxyl groups from apple pectin concentration; 70°C, $[H_2O_2]_0 = 1$ mol/L.

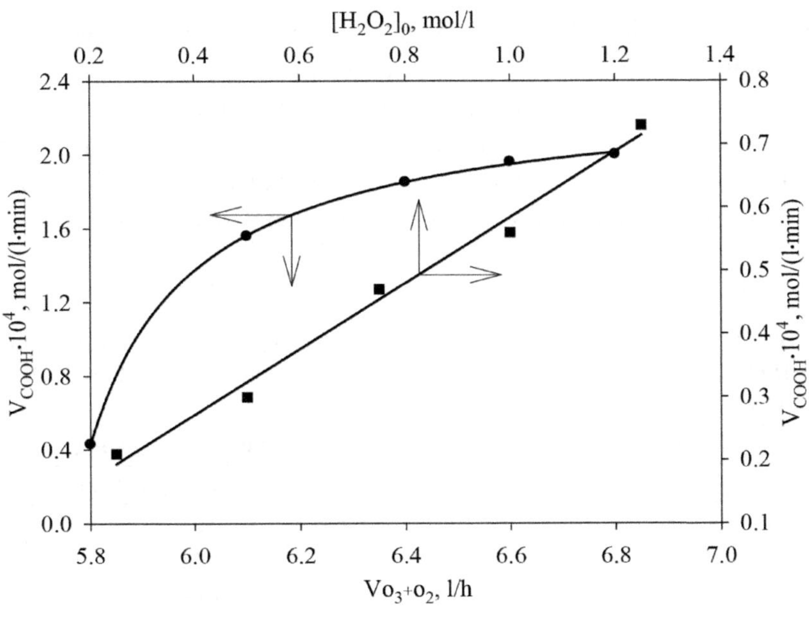

FIGURE 7.3 Dependence of the initial speed of accumulating carboxyl groups from hydrogen peroxide concentration and the speed of bubbling the ozone-oxygen mixture 70°C, $[AP]_0 = 0.5\%$ mass.

bubbling the ozone-oxygen mixture, the oxidation of apple pectin probably takes place in the diffusion mode and is limited by the diffusion of O_3 (and O_2) molecules to polysaccharide ones. If the bubbling speed of the gas mixture is higher than 64 L/h, the oxidative transformations are held in the kinetic mode, which is demonstrated by inconsiderate changes of the initial speed of accumulating carboxyl groups from the gas flow rate.

The results of studying the temperature dependence of the initial speed of accumulating carboxyl groups in the range 60–90°C are presented in Table 7.1.

As seen from the Table 7.1, the temperature rises and V_{COOH} grows either leading to the increase of COOH-groups concentration accumulated to a certain moment of time. It is found out that the temperature dependence of the initial speed of accumulating carboxyl groups in the studied reaction systems is described by the following Arrhenius equations:

$$\text{“AP + H}_2\text{O}_2 + \text{O}_2 + \text{H}_2\text{O” } \lg V_{COOH} = (1.6 \pm 0.5) - (38 \pm 4)/\Theta \ (r = 0.999),$$
$$\text{“AP + O}_3 + \text{O}_2 + \text{H}_2\text{O” } \lg V_{COOH} = (3 \pm 1) - (44 \pm 8)/\Theta \ (r = 0.995),$$

where $\Theta = 2.303RT$ kJ/mol; $R = 8.314$ J/mol·K – a universal gas constant; T – temperature, K; r – correlation coefficient.

It was experimentally found that the overwhelming part of carboxyl groups after pectin oxidation is in a high molecular weight fraction extracted by precipitating with acetone from water solution. In Table 7.2 the distribution of carboxyl groups in high and low molecular weight

TABLE 7.1 Temperature Dependence of V_{COOH} in the Reaction Systems “AP + H_2O_2 + O_2 + H_2O” ($[H_2O_2]_0$ = 1 mol/L) and “AP + O_3 + O_2 + H_2O” ($V_{O_3+O_2}$ = 6.4 L/h); $[AP]_0$ = 0.5% mass

“AP + H_2O_2 + O_2 + H_2O”		“AP + O_3 + O_2 + H_2O”	
T, °C	**$V_{COOH}\times10^4$, mol/(l×min)**	**T, °C**	**$V_{COOH}\times10^4$, mol/(L×min)**
70	0.54	60	0.94
75	0.64	65	1.08
80	0.76	70	1.85
85	0.93	75	1.94
90	1.12	80	2.28

TABLE 7.2 Number* of Carboxyl Groups After AP Oxidation in the Solution in High and Low Molecular Weight Fractions; $[AP]_0 = 0.5\%$ mass., $Vo_{3+}o_2 = 6.4$ L/h, t = 20 min

T, °C	(–COOH)× 10^5, mol	$(–COOH)_{hmwf}$ ×10^5, mol	$(–COOH)_{lmwf}$ ×10^5, mol
60	7.7	5.6	2.1
65	8.1	7.2	0.9
70	8.6	7.4	1.2

*(–COOH) – total amount of COOH-groups in a water solution of the oxidized pectin; $(–COOH)_{hmwf}$ and $(–COOH)_{lmwf}$ – number of carboxyl groups in high and low molecular weight fractions correspondingly. $(–COOH)_{lmwf}$ is received by the difference of (–COOH) and $(–COOH)_{hmwf}$.

fractions after 20-minute oxidation of apple pectin by the ozone-oxygen mixture in different temperature modes are presented.

Thus, the oxidation of apple pectin leads to formation of additional carboxyl groups in the biopolymer. Any changes in the conditions of the oxidative process (initial concentration of the starting reagents and temperature) allow to regulate the concentration of COOH-groups that can be used for improving the properties of the natural biopolymer (as a matrix for next-generation drugs).

7.3.2 OXIDATIVE DESTRUCTION OF APPLE PECTIN

Oxidation of apple pectin under the action of hydrogen peroxide and ozone-oxygen mixture is accompanied by destruction of its molecules. This is proved by lowering the intrinsic viscosity of the biopolymer solution whereas increasing the time of pectin oxidation by hydrogen peroxide or ozone-oxygen mixture (Figure 7.4).

Kinetics of oxidative destruction of apple pectin is studied depending on the initial concentrations of the reagent and temperature. On the example of the system "AP + H_2O_2 + O_2 + H_2O" it is shown that as the concentration of the oxidizing agent increases, the destruction degree of pectin grows as well (Table 7.3).

The study of temperature dependences of intrinsic viscosity provides for some controversial results (Table 7.4). As the temperature of the reaction system "AP + H_2O_2 + O_2 + H_2O" rises, the increase of the polymer destruction is observed. At the same time similar actions with the reaction system "AP + O_3 + O_2 + H_2O" lead to an opposite effect.

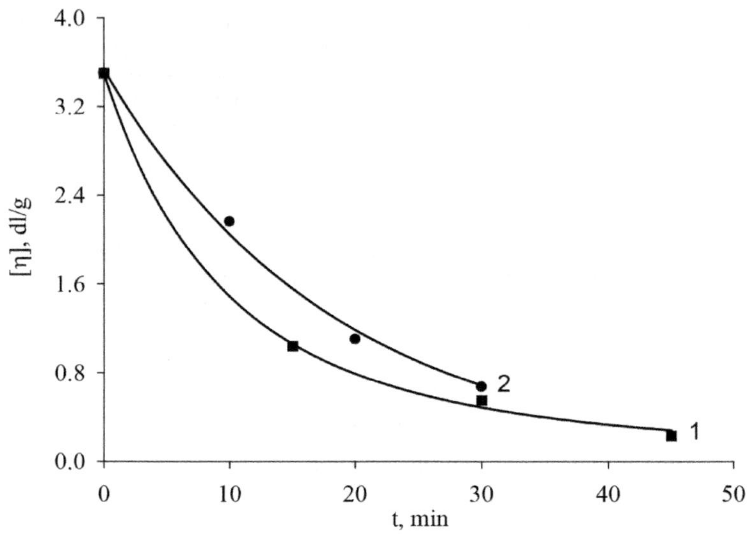

FIGURE 7.4 Kinetics of intrinsic viscosity changes of apple pectin water solution in the systems "AP + H_2O_2 + O_2 + H_2O" (1) ($[H_2O_2]_0$ = 1 mol/L) and "AP + O_3 + O_2 + H_2O" (2) ($VO_{3+}O_2$ = 6.4 L/h); 70°C, $[AP]_0$ = 0.5% mass.

TABLE 7.3 Dependence of Apple Pectin Destruction from the Initial Concentration of Hydrogen Peroxide in the Reaction System "AP + H_2O_2 + O_2 + H_2O"; 70°C, $[AP]_0$ = 0.5% mass

$[H_2O_2]_0$, mol/L	0.25	0.50	0.75	1.00	1.25
$[\eta]_t$, dl/g	0.50	0.37	0.33	0.25	0.22
$(\Delta[\eta]/[\eta]_0)\times 100$, %*	78	84	86	89	90
M, kDa	18	12	11	7.5	6.4

*$\Delta[\eta] = [\eta]_0 - [\eta]_t$, where $[\eta]_0$ and $[\eta]_t$ – are intrinsic viscosities of pectin solutions at the initial moment of time and t = 60 min, accordingly.

The increase of the polysaccharide molecular weight in oxidizing it by ozone-oxygen mixture is probably connected with an association of oxidized macromolecules in the solution resulted in forming additional carboxyl groups. As seen in Table 7.2 a number of COOH-groups in high molecular weight fractions grows as the temperature of the ozonized oxidation increases. Ozone of a stronger oxidative potential as compared with hydrogen peroxide enables to a more intensive accumulation of COOH-groups (Figure 7.1) and more active association of macromolecules. The literature analysis [5–7] shows that both mixtures of separate polymer

TABLE 7.4 Temperature Dependence of Intrinsic Viscosity of Apple Pectin Solutions in the Reaction Systems "AP + H_2O_2 + O_2 + H_2O" ($[AP]_0$ = 0.5% mass., $[H_2O_2]_0$ = 1 mol/L, t = 60 min) and "AP + O_3 + O_2 + H_2O" ($[AP]_0$ = 1.0% mass., $V_{O_3+O_2}$ = 6.4 L/h, t = 30 min)

T, °C	"AP + H_2O_2 + O_2 + H_2O"			"AP + O_3 + O_2 + H_2O"		
	[η], dl/g	$(\Delta[\eta]/[\eta]_0)\times100$, %	M, kDa	[η], dl/g	$(\Delta[\eta]/[\eta]_0)\times100$, %	M, kDa
60	0.45	80	16	0.30	94	9.5
65	0.36	84	12	0.40	92	14
70	0.25	89	7.5	0.58	88	22
75	0.23	90	6.8	0.60	87	23
80	0.22	91	6.4	0.64	86	25

molecules and their aggregates are found in pectin solutions with different degree of esterification. And different degree of esterification of macromolecules directly depends on the number of carboxyl groups not subjected to esterification.

Thus, introducing hydrogen peroxide or ozone-oxygen mixture in the heated apple pectin solutions leads to the oxidative destruction of the biopolymer macromolecules. Varying the conditions of the oxidation process (initial concentrations of the starting reagents and temperature) allows to get oxidized polymer products of the defined molecular weight.

7.3.3 RADICAL NATURE OF APPLE PECTIN OXIDATION

In early works [8, 9] it was shown that the ozonized and peroxide oxidation of citrus pectin (CP) is held according to the radical mechanism. It was established that decomposition of hydrogen peroxide under impurities of transition metals in pectin and redistilled water is the initiation stage of the reaction system "CP + H_2O_2 + O_2 + H_2O":

$$H_2O_2 + M^{n+} \rightarrow HO^{\bullet} + HO^{-} + M^{(n+1)+} \quad (1)$$

In the system "CP + O_3 + O_2 + H_2O" initiation of the radical process is carried out by the abstraction reaction of ozone molecules to hydrogen atoms from C-H-links of polysaccharide (PH):

$$PH + O_3 \rightarrow P^{\bullet} + HO^{\bullet} + O_2 \quad (2)$$

It should be expected that in the researched reaction systems apple pectin is also oxidized by the radical mechanism. The following experiment results received on the example of the reaction system "AP + H_2O_2 + O_2 + H_2O" ground these assumptions.

Adding phenol PhOH (a classic inhibitor of radical oxidation of organic compounds [10] that cuts oxidation chains by the reaction with peroxy-radicals) leads to the inhibition of the oxidation process demonstrated by noticeable decrease of the initial speed of carboxyl groups accumulation (Table 7.5).

Increasing the concentration of ferrous sulfate (II) in the reaction system "AP + H_2O_2 + O_2 + H_2O" leads to a sharp growth of V_{COOH} (Table 7.5). This fact is explained that the speed of the reaction (1) (determining the speed of oxidative transformations of apple pectin) is:

$$V = k\,[H_2O_2][M^{n+}],$$

where k is a speed constant of the reaction of hydrogen peroxide and M^{n+}. Here the first order of the reaction by H_2O_2 in Figure 7.3 becomes quite clear.

In introducing disodium edentate (Trilon B) able to bind transition metal ions [11] in the researched system the slow of the reaction (1) and the oxidative process as a whole takes place (Table 7.5).

The given facts unambiguously state the radical nature of the oxidative transformations of apple pectin in the reaction system "AP + H_2O_2 +

TABLE 7.5 Influence of the Initiating System on the Speed of Carboxyl Group Accumulation; 70°C, $[AP]_0$ = 0.5% mass., $[H_2O_2]_0$ = 1 mol/L, $V_{O_3+O_2}$ = 6.3 L/h, $[FeSO_4]_0$ = 1×10^{-3}mol/L, $[PhOH]_0$ = 1×10^{-5}mol/L, [Trilon B]$_0$ = 1×10^{-5}mol/L

System	$V_{COOH}\times10^4$, mol/(L×min)
"AP + H_2O_2 + O_2 + H_2O"	0.54
"AP + O_3 + O_2 + H_2O"*	1.30
"AP + H_2O_2 + O_2 + PhOH + H_2O"	0.30
"AP + H_2O_2 + O_2 + $FeSO_4$ + H_2O"	14.80
"AP + H_2O_2 + O_2 + Trilon B + H_2O"	0.20

*for comparison.

O_2 + H_2O." It should be supposed that the ozonizing oxidation of pectin will also be of the radical nature as the same regularities (Figures 7.1–7.3) are observed in oxidizing apple pectin in the considered reaction systems "AP + H_2O_2 + O_2 + H_2O" and "AP + O_3 + O_2 + H_2O."

7.4 CONCLUSION

Thus, the oxidation of apple pectin under the action of hydrogen peroxide and ozone-oxygen mixture takes place by the radical mechanism and is accompanied by oxidative destruction of its molecules. In the process of the destructive transformations of polysaccharide low molecular weight polymer products containing additional carboxyl groups as compared with the initial pectin are received. Varying the initial conditions of the process allow to regulate the degree of apple pectin destruction and functionalization of the formed polymer products. So it is quite important in creating new drugs of prolonged action.

The work was supported by RFBR (project 14-03-97026 Povolgie) and project (project code: 299, 2014), executable under the project part of the state task of the Ministry of education and science of the Russian Federation in the sphere of scientific activity.

KEYWORDS

- **apple pectin**
- **hydrogen peroxide**
- **kinetics**
- **oxidation**
- **oxidative destruction**
- **oxidative functionalization**
- **ozone-oxygen mixture**
- **radical mechanism**

REFERENCES

1. Kolmakova, N. P. Pectin and Its Application in Different Food Industries. Food Industry. 2003, № 6, 60–62 (in Rus.).
2. Vendillo, V. G., Emeliyanov, Yu. M., Filippov, Yu. V. Laboratory Setup for Ozone Production. Industrial Laboratory. 1959, V. 25. № 11, 1401–1402 (in Rus.).
3. Rafikov, S. R., Budtov, V. P., Monakov, Yu. B. Introduction to Physics and Chemistry of Polymer Solutions. "Nauka" ("Science," in Rus.) Publishing House. 1978, 328 p. (in Rus.).
4. Goronovskiy, I. T., Nazarenko, Yu. P., Nekryatch, E. F. Brief Guide to Chemistry. "Naukova Dumka" ("Scientific Thought," in Rus.) Publishing House. 1974, 985 p.
5. Round, A. N., Rigby, N. M., MacDougal, A. J., Ring, S. G., Morris, V. J. Investigating the Nature of Branching in Pectin by Atomic Force Microscopy and Carbohydrate Analysis. Carbohydrate Research. 2001, V. 331, 337–342.
6. Yoo, S.-H., Fishman, M. L., Hotchkiss, A. T., Lee, H. G. Viscometric Behavior of High-methoxy and Low-methoxy Pectin Solutions. Food Hydrocolloids. 2006, V. 20, 62–67.
7. Ovodov, Yu. S. Modern Considerations on Pectin Substances. Bioorganic Chemistry. 2009, V. 35. № 3, 293–310 (in Rus.).
8. Timerbaeva, G. R., Zimin, Yu. S., Borisov, I. M. Radical Nature of Pectin Oxidation in Water Medium. Russian Journal of Applied Chemistry. 2007, V. 80. № 11, 1890–1893 (in Rus.).
9. Zimin, Yu. S., Timerbaeva, G. R., Borisov, I. M. Kinetics of Citrus Pectin in Water Medium. "Izvestia Vuzov," Series "Chemistry and Chemical Technology." 2009, V. 52. № 4, 79–84 (in Rus.).
10. Roginskiy, V. A. Phenol Antioxidants. Reaction Ability and Effectiveness. "Nauka" ("Science," in Rus.) Publishing House. 1988, 247 p.
11. Kharitonov, Yu. Ya. Analytical Chemistry. V.1. "Vysshaya shkola" ("Higher School," in Rus) Publishing House. 2001, 615 p.

CHAPTER 8

ULTRATHIN POLY (3-HYDROXYBUTYRATE) FIBERS

S. G. KARPOVA,[1] A. L. IORDANSKII,[2] M. V. MOTYAKIN,[2] A. A. OL'KHOV,[2,3] O. V. STAROVEROVA,[2] S. M. LOMAKIN,[1] N. G. SHILKINA,[2] S. Z. ROGOVINA,[2] and A. A. BERLIN[2]

[1]*Emanuel Institute of Biochemical Physics, Russian Academy of Sciences, ul. Kosygina 4, Moscow, 119991 Russia*

[2]*Semenov Institute of Chemical Physics, Russian Academy of Sciences, ul. Kosygina 4, Moscow, 119991 Russia*

[3]*Plekhanov Russian University of Economics, Stremyannyi per. 36, Moscow, 117997 Russia*

CONTENTS

ABSTRACT

Structural-dynamic analysis based on combined thermophysical and molecular-mobility measurements via spin-probe ESR spectroscopy has been applied to films and fibrous matrixes based on poly(3-hydroxybutyrate). The dynamic behaviors of partially crystalline samples during deformation under conditions of electrospinning and cold rolling have been compared. The comparative results of the complex investigation of films and ultrathin fibers in the poly(3-hydro-xybutyrate) matrix have shown that the electromechanical action leads to additional crystallization of the crystalline regions, spherulites, and lamellas in the polymer. The changes in the crystalline phase of the polymer are accompanied by an increase in the packing density of macromolecules in the intercrystalline space. With the use of the spin-probe ESR method, the effect of water and the oxidant ozone on the morphology of the amorphous phase of poly(3-hydroxybutyrate) ultrathin fibers has been determined. The measurements of the dynamics of spin-probe rotation in samples before and after cold rolling have shown that the additional orientation of poly(3-hydroxybutyrate) spherulites in a mechanical field results in stabilization of amorphous regions more resistant to the aggressive effects of ozone.

8.1 INTRODUCTION

Ultrathin fibrillar structures (fibers, fibrils, nets, and porous fibrous matrixes) are of great interest as advanced functional materials with specific properties owing to the high surface–volume ratio for an individual filament, the capacity for surface modification, and the controllable mechanical characteristics and diffusion transport features [1–3]. At present, ultrathin fibers and the related products have found wide application in biomedicine, cell engineering, separation and filtration processes, the production of reinforced composites, electronics, analytics, sensor diagnostics, and a number of other innovative areas [4–6].

The formation of fibers and filaments of submicron and nano diameters via conventional spinning technologies is accompanied by significant difficulties [7].

At present, the alternative to ultrathin-fiber formation is an original technology that is based on electrospinning and allows the production of fibers and related unwoven materials. In regards to the equipment, electrospinning is a technologically simple, but multiparameter process of formation of micro- and nano fibers 10 nm–10 μm in diameter [8]. The electrospinning process is based on the combined action of mechanical and electrostatic forces applied to a polymer solution or a melt oriented in an electric field [9]; thus, some parameters—such as the potential at the electrodes, the distance between the electrodes, conductivity, and the density and viscosity of the spinning polymer system—affect the morphology, surface properties, functionality, porosity, and geometry of fibers [10].

In previous works [11–13], for the examples of the natural polyester poly(3-hydroxybutyrate) (PHB) and a number of its compositions with other polymers, the physicochemical, dynamic, and transport characteristics of macroscopic biodegradable matrixes and microparticles of PHB that are prospective for prolonged and controlled drug delivery were studied [14].

The high biocompatibility, controlled biodegradation, and satisfactory mechanical characteristics make it possible to consider this polymer one of the most promising for biomedical use. PHB is not only used therapeutically but also widely applied for the development of bone implants, nerve fiber connections, matrixes for cell and tissue cultivation, and filters and ultrafiltration membranes as well as for the design of artificial heart elements and vascular prostheses [11, 15].

At present, there are a number of methods for obtaining oriented polymer structures: for example, blow molding, hot molding, compression stretching, and die extrusion of fibers [16]. These methods are widely used in manufacturing; however, sometimes they are accompanied by an undesirable effect, such as the formation of micro- and macrocavitation defects, which deteriorate the diffusion and strength characteristics of fibers and films [17]. Fibers prepared via electro-spinning are less prone to defect formation.

Moreover, to improve the orientation of ultrathin fibers, the cold rolling technique (related to compression methods), which minimizes the formation of defect voids in the fibrous matrix, has been employed [18]. In spite of some limitations, the above method has a low energy consumption and, hence, remains among most economical processes of polymer-material production.

The transfer from film and matrix systems to fibers and related unwoven materials is accompanied by significant changes in their physicochemical and transport characteristics, a circumstance that is related to the space limitations arising in ultrathin fibers and affecting the dynamics and set of macromolecular conformations [19].

In this study, some structural–dynamic characteristics of PHB matrixes formed from ultrathin fibers were compared at the molecular level to reveal features of the diffusion mobility and degradation processes occurring during space limitations. With the use of spin-probe ESR spectroscopy, the influence of water and the oxidant ozone on the structure of the fibrous PHB matrix was demonstrated, a circumstance that made it possible to estimate the changes in the segmental mobility of macromolecules at the early stages of their interaction with aggressive media.

8.2 EXPERIMENTAL PART

For film preparation, the natural biodegradable polymer PHB 16F (Biomer®, Germany), obtained via microbiological synthesis, was used. The pristine polymer is a white finely dispersed powder with a particle size from 5 to 20 μm. The viscosity-average molecular mass of the polymer was $M = 2.06\times10^5$, the density was $\rho = 1.248$ g/cm^3, the melting temperature was $T_m = 177$°C, and the crystallinity was 65%. Ultrathin fibers of PHB were prepared via electrospinning [20]. The fibers were formed from 5 and 7% PHB solutions in chloroform. The process of spinning ultrathin fibers of PHB was described in more detail in [21]. The physical modification of the structure of PHB-based unwoven materials was performed via rolling between two metal rollers rotating at a rate of 30 rpm and providing a pressure of 20 MPa at 22 ± 1°C (so-called "cold" rolling [22]). As a result of the rolling, the effective thickness of the fibrous matrix was reduced from 120 ± 15 to 95 ± 10 μm.

The diameters of PHB fibers and their size distribution were studied via electron microscopy on a Hitachi TM-1000 scanning electron microscope (Japan). The statistical analysis of the fiber-diameter distribution was performed via mathematical treatment of images recorded with the scanning electron microscope.

The molecular mobility was estimated via the spin-probe technique on an EPR-V automated EPR spectrometer (Semenov Institute of Chemical Physics, RAS, Moscow). As a probe, the stable nitroxyl radical 2,2,6,6-tetramethylpiperi-din-1-oxyl (TEMPO) was used. The radical was incorporated into films from vapor at 60°C to a concentration not exceeding 10^{-3} mol/L. The ESR spectra were recorded in the absence of saturation, a condition that was verified by the dependence of the signal intensity on the microwave field power. The correlation times of probe rotation, τ, were calculated from the ESR spectra according to the following formula [23]:

$$\tau = 6.65 \times 10^{-10} \Delta H^{+} \left(\sqrt{I^{+} / I^{-}} - 1 \right)$$

where ΔH^{+} is the width of the low-field component of the spectrum and I^{+}/I^{-} is the ratio of intensities of the low- and high-field components. The measurement error for τ was ±7%.

The ozone oxidation of samples was conducted in an ozone–oxygen mixture with a partial ozone concentration of 5×10^{-4} mol/L. The DSC measurements were performed on a Netzsch DSC 204 F1 calorimeter in argon at a heating rate of 10 K/min. The average measurement error for heat effects was approximately ±3%.

8.3 RESULTS AND DISCUSSION

During electrospinning, the characteristics of the initial solution have an effect not only on the morphology of the resulting ultrathin fiber but also on such an important parameter as the cross section (Figure 8.1). The microscopic images show the change in the fiber diameter distribution and the position of its maximum with a relatively weakly increase in the PHB concentration in the spinning solution from 5 to 7 wt. %. The requirements on the fiber geometry may vary with the application conditions. The development of sensors—for example, fiber-optic sensors—requires, as a rule, a decrease in the fiber diameter to the nano level to improve the sensitivity, whereas, in the case of biomedical fiber materials used, in

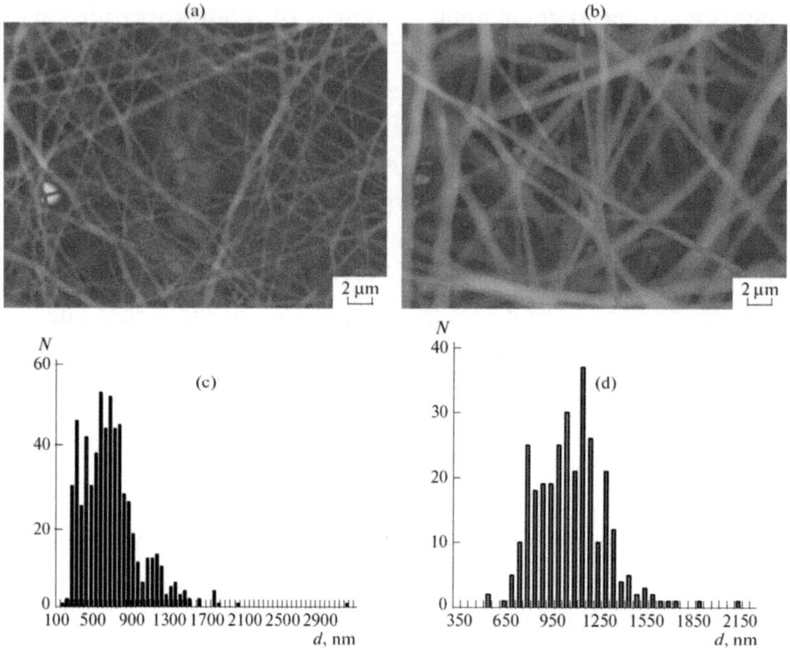

FIGURE 8.1 (a,b) SEM images of PHB fibers prepared via electrospinning of (a) 5 and (b) 7% polymer solutions; (c,d) the corresponding fiber-diameter distributions.

particular, as matrixes for cell engineering, nano-fibers result in deterioration of cell adhesion and a reduced cell-growth rate [24].

The possibility to control ultrathin-fiber geometry through variation in the physicochemical characteristics of a solution has been described in a number of recent works [25]. In our case, a 7% PHB spinning solution in chloroform was chosen as a basic variant mainly because of the good reproducibility of the physicochemical characteristics and the geometry of fibers prepared from this solution. As was shown in our previous work, matrixes based on fibers ~1 μm in diameter have demonstrated high biocompatibility relative to that of fibers of smaller diameters [24]. On the basis of the above results, further study was performed with matrix fiber systems characterized by a size distribution maximum near 1 μm.

The results of measurement of segmental mobility were considered along with DSC measurements of the temperature transitions of PHB. With the use of this method, the regions of melting of the crystalline phase in fiber samples were determined.

The melting of PHB films was studied in previous studies [26–28]. The DSC curves for all samples before and after rolling have similar patterns, but their melting maxima fall in different temperature ranges (Figure 8.2). The discrepancy in the melting temperatures becomes more pronounced for samples subjected to a heating–cooling cycle (second temperature scan). First, for all samples after the repeated heating, the single melting

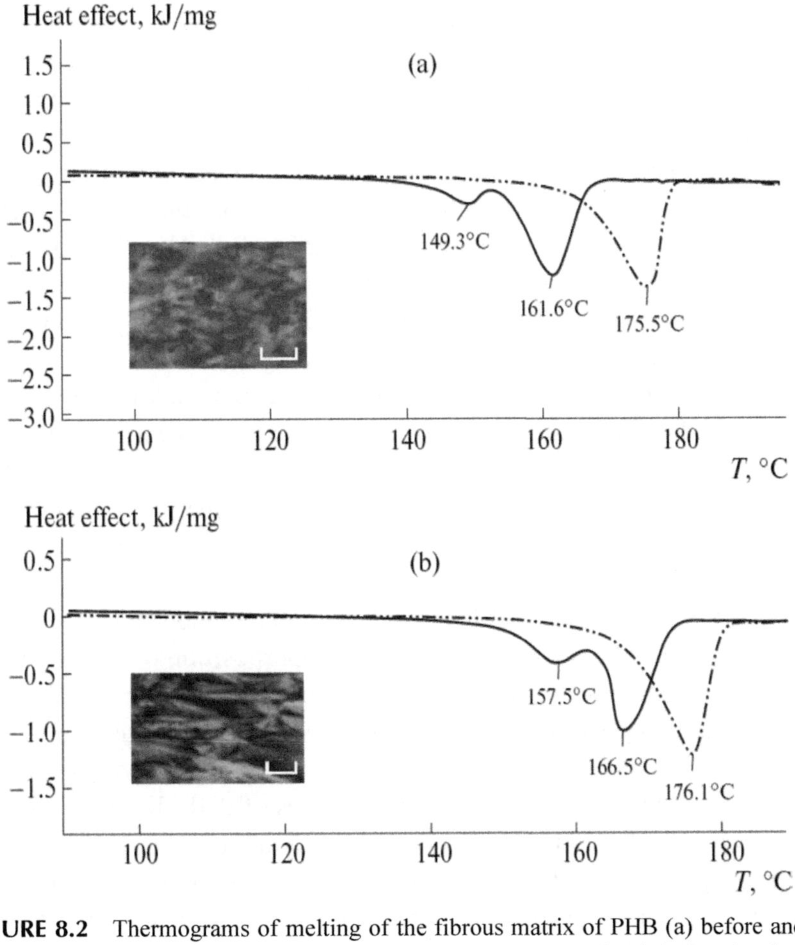

FIGURE 8.2 Thermograms of melting of the fibrous matrix of PHB (a) before and (b) after mechanical treatment via rolling: The single maximum to the right is related to the first temperature scan; the binary maximum corresponds to the second temperature scan. Inserts: SEM images of crystalline regions of PHBs in the form of (a) a fibrous matrix, and (b) a fibrous matrix adter cold rolling. The markers indicate 0.2 μm.

maxima change to maxima of the bimodal shape at 161.6 and 149.3°C for samples before rolling and at 166.5 and 157.5°C for the same samples after rolling. Second, the high-temperature melting peaks during the second heating decrease by more than 10°C, a phenomenon that indicates less than perfect organization of the crystalline phase of samples that is due to their melt cooling to room temperature.

The thermograms of PHB after rolling show higher low and high temperature maxima of melting (see above) with no change in the high crystallinity of samples (~79%). The basic conclusions made from the quantitative estimation of thermograms of PHB fibrous matrix melting are as follows. The mechanical treatment via rolling and single heating results in the bimodal separation of the crystalline phase into closer to and further from perfect crystalline regions, for example, regions with different numbers of defects. The same mechanical action favors the improvement of the crystalline structure, a circumstance that is reflected by higher melting maxima on the DSC curves. The presence of two crystalline fractions may lead to the heterogeneity of the amorphous intercrystalline phase of PHB, which is denser and more ordered or less ordered.

The latter conclusion is less evident and requires additional corroboration, which will be attained via microprobe ESR spectroscopy, which makes it possible to estimate both the dynamic characteristics of the motion of PHB segments and the differences in density of its amorphous phase considered as a combination of more ordered and less ordered regions [29].

The ESR spectra of the TEMPO radical in a film and in a matrix formed on the basis of PHB ultrathin fibers is the superposition of two individual spectra of radicals with different correlation times τ_1 and τ_2 (Figure 8.3). Here, τ_1 and τ_2 reflect the segmental mobilities in amorphous regions with higher density and lower density, respectively. The presence of at least two correlation times in the amorphous regions of PHB fibrous matrixes is indicative of the heterophasic structure of intercrystalline polymer regions and is in agreement with the modern model of the bimodal structure of the amorphous phase of partially crystalline polymers, such as PHB, polylactide, and PET [27].

The ESR spectra were quantitatively processed with the use of the equation as well as via calculation of the intensity ratios of low-field peaks

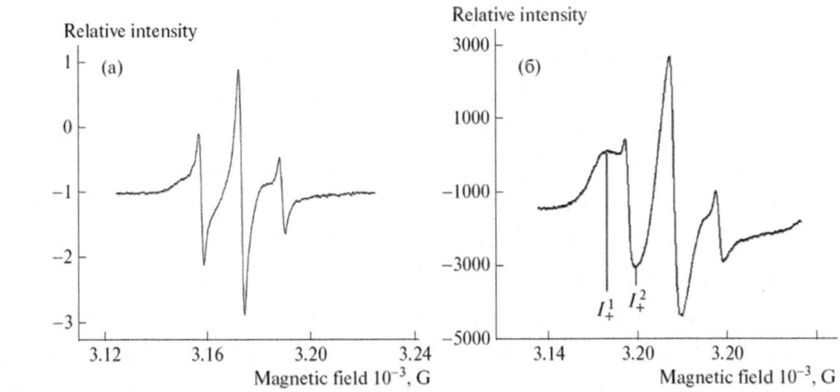

FIGURE 8.3 ESR spectra of TEMPO in (a) the PHB film and (b) the fibrous matrix at 25°C that demonstrate the superposition of two simple spectra of radicals with different correlation times (rapid and slow modes).

due to radicals with a retarded-motion mode and a more rapid motion mode, and, respectively (Figure 8.3). The distinctions between the two states of the radical in the PHB film are more pronounced than those in the fiber system, as exemplified by the comparison of ratios. In the film, this ratio is higher than that in the fiber (0.52 and 0.37). Moreover, the effective correlation times corresponding to the rotational mobility of the radical in the fiber before and after rolling (0.92×10^{-9} and 1.8×10^{-9} s, respectively) are significantly lower than the analogous characteristic in the film (4.5×10^{-9} s), an outcome that suggests more rapid rotation of the radical in the amorphous phase of PHB fibers.

In addition to the dynamic measurements, the concentrations of the radical sorbed in identical weighed portions of film, the initial fiber, and the fiber after cold rolling were determined experimentally. The calculations based on the integration of the ESR spectra showed that the above concentrations were as high as 4.6×10^{18} spin/g (9.5×10^{-4} mol/L) in the film, 2.9×10^{18} spin/g (6.0×10^{-4} mol/L) in the initial film, and 2.6×10^{18} spin/g (5.4×10^{-4} mol/L) in the fiber after rolling. With consideration for the fact that the radical sorption was conducted under identical conditions at a lower PHB crystallinity (~79%), these results are indicative of a decrease in the specific volume of the polymer capable of accommodating the radical, for example, an increase in the fraction of the denser amorphous phase of fibers due to cold rolling. Thus, the comparative analysis of the ESR spectra showed that the pristine samples in the form of a continuous matrix or a matrix formed by fibers substantially differ in the rotational mobility of the radical and the organization of the intercrystalline phase in PHB films and fibers. The above effect supposedly arises as a result of the turning of spherulites and their orientation along the stretching (rolling) direction (micrographs in Figure 8.2), a situation that affects the intercrystalline space without significant changes in the polymer crystallinity.

The temperature dependences of the correlation time in conventional semilogarithmic coordinates allowed the determination of the activation energy of the rotational mobility of the radical in the studied polymer samples (Figure 8.4a). The related calculations showed that activation energies in films, fibers, and fibers after rolling are 18, 42, and 55 kJ/mol, respectively. The successive growth of activation energy after moving from the film to the oriented fiber matrix suggests that there is an increasing energy barrier during radical rotation owing to the hindrance of the segmental mobility in dense regions of the amorphous phase of the polymer in the above-given sequence during consideration of the radical concentration: in the film < in the fiber < in the fiber after rolling.

The temperature dependence of the correlation time for the fiber matrix after rolling demonstrated anomalous behavior (Figure 4b) observed in the moderately high temperature range from 72°C or lower as a deviation from the semilogarithmic dependence. This phenomenon may be explained by the bimodal distribution of the radical in the

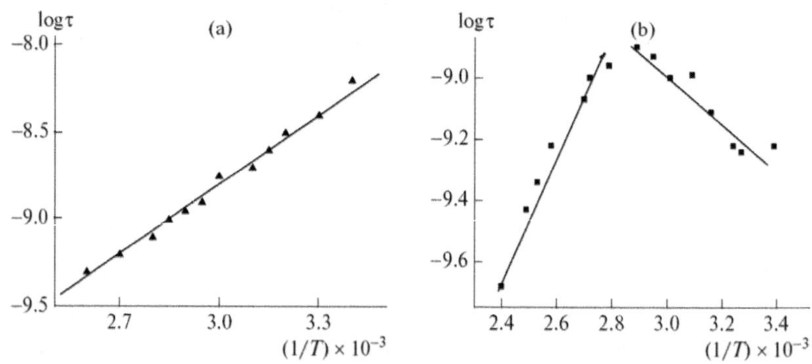

FIGURE 8.4 Dependences of correlation time on inverse temperature for (a) the fibrous matrix and (b) the same matrix after cold rolling.

partially oriented amorphous phase and by the related difference in the rates of radical rotation. The changing rate of radical rotation results from the redistribution of dense and less dense regions of the amorphous phase of PHB. The observed decrease in the probe mobility is probably related to the increase in the fraction of the dense amorphous phase, a finding that correlates with the measurements of heat effects via the DSC method [28]: During heating of PHB samples, the segmental mobility and the probability of a denser packing of macro-molecules in the intercrystalline space increase simultaneously. At temperatures above room temperature but below 70°C, the content of the dense amorphous phase increases, resulting in reduced effective mobility of the radical in PHB.

Note that, in a thorough study by M.L. Di Lorenzo, et al. [27], it was shown that, during the heating of PHB samples accompanied by cold crystallization (a broad exothermic peak above 60°C), regions of a denser amorphous phase characterized by low mobility and significant segmental orientation are formed. In our case, during dynamic ESR measurements, this transition is observed at ~70°C or below (it corresponds to the beginning of the anomalous portion on the curve of inverse temperatures (Figure 8.4)), for example, in the temperature range preceding cold crystallization.

The effect of water, which is known to be a moderate plasticizer of PHB [30], first results in an increase in macromolecular mobility and then favors the formation of more dense regions, where the mobility of the radical decreases. Figure 8.5 shows the dependences of effective correlation times on the times of contact of the fibrous matrix with water at 40°C, for example, in the physiological temperature range.

As can be seen from Figure 8.5, the correlation time of rotational diffusion of the radical (hereinafter, it will be called the *correlation time* for short) increases with the time of contact of the fiber with water, a result that suggests a reduced segmental mobility of PHB, most probably, due to the same process described in the preceding series of temperature-dependence measurements. During the effect of water, the segmental mobility grows, thereby resulting in easier formation of ordered structures, as in the case of the preliminary crystallization of plasticized polymers. Finally, the fraction of the ordered denser

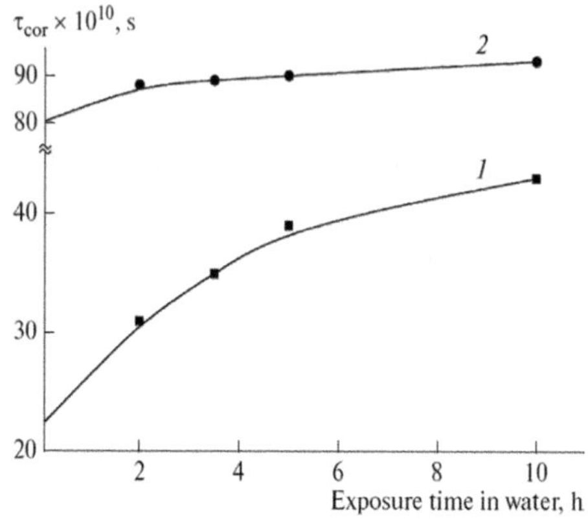

FIGURE 8.5 Correlation time vs. time of contact with the aqueous medium: (*1*) pristine fibrous matrix and (*2*) the same matrix after mechanical treatment.

amorphous phase of PHB increases. By comparing the correlation times in fibrous matrixes before and after rolling, we can easily see that, as a consequence of the rolling, the correlation times of oriented samples increase, for example, the segmental mobility decreases (Figure 8.5). The fiber orientation in the course of rolling is in fact responsible for the depleted set of extended-chain conformations, a circumstance that results in a drop in the segmental mobility and, hence, increased τ values for treated fibers.

The behavior of the PHB samples in ozone, which is a strong oxidant affectting the structure and mobility of PHB molecules, was estimated via investigation of the rotational dynamics of the TEMPO probe. Figure 8.6 shows the dependence of correlation time on the time of ozone oxidation. Within the first hour, a sharp drop of τ values occurs, and then (up to 4 h later) the process is strongly retarded and the plot of τ versus oxidation time flattens out. As the action of ozone continues, the PHB samples demonstrate a further increase in segmental mobility (a decrease in correlation time). Note that the dynamics of the rotation of the radical in pristine fibrous PHB samples and in the same samples after rolling remains the same.

During the action of moisture and temperature, the samples modified via rolling demonstrated lower mobility of the radical at all ozonization times, as previously. On the basis of the preceding measurements of the

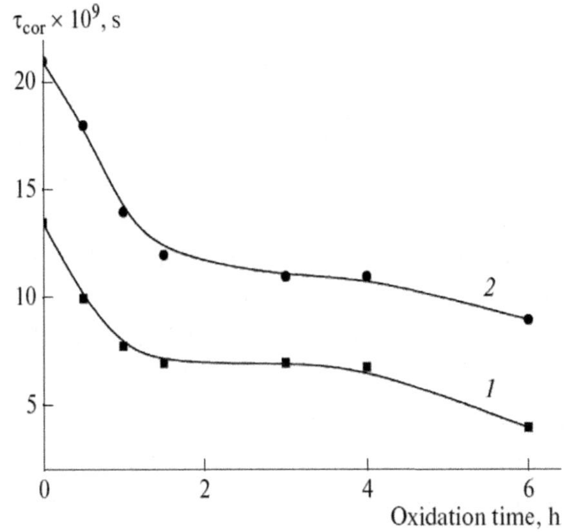

FIGURE 8.6 Correlation time of TEMPO vs. time of ozone oxidation in: (*1*) the pristine fibrous matrix and (*2*) the same matrix after cold rolling.

crystalline structure and probe mobility in ozonized PHB samples [30], it may be proposed that, at the initial stage of oxidation, ozone initiates the partial degradation of macrochains and, consequently, their molecular mobility increases. At this stage, the most accessible and imperfect polymer molecules, which are located in less dense regions of the amorphous phase, participate in the reaction. With their oxidation completed, the process is stabilized, a circumstance that is reflected by the 2- to 3-h-long induction period (Figure 8.6). Further oxidation with ozone proceeds more intensely and involves not only the "loose" amorphous region of the polymer but also the more ordered dense amorphous zones.

Note that, in fibers after rolling, the jump-wise increase of mobility is less pronounced than that of the initial fibers. With consideration for the above results, this effect makes it possible to assume that the rolling leads to the formation of denser amorphous regions in PHB fibers and thereby improves their stability against oxidation by ozone.

Thus, the structural–dynamic analysis combining DSC measurements and molecular-mobility determination via the spin-probe method has shown that, during the action of an external mechanical field, not only a change in the morphology of PHB and turning of its spherulites occur but also, according to the ESR data, the redistribution of the radical involving well-oriented chains in the amorphous region of polymer is observed.

According to the DSC data, the changes in the crystalline phase are related to the redistribution of macromolecular segments in the amorphous phase. To allow for molecular mobility, a model of the binary distribution of segments in less and more dense intercrystalline regions responsible for more and less rapid rotation of the TEMPO radical, respectively, was used [30]. The action of temperature and the exposure time in an aqueous medium results in redistribution of the ratio between the "loose" phase and the dense phase of PHB in favor of the dense phase, a circumstance that is reflected in the rotational dynamics of the probe encapsulated in ultrathin PHB fibers. The oxidation of fibrous matrixes of PHB via ozonization occurs in three stages: the reaction of ozone with macromolecules in the "loose" region; a latent period of 2–3 h without a significant change in the segmental mobility; and the reaction of ozone in denser regions of the polymer amorphous phase, which results in higher rotational mobility of the ESR probe. The comparative measurements of the rotational dynamics of the probe before and after rolling showed that the additional orientation of the PHB spherulites in a mechanical field leads to stabilization of the amorphous regions more resistant to the aggressive effect of ozone.

The proposed approach to the investigation of ultrathin fibers in the PHB matrix makes it possible to analyze the changes in the segmental mobility of macromolecules at early stages of their interaction with aqueous and oxidative aggressive media.

ACKNOWLEDGMENTS

This work was supported by the Russian Foundation for Basic Research (project no. 14–03–00405-a) and the Program of Basic Research of the Division of Chemistry and Materials Sciences, Russian Academy of Sciences, "Development of Macromolecular Structures of a New Generation" (OKhNM-1–2014).

KEYWORDS

- **crystalline phase**
- **molecular-mobility**
- **poly(3-hydroxybutyrate)**
- **stabilization**
- **thermophysical**
- **ultrathin fibers**

REFERENCES

1. Greiner, A., Wendorff, J. H. Adv. Polym. Sci. 219, 107 (2008).
2. Agarwal, S., Greiner, A., Wendorff, J. H. Prog. Polym. Sci. 38 (6), 963 (2013).
3. Kim, I. D. Macromol. Mater. Eng 298, 473 (2013).
4. Sill, T. J., von Recum, H. A. Biomaterials 29, 1989 (2008).
5. Filatov, Y., Budyka, A., Kirichenko, V. *Electrospinning of Microand Nanofibers: Fundamentals in Separation and Filtration Processes* (Begell House Inc., New York, 2007).
6. Banica, F. G. *Chemical Sensors. Biosensors, Fundamentals and Applications* (Wiley, New York, 2012).
7. Kulkarni, A., Bambole, V. A., Mahanwar, P. A. Polym. Plast. Technol. Eng. 49 (5), 427 (2010).
8. Dvir, T., Timko, B. P., Kohane, D. S., Langer, R. Nature Nano-technol. 6 (1), 13 (2011).
9. Raghavan, P., Lim, D. H., Ahn, J. H. Nah, Ch., Sherrington, D. C., Ryu, H. S., Ahn, H. J. React. Funct. Polym. 72, 915 (2012).
10. Baji, A., Mai, Y. W., Wong, S. C., Abtahi, M., Chen, P. Comp. Sci. Technol. 70, 703 (2010).
11. Iordanskii, A. L., Bonartseva, G. A., Yu. Pankova, N., Rogovina, S. Z., Gumargalieva, K. Z., Zaikov, G. E., Berlin, A. A., *Current State-of-the Art on Novel Materials*, Ed. by D. Balkцse, Horak, D., L. Šoltй (Apple Academic Press, New York, 2014), Vol. 1, Chap. 12, p. 122.
12. Karpova, S. G., Iordanskii, A. L., Chvalun, S. N., Shcherbina, M. A., Lomakin, S. M., Shilkina, N. G., Rogovina, S. Z., Markin, V. S., Popov, A. A., Berlin, A. A., Dokl. Phys. Chem. 446 (2), 176 (2012).
13. Bonartsev, A. P., Livshits, V. A., Makhina, T. A., Myshkina, V. L., Bonartseva, G. A., Iordanskii, A. L., eXPRESS Polym. Lett. 1 (12), 797 (2007).
14. Ivantsova, E. L., R. Yu. Kosenko, Iordanskii, A. L., Rogovina, S. Z., Prut, E. V., Filatova, A. G., Gumargalieva, K. Z., Berlin, A. A., Novikova, S. P., Polym. Sci., Ser. A 54 (2), 87 (2012).

15. Ueda, H., Tabata, Y., Adv. Drug Delivery Rev. 55, 501 (2003). 138.
16. Kulezneva, V. N. Foundations of Plastic Processing Technology (Khimiya, Moscow, 1995) [in Russian].
17. Galeski, A., Prog. Polym. Sci. 28, 1643, 2003.
18. Pluta, M., Bartczak, Z., Galeski, A., Polymer 41, 2271, 2000.
19. Ma, Q., Mao, B., Cebe, P., Polymer 52, 3190, 2011.
20. Filatov, Yu. N. *Electrospunning of Fibrous Materials (ESF processes)*, Ed. by V. N. Kirichenko (Neft' i gaz, Moscow, 1997) [in Russian].
21. Staroverova, O. V., A. A. Ol'khov, G. M. Kuz'micheva, Domoroshchina, E. N., Vlasov, S. V., Yu. Filatov, N., Vestn. MITKHT 6 (6), 120 (2011).
22. Vlasov, S. V., A. A. Ol'khov, Plast. Massy, No. 6, 40 (1996).
23. Buchachenko, A. L., Vasserman, A. M., *Stable Radicals* (Khimiya, Moscow, 1973).
24. Staroverova, O. V., Shushkevich, A. M., G. M. Kuz'micheva, A. A. Ol'khov, Voinova, V. V., A. A. Ol'khov, Zharkova, I. I., Shaitan, K. V., Sklyanchuk, E. D., V. V. Gur'ev, Tekhnol. Zhivykh Sist. 10 (8), 74 (2013).
25. Rutledge, G. C., Fridrikh, S. V., Adv. Drug Delivery Rev. 59 (14), 1384 (2007).
26. Kamaev, P. P., Aliev, I. I., Iordanskii, A. L., Wasserman, A. M., Polymer 42 (2), 515 (2000).
27. Di Lorenzo, M. L., Gazzano, M., Righetti, M. C., Macromolecules 45, 5684 (2012).
28. Di Lorenzo, M. L., Righetti, M. C., J. Therm. Anal. Calorim. 112 (3), 1439 (2013).
29. Iordanskii, A. L., Kamaev, P. P., Polym. Sci., Ser. B 40 (1–2), 8 (1998).
30. Karpova, S. G., Iordanskii, A. L., Popov, A. A., Shilkina, N. G., Lomakin, S. M., Shcherbin, M. A., Chvalun, S. N., Berlin, A. A., Russ. J. Phys. Chem. B 6 (1), 72 (2012).

CHAPTER 9

ACTIVATED CARBON NANOFIBERS IN FUEL CELLS: AN ENGINEERING PERSPECTIVE

SAEEDEH RAFIEI, MARYAM ZIAEI, and A. K. HAGHI

University of Guilan, Rasht, Iran

CONTENTS

9.1 INTRODUCTION TO HYDROGEN STORAGE AND FUEL CELL SYSTEMS

Fuel cells are an electrochemical energy conversion technology to directly convert the chemical energy of fuels, such as hydrogen, methanol and natural gas, to electrical energy, which was invented by Sir William Grove [1]. They inherently have a significantly higher efficiency than that of conventional energy conversion technologies, such as an internal combustion engine (ICE). Fuel cells are considered to be the most efficient and least polluting power-creating technology and are a potential and practical candidate to moderate the fast increase in power requirements and to minimize the impact of the increased power consumption on the environment.

During last decades, fuel cells have received enormous attention worldwide as suitable electrical energy conversion systems because of their huge potential for power generation in different aspects of industry. One of the most important causes that have influenced the development of fuel cell technologies recently, is the increased public awareness of the finite supplies of fossil fuels and the environmental consequences of the increasing consumption of fossil fuels in electricity production and transportation section. The fuel cells operation is basis on natural gas storage specially hydrogen.

The preferred fuel for a fuel cell is hydrogen due to unique properties of it, such as nontoxicity, plentifulness and odorless, colorless, tasteless nature of hydrogen. On the other hand, most fuel cells can work only with fairly pure hydrogen, and those that can use other fuels still typically work best with hydrogen.

The importance of hydrogen as a future energy carrier is generally recognized. The advantages of a hydrogen-based economy would be its sustainable and environmentally friendly character. Hydrogen can be produced from renewable energy sources and then be conveniently converted into electricity, particularly through the use of fuel cell technology [2–3]. Apart from factors concerning the energy-efficient production of hydrogen as well as cost-effective use of hydrogen in fuel cells, the application of hydrogen as an energy carrier is limited by hydrogen storage problems Storage of hydrogen is expensive, heavy, takes too much volume and unsafe. Therefore, discovering materials with capability of hydrogen storage is necessary field for worldwide research [4–5].

Storage of hydrogen in new carbon nanomaterials such as activated carbons (AC), carbon nanotubes (CNT), carbon nanofibers (CNF) and activated carbon nanofibers (ACNF) attracted global interest due to their porosity characteristics, low cost and low weight. It has been proposed that hydrogen can be absorbed onto carbon materials by physisorption and/or chemisorption. Physisorption occurs when hydrogen maintains its molecular structure and "is trapped" in the carbon material by van der Waals forces. In chemisorption, atoms of hydrogen create chemical bonds with the carbon [2, 5–7].

CNFs are non-hazardous, processed, and carbonaceous materials, having a porous structure, a high adsorption capacity and rate, and a large specific surface area. The advantages of ACNFs are smaller fiber diameter, more concentrated pore size distribution and excellent adsorption capacity at low concentration of absorbents compared with conventional activated carbons. The porosity of ACNFs is developed during activation and it is influenced by many factors, such as the degree of activation and the conditions used for carbonization [8]. As it mentioned above, hydrogen can be converted into electricity energy by applying fuel cells, so fuel cells history and operation will be discussed in subsequent sections.

This chapter presents a general introduction about hydrogen storage in fuel cell technologies. The general issues related to the various types of fuel cells will also be briefly discussed. However, specific challenges and advances in various applications of CNFs and CNTs in fuel cells will be critically reviewed and discussed.

9.2 DIFFERENT TYPES OF FUEL CELLS

Several types of fuel cells have evolved in the past decades. They are called after their electrolyte, the substance that transports the ions. The electrolyte dictates the operating temperature of a fuel cell type. Depending on the operating temperature, a specific catalyst is chosen to oxidize the fuel. Various types of fuel cells operate a bit differently. But in general terms, hydrogen atoms enter a fuel cell at the anode where a chemical reaction strips them of their electrons.

The hydrogen atoms are now "ionized," and carry a positive electrical charge. The negatively charged electrons provide the current through wires to do work. If alternating current (AC) is needed, the DC output of the fuel cell must be routed through a conversion device called an inverter. Some cells need pure hydrogen, and therefore demand extra equipment such as a "reformer" to purify the fuel. Other cells can tolerate some impurities, but might need higher temperatures to run efficiently. Liquid electrolytes circulate in some cells, which require pumps. Most important types are described briefly as below:

9.2.1 ALKALI FUEL CELLS (AFC)

In this kind of cells a solution of KOH in water is used as electrolyte. Efficiency is about 70%, and operating temperature is 150–200°C. They were used in Apollo spacecraft to provide both electricity and drinking water and they require pure hydrogen fuel.

9.2.2 MOLTEN CARBONATE FUEL CELLS (MCFC)

They apply high-temperature compounds of salt carbonates as the electrolyte. Their efficiency ranges from 60 to 80%, and operating temperature is about 650°C. Their high operation temperature and expensive nickel electrodes make them improper for common usages.

9.2.3 PHOSPHORIC ACID FUEL CELLS (PAFC)

They use phosphoric acid as the electrolyte. Efficiency ranges from 40 to 80%, and operating temperature is between 150 to 200°C. Platinum electrode-catalysts are needed, and internal parts must be able to withstand the corrosive acid.

9.2.4 PROTON EXCHANGE MEMBRANE FUEL CELLS (PEM)

They work with a polymer electrolyte in the form of a thin, permeable sheet. Efficiency is about 40–50%, and operating temperature is about 80°C. The solid, flexible electrolyte will not leak or crack and these cells operate at a low enough temperature to make them suitable for homes and cars. But their fuels must be purified, and a platinum catalyst is used on both sides of the membrane, raising costs.

9.2.5 SOLID OXIDE FUEL CELLS (SOFC)

They use a hard, ceramic compound of metal oxides as electrolyte. Efficiency is about 60%, and operating temperatures are about 1,000°C. The high temperature limits applications of SOFC units and they tend to be rather large. While solid electrolytes cannot leak, they can crack. The operation of PEM and SOFC fuel cells are illustrated in Appendix C.

9.2.6 ADVANTAGES OF DIFFERENT KIND OF FUEL CELLS

9.2.6.1 Molten Carbonate Fuel Cell (MCFC)

Advantages

- They allow spontaneous internal reforming fuel.
- Generate a lot of heat.
- High-speed reactions.
- High efficiency.
- No need for noble metal catalyst (cost reduction).

Disadvantages

- For further development, they need to be designed using materials resistant to corrosion and dimensionally stable and resistant. The catalyst of nickel oxide cathode can be dissolved in the electrolyte, causing a malfunction. Dimensional instability can cause distortion, changing the active area of the electrodes.
- They have a high intolerance to sulfur. In particular, the anode does not tolerate more than 1.5 ppm of sulfur particles in the fuel. Otherwise, the fuel cell will suffer a significant deterioration in their functioning.
- They have a liquid electrolyte, with the corresponding handling problems.
- They require preheating before starting work.

9.2.6.2 Solid Oxide Fuel Cell (SOFC)

Advantages

- They allow spontaneous internal reforming fuel. Because the oxide ions travel through the electrolyte, fuel cell can be used to oxidize any combustible gas.
- Generate a lot of heat.
- Chemical reactions are very fast.
- They have a high efficiency.
- You can work at current densities higher than molten carbonate fuel cells.
- The electrolyte is solid. Avoids the problems of liquid handling.
- No need noble metal catalysts.

Disadvantages

- For total market penetration, they need to develop materials that have sufficient conductivity, which remain solid at temperatures of operation, which are chemically compatible with other components of the cell, which are dimensionally stable and have high resistance.
- They have a moderate intolerance to sulfur (50 ppm).
- It is not a mature technology.

9.2.6.3 Alkaline Fuel Cell (AFC)

Advantages

- They can work at low temperature.
- They have a fast start.
- They have a high efficiency.
- They use very little amount of catalyst, and thereby lowers costs.
- They do not have corrosion problems.
- They have a simple operation.
- They have low weight and volume.

Disadvantages

- They are extremely intolerant to CO_2 (up to 350 ppm) and show certain intolerance to CO. This limits both the type of oxidant and fuel. Oxidant must be pure oxygen or air free of CO_2. The fuel must be pure hydrogen.

- Employ a liquid electrolyte, resulting in handling problems.
- They require an evacuation of the water treatment complex.
- They have a relatively short lifetime.

9.2.6.4 Proton Exchange Membrane Fuel Cell (PEMFC)

Advantages

- Thanks to the separator of anode and cathode is a solid polymer film (planar structure) and that the cell operates at relatively low temperatures, aspects such as handling, assembly or tightness are less complex than in most other types of cells.
- They use a non-corrosive electrolyte. They remove the need to handle acid or any other corrosive, increasing security.
- They are tolerant of CO_2; so they can use the atmospheric air.
- They employ a solid and dry electrolyte so it eliminates the handling of liquids and the problems of resupply.
- They have high voltage, current and power density.
- They can work at low pressure (1 or 2 bars), which adds security.
- They have a good tolerance to the difference of pressure of the reactants.
- They are compact and robust.
- They have a simple mechanical design.
- They use stable building materials.

Disadvantages

- They are very sensitive to impurities of hydrogen, which have developed a number of reforming units to be able to use conventional fuels such fuel cells. PEM fuel cells that directly use methanol as fuel without being reforming are a variant of the direct methanol fuel cell (DMFC).
- They do not tolerate more than 50 ppm of CO and have a low tolerance to sulfur particles.
- They need humidification units of reactive gases. If water is used for humidification of gases, the operating temperature of the fuel cell must be less than the boiling water, restricting the potential for cogeneration.

- They use a catalyst (platinum) and a membrane (solid polymer) very expensive.

9.2.6.5 Phosphoric Acid Fuel Cell (PAFC)

Advantages

- They tolerate up to 30% CO_2, therefore, these fuel cells may use air directly from the atmosphere.
- While working at medium temperature, they can use the waste heat for cogeneration.
- They use an electrolyte with stable characteristics, low volatility even for temperatures above 200°C.

Disadvantages

- They have a maximum tolerance of 2% CO.
- They utilize liquid electrolyte, witch is corrosive to average temperatures, which involves handling and safety problems.
- They allow the entry of water that can dilute the acid electrolyte.
- They are big and heavy.
- They cannot auto reform fuel.
- They need to reach a certain temperature before starting to work, for example, they have an operating temperature.

9.2.6.6 Direct Methanol Fuel Cell (DMFC)

Advantages

- They use a liquid fuel. The size of the deposits is less and can take advantage of existing infrastructure provision.
- They do not need any reforming process.
- The electrolyte is a proton exchange membrane (similar to the PEM fuel cell type).

Disadvantages

- They have low-efficiency with respect to the hydrogen cells.
- They need large amounts of catalyst (noble metal) to the electro-oxidation of methanol at the anode.

9.3 HYDROGEN STORAGE TECHNOLOGY

9.3.1 INTRODUCTION TO THE PROBLEM

Up to now we have mainly considered the production of hydrogen from fossil fuels on an 'as needed' basis. However, there are times when it is more convenient and efficient to store the hydrogen fuel as hydrogen. This is particularly likely to be the case with low-power applications, which would not justify the cost of fuel-processing equipment. It can also be a reasonable way of storing electrical energy from sources such as wind driven generators and hydroelectric power, whose production might well, is out of line with consumption. Electrolysers might be used to convert the electrical energy to hydrogen during times of high supply and low demand.

A small local store of hydrogen is also essential in the use of fuel cells for portable applications, unless the direct methanol fuel cell is being used. As a result of its possible importance in the world energy scene as a general-purpose energy vector, a great deal of attention has been given to the very difficult problem of hydrogen storage. The difficulties arise because although hydrogen has one of the highest specific energies (energy per kilogram) – which is why it is the fuel of choice for space missions – its density is very low, and it has one of the lowest energy densities (energy per cubic meter). This means that to get a large mass of hydrogen into a small space very high pressures have to be used. A further problem is that, unlike other gaseous energy carriers, it is very difficult to liquefy. It cannot be simply compressed in the way that LPG or butane can. It has to be cooled down to about 22 K, and even in liquid form its density is quite low at 71 kgm−3.

Although hydrogen can be stored as a compressed gas or a liquid, there are other methods that are being developed. Chemical methods can also be used. The most common methods of storing hydrogen are as follows:

I. Compression in gas cylinders;
II. Storage as a cryogenic liquid;
III. Storage in a metal absorber – as a reversible metal hydride;
IV. Storage in carbon nanofibers.

None of these methods are without considerable problems, and in each situation their advantages and disadvantages will play differently. However, before considering them in detail we must address the vitally important issue of safety in connection with storing and using hydrogen.

9.3.2 SAFETY OF HYDROGEN

Hydrogen is a unique gaseous element, possessing the lowest molecular weight of any gas. Hydrogen has several properties that differ strongly from natural gas or methane. It has the highest thermal conductivity, velocity of sound, mean molecular velocity, and the lowest viscosity and density of all gases. Such properties lead hydrogen to have a leak rate through small orifices faster than all other gases. Hydrogen leaks 2.8 times faster than methane and 3.3 times faster than air. In addition, hydrogen is a highly volatile and flammable gas, and in certain circumstances hydrogen and air mixtures can detonate. The implications for the design of fuel cell systems are obvious, and safety considerations must feature strongly [9, 10].

Hydrogen is being used experimentally as a vehicle fuel, not only because it oxidizes to harmless water, but also because it has a higher energy density per unit of weight than CNG or methane. One of the other positive characteristics of hydrogen is that it disperses very quickly, meaning that hydrogen concentrations under normal pressure dissolve to incombustible levels very quickly. This also means that under ambient air pressure hydrogen has very little energy density per unit of volume compared to other vehicle fuels. Hydrogen also rises very quickly and therefore is less of a threat outdoors. Hydrogen therefore needs to be handled with care. Systems need to be designed with the lowest possible chance of any leaks and should be monitored for such leaks regularly. However, it should be made clear that, all things considered, hydrogen is no more dangerous, and in some respects it is rather less dangerous than other commonly used fuels [11, 12].

Hydrogen can be stored in four main ways: compressed, as a liquid, in hydride form bound to metals, or on the surface of solid porous materials such as carbon nanofibers and carbon nanotubes. The forms of storage that are currently being tested are compressed hydrogen and liquid hydrogen.

Compressed hydrogen is stored at 3,600 psig, 5,000 psig or 10,000 psig. The first two of these pressures are currently being tested. In contrast with compressed natural gas (CNG), compressed hydrogen cannot be stored in steel tanks, as the hydrogen molecules would embrittle the metal. Also the pressure in these tanks is much higher than in tanks for CNG. Even though these tanks were and are being designed with extreme care, the possibility of a leak still exists. This again might lead to the accumulation of flammable or even explosive hydrogen air mixtures.

The rupture of the pressure tank can create very high concentrations of hydrogen to form in the vicinity of the vehicle. Also hydrogen disperses very quickly; this emission will cause a combustible mix to form for a short period in the open. Enclosed areas could collect enough hydrogen-air mixture for a large explosion. Liquid hydrogen (LH_2) has to be stored at temperatures lower than 20° Kelvin or −435° F. Even well-insulated storage tanks cannot keep this low a temperature without relying on outside cooling, which is preventing for passenger vehicles from an energy-needs standpoint. The result of this is the expected leakage of 1%-3% of the hydrogen contained in the LH_2 tank per day, depending on the use and build of the vehicle. This controlled emission of hydrogen over time is not considered overly dangerous, as controlled oxidization by catalysts or dispersion is possible. Horrible ruptures of the LH_2 tank cannot only endanger people due to the extreme temperature of the liquid, but can also lead to very high hydrogen concentrations in the surrounding air, especially in confined places such as tunnels.

Solid porous materials such as carbon nanofibers and carbon nanotubes are still in the early stages of development. While their properties appear to be close to metal hydride storage, they have other problems related to the porous material, the high volume and weight of the material per weight of hydrogen carried, which are still unsolved. The carbon nanotubes and nanofibers are now still prohibitively expensive and also have the potential problem of flammability.

9.4 HYDROGEN STORAGE OF CARBON NANOFIBERS

Due to the exhaustion of gasoline or diesel fuel, new energy sources are necessary to be developed as an assistant or alternative energy. Among them, hydrogen gas is an attractive possibility to provide new solutions for ecological and power problem [13]. As mentioned Hydrogen is a promising substitution energy source for fossil fuels because of its non-limited and non-polluting energy. Recently, the requirement for a safe, low-cost, car-carry type of hydrogen gas generator has become stronger because the use of fuel cells will soon be practical [14]. Since it is difficult to store hydrogen, its use as a fuel has been limited [15–17]. The possibility of developing hydrogen into an environmentally friendly, convenient fuel for transportation has lead to the search for suitable materials for its storage. The suitable media for hydrogen storage have to be light, industrial, and in compliance with national and international safety laws. An important condition for the wide usage of hydrogen as a future fuel, especially for electrical vehicles, is the development of a safe, cheap, and simple storage method [14]. In the last few years, researchers have paid much attention on hydrogen adsorption storage in nanostructured carbon materials [18], such as activated carbon [14, 19], carbon nanotubes [17], and carbon nanofibers. Carbon nanomaterials due to their high porosity and large surface area has been suggested as a promising material for hydrogen storage [20].

In 1998, a paper was published on the absorption of hydrogen in carbon nanofibers [21]. The authors presented results suggesting that these materials could absorb in excess of 67% of hydrogen. This amazingly large figure took the academic world by storm, as it offered the prospect of levels of hydrogen storage in material at ambient temperatures and pressures, which would ensure hydrogen a certain future as a fuel for vehicles. Using the new graphitic storage material, it was predicted that a fuel cell car with hydrogen stored in carbon nanofibers would run for 5000 miles on one charge of fuel. Before long several groups around the world were carrying out research on hydrogen storage in carbon nanofibers. The initial euphoria became tempered as various groups tried to repeat the work and found that they could not – only relatively small amounts of hydrogen were absorbed. Nevertheless, the US Department of Energy set a benchmark for hydrogen storage in Fives of 62 kg H_2/m^3 and 6.5% weight (i.e., the percentage weight of stored hydrogen to the total system weight). This would provide enough hydrogen in a car for a journey of some 300 miles. The DOE target continues to stimulate a lot of research in this area but so far no carbon nanoscale material has reproducibly come near this figure.

The story of carbon nanofibers really started with the discovery of fullerenes in the early 1980s [22] when it was found that new molecular structures could be synthesized at the nanoscale, which have fascinating and useful properties. By nanoscale we mean of the dimension of nanometers (1 nm is 10^{-9} m). So we are talking about the sorts of units you need to describe the sizes of individual atoms/molecules. Carbon nanomaterials can be made by a variety of processes and can be examined under the electron microscope when their structure becomes apparent.

Amorphous and graphitic carbons have been used as an absorbent for some materials for many years. 'Activated carbon' has been available as a storage and filter medium industrially for decades. However, activated carbon is really ineffective for storing hydrogen because the hydrogen molecules only weakly absorb on the carbon surface at room temperatures, despite such materials having high surface areas (2000 m^2 g^{-1}). On the other hand, carbon whose filaments are at the nanometer scale would be expected to have a larger surface area and porosity, and therefore be able to store much more hydrogen than normal graphite. The question is whether carbon nanofibers can live up to their promise Molecular simulations and quantum-mechanical calculations tell us that the very high storage densities that were first reported are inconsistent with theory. Later work [23, 24] has suggested that some of the high absorption figures were a result of the presence of water vapor in the samples, which tended to expand the graphitic layers in the carbon structure so that multiple layers of hydrogen atoms would absorb. Meticulous care is needed in measuring hydrogen uptake on very small samples of materials.

Without going into too much detail, we should point out that there are really three types of carbon nanofibers that are now being investigated for hydrogen storage: graphitic nanofibers, single-walled carbon nanotubes, and multi-walled carbon nanotubes. The essential structures of these materials are shown in Figure 9.1. Graphitic nanofibers are made by decomposing hydrocarbons or carbon monoxide over suitable catalysts.

The catalyst influences the arrangement of the graphitic sheets that are produced, and many different types of structures have been observed, with the most common being described as platelets, tubular, and a herringbone arrangement (see Figure 9.1). Graphitic nanofibers vary from around 50–1000 nm in length and 5–100 nm in diameter [21]. There is a wide range of reported hydrogen absorption values on graphitic nanofibers [25], and a consensus has yet to be reached on the practicality of using these materials in real gas storage systems. There is a schematic representation of carbon nanofiber structure containing graphitic sheets in appendix B.

Carbon nanotubes, either single walled (typical diameter 1–2 nm), or multi-walled (diameter 5–50 nm), are an alternative to the graphitic nanofiber systems. They are special because of their small dimensions, each tube possessing a smooth and highly ordered graphitic surface. Single-walled nanotubes (SWNT) (Figure 9.2b) can be thought of as a one-dimensional Buckminster fullerene, or as a tube made up of hexagonal graphite plates linked together. Carbon nanotubes were first identified by Iijima [26] who prepared them by accident, using an electric arc drawn between two carbon electrodes. Nanotubes are still produced by electric arcs, but more recently other methods have also been employed, such as laser ablation, and chemical vapor deposition [27]. Unfortunately, there is a cost penalty associated with using such high-tech methods of preparation.

As with graphitic nanofibers, there is yet to be a consensus reached on the efficacy of carbon nanotubes for hydrogen storage. Experimental

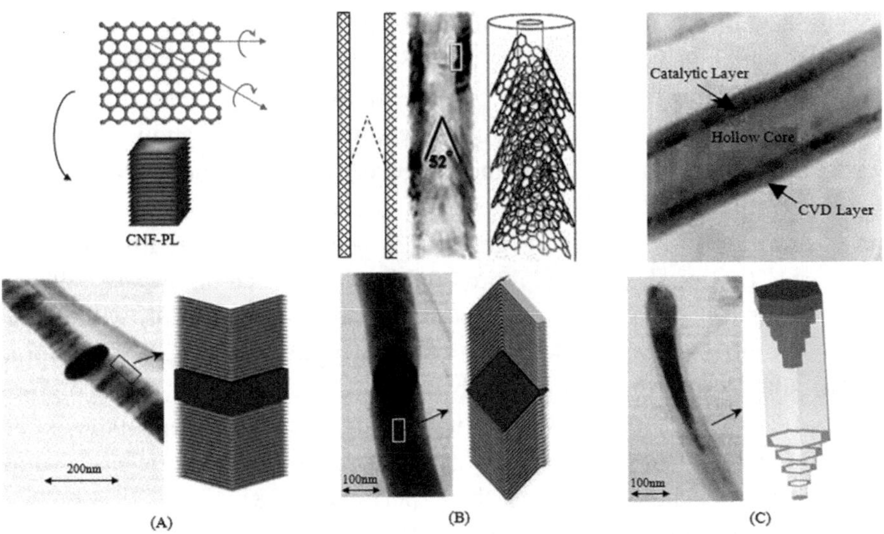

FIGURE 9.1 Schematic representation of different types of carbon nanofibers (A) platelet structure, (B) herringbone structure, and (C) tubular structure.

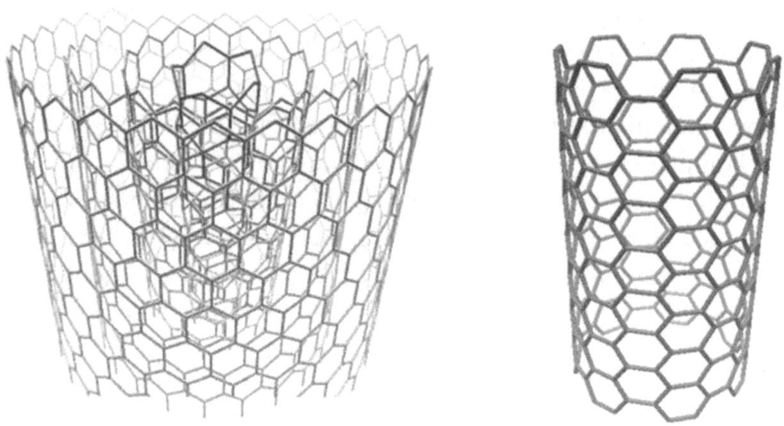

(a) (b)

FIGURE 9.2 Schematic representation of (a) multi wall carbon nanotube, (b) single wall carbon nanotube.

results are often difficult to interpret, and issues of purity of materials and subtleties in experimental techniques (gravimetric vs. volumetric methods) make it difficult to draw firm conclusions. The experimentalist is often trying to measure very small changes in weight or volume, for example, Dillon et al. [28] reported absorptions of ca 0.01 wt.% H_2 on SWNT, which had a purity of only 0.01%, from which he extrapolated that 5 to 10 wt.% could be absorbed on pure materials.

Before long, Dillon and some others were able to show absorption capacities of a few percent were indeed measurable on some purer materials. However, these findings were later contradicted by Hirscher et al. [29], and even more recently by Dillons group [28] who showed that absorptions of only ca. 1 to 2 wt.% H^+ could reproducibly be absorbed on impure SWNTs.

Some success has been reported using alkali-doped multi-walled carbon nanotubes (MWCNT). Values of 20% uptake have been reported at 20 to 400°C at ambient pressure by Chen [30]. Yet again, lower figures (2.5 wt. %) were obtained when other researchers tried to repeat the experiments. Most recent work has, however, started to demonstrate a reproducibility and consistency amongst the experiments [28].

This is giving grounds for hope that carbon nanofibers will feature in real hydrogen storage systems in the future. Certainly the fabrication of nanofibers is an interesting and fascinating area of research. It is somewhat

TABLE 9.1 Summary of Reported Hydrogen Storage Capacity of Carbon Nanofiber

Sample	Purity%	T(K)	P(Mpa)	H2 (Wt %)	Ref.
GNF	—	77–300	0.8–1.8	0.08	Ahn et al. [16]
GNF herringbone	88	298	11.35	6.6	Browning et al. [31]
GNF platelet	92	298	11.35	6.5	Browning et al. [31]
Vapor grown carbon fiber	98	298	3.6	<0.1	Tibbetts et al. [17]
CNF	—	77	12	12.38	Rzepka et al. [32]
ACNF with CO_2	78	77	11	0.33	Blackman et al. [33]
ACNF with KOH	78	77	10	0.42	Blackman et al. [33]

controversial and the jury is still out on the practical feasibility of using these materials for cost-effective hydrogen storage.

Table 9.1 shows a summary of hydrogen storage capacity in different forms of carbon nanofibers.

Rodriguez and Baker investigated that the hydrogen storage capacities of CNF at room temperature and pressures up to 140 bars were quantified independently by gravimetric and volumetric methods, respectively [16, 31, 32]. Ji Sun Im and Soo-Jin Park [34] studied the relation between pore structure and the capacity of hydrogen adsorption, textural properties of activated CNFs with micropore size distribution, specific surface area, and total pore volume by using BET (Brunauer–Emmett–Teller) surface analyzer apparatus and the capacity of hydrogen adsorption was evaluated by PCT (pressure–composition–temperature) hydrogen adsorption analyzer apparatus with volumetric method.

They indicated that Even though specific surface area and total pore volume were important factors for increasing the capacity of hydrogen adsorption, the pore volume which has pore width (0.6–0.7 nm) was a much more effective factor than specific surface area and pore volume in PAN-based electrospun activated CNFs [34].

Chemically activated carbon nanofiber with NaOH and KOH, were evaluated by Figueroa-Torres [18] for hydrogen adsorption at 77 K and atmospheric pressure. Hydrogen adsorption reached values in the order of 2.7 wt. % for KOH activated carbon. The mechanism of formation of the porous nanostructures was found to be the key factor in controlling the hydrogen adsorption capacity of chemically activated carbon [35].

Loading of CNF with metallic particles can enhance the hydrogen storage capacity of it. The hydrogen storage behaviors of porous carbon nanofibers decorated by Pt nanoparticles were investigated, It was found that amount of hydrogen stored increased with increasing Pt content to 3.4 mass%, and then decreased [36].

The main reason suggested for improved hydrogen adsorption was that electrospun activated carbon nanofibers might be expected to have an optimized pore structure with controlled pore size. This result may come from the fact that the diameters of electrospun fibers can be controlled easily, and optimized pore sizes can be obtained with a highly developed pore structure. To find the optimized activation conditions, carbon nanofibers were activated based on varying the chemical activation agents, reaction time, reaction temperature, and the rate of inert gas flow [37] (Figure 9.3).

Porous carbon-nanofiber-supported nickel nanoparticles also can be used as a promising material for hydrogen storage. It was found that the amount of hydrogen stored was enhanced by increasing nickel content [38] (Figures 9.4–9.5).

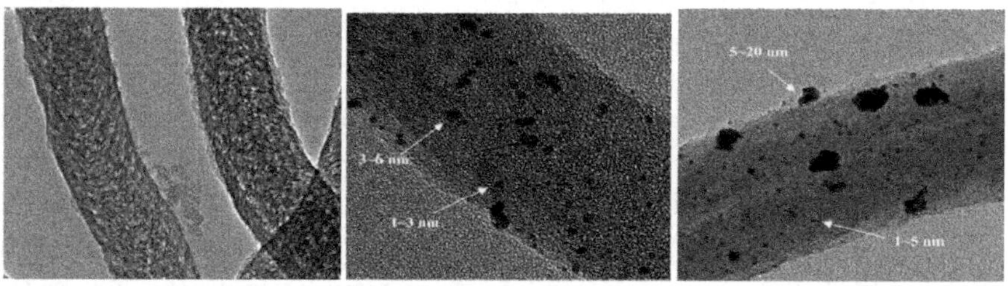

FIGURE 9.3 TEM images of porous graphite nanofibers: (a) PGNF, (b) Pt-5/PGNF, and (c) Pt-10/PGNF [36].

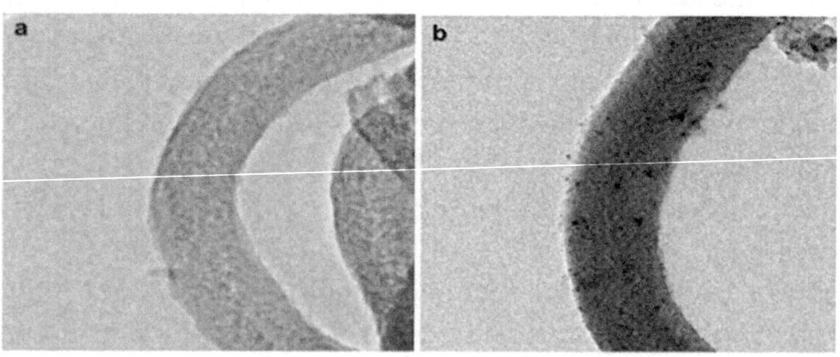

FIGURE 9.4 TEM images of Ni-plated porous carbon nanofibers: (a) as-received, (b) Ni-plated CNF [38].

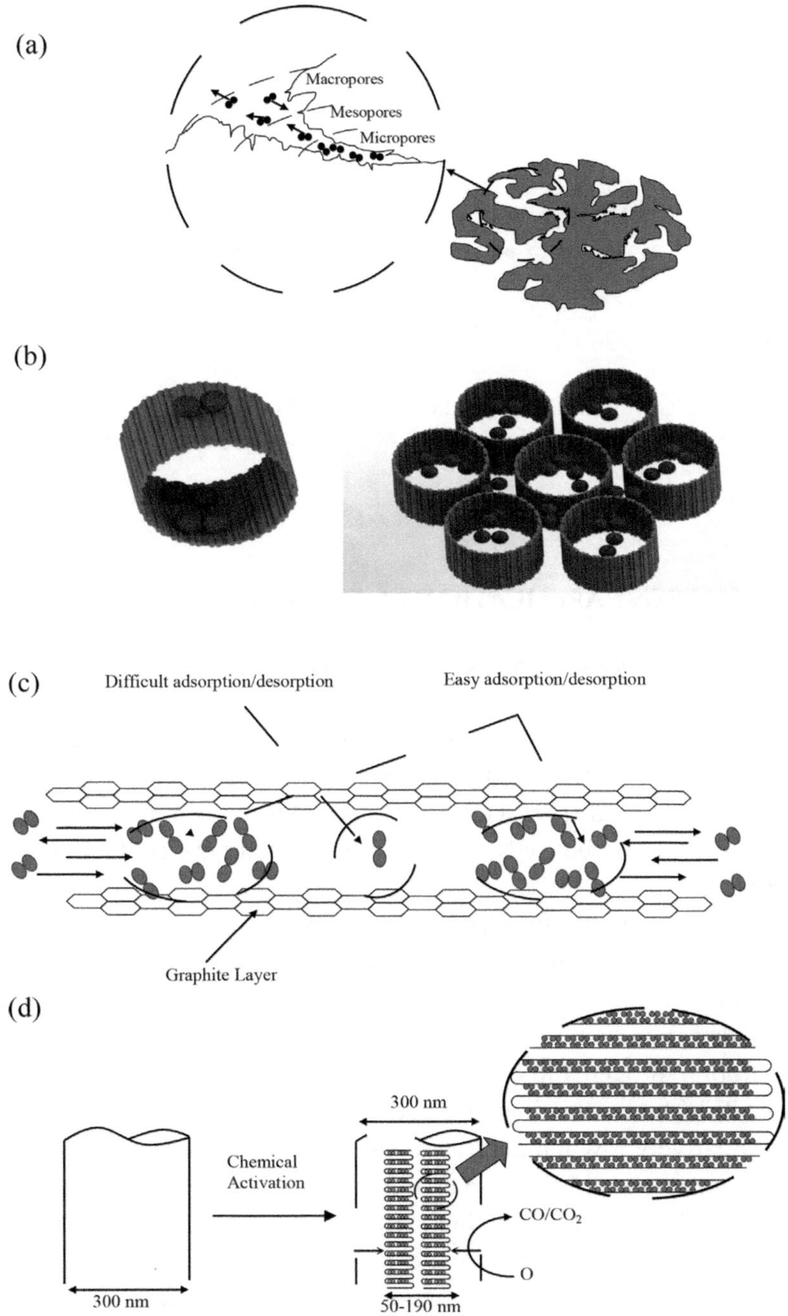

FIGURE 9.5 Continued

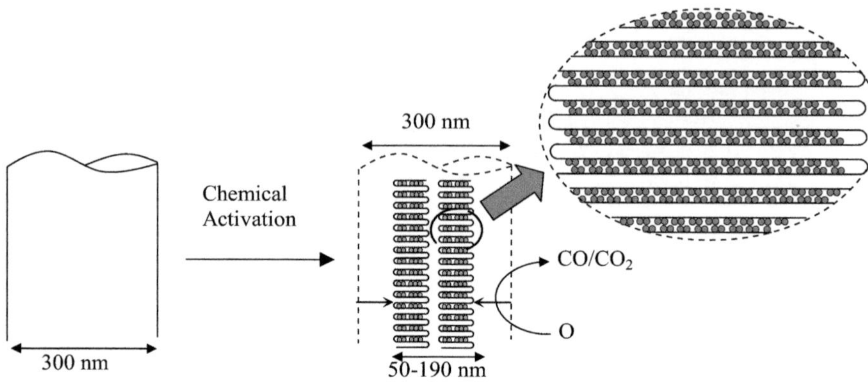

FIGURE 9.5 The mechanism of hydrogen adsorption using various carbon materials; (a): activated carbon, (b): single walled carbon nanotube, (c): graphite, (d): electrospun activated carbon nanofibers.

9.5 DIFFERENT APPLICATIONS OF CNFS IN FUEL CELLS

9.5.1 THE CARBON NANOFIBER HYDROGEN FUEL TANK

The principle application of a carbon nanofiber hydrogen storage medium is in a fuel tank for an integrated on-board fuel cell system with a polymer electrolyte membrane (PEM) fuel cell stack at its core and a hydrogen supply stored as adsorbed hydrogen in a pressurized tank containing carbon nanofibers.

In a PEM fuel cell, two half-cell reactions take place simultaneously, an oxidation reaction (loss of electrons) at the anode and a reduction reaction (gain of electrons) at the cathode. These two reactions make up the total oxidation-reduction (redox) reaction of the fuel cell, the formation of water from hydrogen and oxygen gases.

The PEM fuel cell is ideal for automotive applications because it operates at relatively low temperatures and can vary its output to meet varying power demands [39]. As in an electrolyzer, the anode and cathode are separated by an electrolyte, which allows ions to be transferred from one side to the other (Figure 9.6). The electrolyte in a PEM fuel cell is a solid acid supported within the membrane. The solid acid electrolyte is saturated with water so that the transport of ions can proceed.

PEM fuel cell reactions are as below:

Anode reaction: $H2 \rightarrow 2H+ + 2e-$
Cathode reaction: $½O2 + 2e- + 2H+ \rightarrow H_2O$
Overall reaction: $H2 + 1/2\ O2 \rightarrow H_2O$

In PEM systems, the hydrogen storage tank consists of a steel or composite tank or canister filled with carbon nanofibers with adsorbed hydrogen present. The unused tank is kept pressurized at about 100–120 atm to maintain the adsorbed state of the hydrogen. Hydrogen's low molecular weight makes storage as a compressed gas less effective than for other fuel gases, such as methane. This fact drives compressed hydrogen storage pressures to very high values, thus requiring expensive storage systems to reach reasonable gravimetric Capacities [40].

The tank is connected to the fuel cell via a regulated pressure nozzle assembly controlled by the onboard computer that monitors the system. As the fuel cell demands hydrogen through the normal operation of the vehicle, the pressure in the fuel tank is decreased and gaseous hydrogen

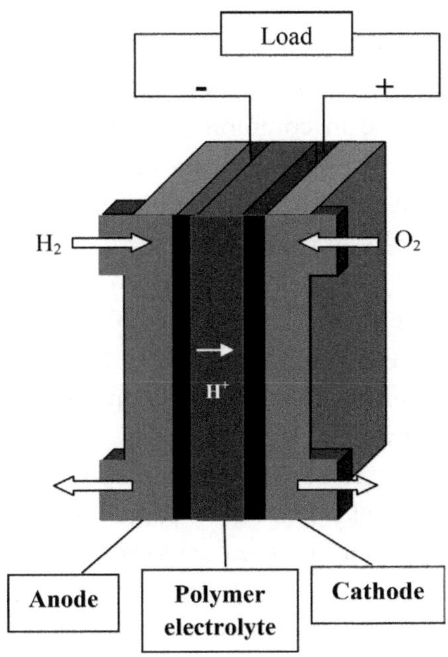

FIGURE 9.6 Diagram of a single PEM fuel cell.

is released through the nozzle assembly and directed to the fuel cell cathode for catalysis and subsequent current generation. The rate of release is variable according to energy demands and can vary from zero to the maximum desorption rate available from the carbon nanofibers via incremental depressurization [41].

The D.O.E.'s target benchmarks for on-board hydrogen storage are based on a model hydrogen fuel cell powered vehicle to be able to travel 500 km without refueling and the metric that 3.1 kg of hydrogen would be required for a fuel cell powered car to travel those 500 km. Based on the 10–15 wt. % storage capability that has been demonstrated with properly prepared carbon nanofibers, as illustrated earlier in this chapter, this would result in a 10 full tank of hydrogen adsorbed carbon nanofibers weighing between 21 and 31 kg with perhaps some additional weight required for the pressure valve and protective covering [42].

9.5.2 CARBON NANOFIBERS FOR ELECTRODE AND CATALYST SUPPORT IN FUEL CELLS

Low temperature fuel cells such as polymer electrolyte fuel cells (PEFC) and alkaline fuel cells (AFC) are more interested in recent decade due to their capability to use in common applications. Cathodes with silver and platinum as catalysts supported on carbon nanofibers for the oxygen reduction reaction were investigated with an alkaline electrolyte by Hacker et al. successfully [43].

CNF and activated form of it can be used as catalyst support or electrodes in Polymer Electrolyte Membrane (PEM) fuel cells [44]. PEM fuel cell electrodes have been traditionally fabricated by deposition of finely divided platinum catalyst particles on a support material such as activated carbon, carbon black and recently carbon fibers or nanofibers. The use of CNF as catalyst support in PEMFC electrodes compared to activated carbon blacks has the advantages of an increase of catalyst activity and selectivity of the catalysts for the hydrogen oxidation reaction [45].

Understanding the long-term stability of the PEM fuel cell is of crucial importance as this technology approaches its commercialization platform.

So far, the operational life-time for real life applications does not satisfy the requirements for state-of-the-art technologies, for example, 5000 h for cars, 20,000 h for buses, and 40,000 h for stationary applications [46]. Therefore, performance degradation of PEM fuel cells and the degradation of the component materials have attracted extensive attention in recent years. In the past few years, numerous papers have been published that focuses on the degradation issues of PEM fuel cells. These research progresses on PEM fuel cell durability and degradation have been reviewed from many different perspectives.

The review of Borup et al. [47] concentrated on the fundamental aspects of PEM fuel cell degradation mechanisms. Shao et al. [48] and Zhang et al. [49] paid attention to the material challenges to developing durable high temperature PEM fuel cells, including electrocatalysts, carbon supports, membranes, polymers, and bipolar plates. While in Wu et al. review [46], the existing strategies for improving the lifetime of the fuel cell components were summarized. It has been found that the degradation of the fuel cell performance is primarily due to the decay of the membrane-electrode- assembly (MEA) [50–52]. Among the MEA components, catalyst layer or electrocatalyst degradation is one of the most critical factors and increasing catalyst layer or electrocatalyst durability is a major challenge and a growing focus of research attention in PEM fuel cell durability studies.

As mentioned before, among the several existing types, polymer electrolyte fuel cells (PEMFC and DMFC) have favorable advantages in transport, portable and micro power applications in terms of high efficiency, high energy density, quick start-up and zero or low emission. Although great progress has been made in the last years, the commercialization of this technology has several technical and economical limitations such as cost, durability and performance. The main reason for this problem is the need to platinum as catalyst, which is costly and extinct. So the challenge remains in decreasing Pt loading while maintaining or even increasing the performance and durability [53].

Carbon nanofibers have an excellent combination of chemical and physical properties due to their unique structure, and blend two properties that rarely coexist in a material: high surface area and high electrical conductivity which make them suitable support for catalysts. Studies indicated that CNF can be applied as catalyst support in polymer electrolyte fuel cells successfully which caused less need for platinum catalyst [54, 55]. Carbon nanofibers are proved to be appropriate catalyst support in fuel cell because of their graphitic structure which improve their corrosion resistance; their textural properties, which might provide benefits for mass and electron transport, as well as, the strong interaction with the metal nanoparticles, which increases the catalyst activity and durability. In addition, carbon nanofiber functionalization influences the amount and dispersion of the deposited nanoparticles [56].

Electrode structure with carbon nanofiber supported Pt in PEM fuel cells has three layers system constituted by a conductive porous support, a diffusion layer and the catalyst itself. The performances of the electrodes depend on many parameters: (i) type of macroscopic carbon support (carbon paper, carbon cloth, etc.) and its characteristics (porosity and thickness); (ii) type of catalyst (metal, metal amount, particles size, type of catalyst support); (iii) thermal treatment; (iv) thickness of diffusion and catalytic layers; and v) fabrication process [53, 56].

9.6 CARBON NANOFIBERS IN DIRECT METHANOL FUEL CELLS

9.6.1 INTRODUCTION TO DIRECT METHANOL FUEL CELLS

In the past two decades, proton exchange membrane fuel cells (PEMFCs), including direct methanol fuel cells (DMFCs), have been demonstrated to be feasible for a large variety of technical areas, such as portable devices, automotive, and stationary power systems. However, their commercial applications have been limited by their high cost and insufficient durability. In particular, DMFCs using liquid methanol rather than hydrogen gas as fuel have been demonstrated to be suitable technologies for low-power portable electronic device applications due to their high theoretical energy density and low working temperatures. The use of methanol as fuel has several advantages in comparison to hydrogen: it is a cheap liquid fuel that is easily handled, transported, and stored [57–59].

The operating principle of DMFCs involves the anode electrooxidation of methanol catalyzed by carbon-supported Pt–Ru alloy catalyst and the cathode electroreduction of oxygen catalyzed by carbon supported platinum catalysts [60]. Unfortunately, both catalyzed methanol electrooxidation and oxygen electroreduction are kinetically slow, resulting in relatively lower performance. As one of the main factors that limit and diminish the practical performance of the DMFCs, the sluggish methanol oxidation and the poisoning of the anode catalyst have attracted a lot of efforts in catalyst development. Platinum (Pt), the most practical catalyst for fuel cells, is a good catalyst for methanol oxidation. However, it can easily suffer from a quick poisoning by the CO, a major reaction intermediate [61–64]. The remedy has been to use Pt-based binary or ternary eletrocatalysts [65–67]. Some improved catalytic activities have been reported when using Pt-based alloys, such as Pt–Ru, Pt–Mo, Pt–Sn, Pt–Os, Pt–Ru–Os [68]. This enhancement effect of alloying has been explained by models, such as the "bi-functional mechanism" [69] and/or by the "electronic effect" [26–28], which indicates a promotional effect of the alloyed metal on Pt. Particularly, Pt–Ru has been the most investigated binary system and has shown the best catalytic activity [70, 71].

The catalytic activity of Pt-based catalysts may be influenced by many factors, among which the catalyst-supporting material, such as a carbon particle, plays an important role in promoting catalyst activity. Carbon black particles are widely used as catalyst supports because of their relative stability in both acidic and basic media, good electric conductivity, and high specific surface area. However, the mesopores on carbon particles can result in part of the Pt nanoparticles getting buried deeply inside the pores and hence becoming inaccessible for the electrochemical reaction at the triple phase boundary. Further, the carbon particle can undergo corrosion (more rampant under peroxide intermediate formation conditions at fuel cell cathodes), resulting in the aggregation and dissolution as well as isolation of Pt nanoparticles [72–74]. To overcome these challenges in carbon support, many efforts have been made to search for new catalyst supports [47].

In this regard, nanomaterials, such as carbon nanotubes (CNTs) and carbon nanofibers (CNFs) have been explored as support materials for catalysts, in particular for fuel cell catalysts in the last few years. For example, a large amount of studies have shown that Pt (or Pt alloys) supported on CNTs and CNFs could exhibit better performance for the electro-oxidation of methanol [75, 76], oxygen reduction reaction [77–79] and higher durability than that on Vulcan XC−72 carbon particles [80, 81].

This chapter begins by briefly describing the processing techniques used to synthesize the carbon nanotubes and carbon nanofibers. Next, the methods by which metal catalysts can be deposited onto the CNTs and CNFs to form supported catalysts will be discussed along with the surface functionalization. Then, this is followed by a review of fabrication processes of DMFC membrane electrode assemblies (MEAs), which are containing CNT- or CNF-supported catalysts. Finally, we will discuss the challenges of using of nanocatalysts in DMFCs.

9.6.2 COMMONLY USED CATALYST CARBON SUPPORTS

Traditionally, carbon black is the most popular material used as support for fuel cell catalysts due to its advantages as follows: (1) high electrical and thermal conductivities; (2) available with a wide range of surface areas and porosities; (3) high stability in acid environments; (4) relatively stable in reducing and reasonably oxidizing environments; (5) available with low impurity levels; and (6) less expensive (less than or equal to a few dollars per kilogram) and available in at least multikilogram quantities.

A wide range of carbon blacks is available from a number of suppliers (e.g., Cabot, Columbian Chemicals, Azko Nobel, Denka, Timcal, Degussa, and Mitsubishi). These types of carbon blacks are characterized by particularly high surface area–to-volume ratios and have much lower impurity levels. However, despite the many black products available, mainly three types of carbon materials, such as Vulcan XC-72, BP2000, and Ketjen EC300J, were reported for fuel cell usage.

Within a DMFC electrode catalyst layer, in order to facilitate the electrocatalytic reaction, such as methanol oxidation, an ideal catalyst carbon support should simultaneously possess high specific surface area, good electric conductivity, suitable pore size, and good corrosion resistance and can also support necessary surface functional groups to guarantee the reaction occurring continuously without significant activity degradation. Normally, good electronic conductivity (graphitization degree) to reduce the electrode resistance and also to protect the carbon support from being oxidized is important for the electrocatalytical reaction. Suitable surface functional groups on the carbon support are also needed to act as anchors for the noble metal particles during catalyst preparation, which would improve the lifetime of the catalyst and the dispersion of the active components. In this regard, the commonly used activated carbon XC−72R seems still not to satisfy meeting these demands for an ideal electrocatalyst support even though it is the best commercial support currently. Therefore, the development of new carbon materials and modification of existing ones are necessary. Actually, the requirements for carbon support materials to have high conductivity and high specific surface area as well as a large amount of mesopores or macropores is contradictive to each other, there exists a trade-off. In order to meet these requirements as much as possible, since 1998, there have been a variety of new carbon materials, such as CNTs and CNFs that have been investigated as electrocatalyst supports in DMFCs and some impressive results have been achieved.

9.6.3 CNT AND CNF-BASED CATALYST SUPPORTS

9.6.3.1 Significance of CNTs and CNFs

With the continuous progress of nanotechnology in materials science, different types of carbon nanomaterials have attracted considerable attention, especially carbon nanotubes (CNTs) and carbon nanofibers (CNFs) and their applications, particularly as fuel cell catalyst supports. The potential benefits of using CNTs and CNFs as fuel cell catalyst supports that have been suggested include higher utilization of active metal due to the lack of smaller porosity and higher corrosion resistance due to the (theoretically) inert surfaces. The use of CNTs for fuel cell catalyst applications has been recently experimented and largely reviewed [55]. It is believed that when catalyst is supported on nanotubes, the reactant accessibility could be improved, leading to higher catalyst utilization [82]. Better performance of a carbon nanotube–based catalyst electrode with respect to a carbon black–based one has also been correlated to the higher electronic conductivity of the former (104 S cm^{-1}) when compared to the latter (4.0 S cm^{-1}) in the case of VulcanXC-72 [83, 84]. Regarding the stability of CNT as a catalyst support, Li et al. carried out some experiments to evaluate their electrochemical durability [85]. The comparison with carbon black Vulcan XC−72 demonstrated that CNTs are more resistant to electrochemical oxidation.

Other carbon nanostructures, in particular graphite nanofibers, have been largely studied as support for catalysts [86, 87]. For example, Bessel et al. [88] reported about the deposition of Pt on various types of nanofibers, and their electrochemical activity toward methanol oxidation reaction was studied. The authors ascribed the enhanced performance of these catalysts compared to carbon black–supported catalysts to a specific crystallographic orientation of the metal particles deposited on highly tailored graphite nanofibers. Literature results suggest that the use of carbon nanostructures, which are different from carbon black, as catalyst supports, is quite promising to prepare electrodes for both fuel cell anodes and cathodes.

9.6.3.2 Carbon Nanotubes (CNTs)

Among several new nanostructured carbon materials used for catalyst supports, such as CNTs [89], CNFs [90], ordered mesoporous carbon [91], carbon aero, and xerogels, CNTs and CNFs are the most well-known nanostructured carbon, which have shown promising results as catalyst supports for fuel cell applications due to their unique electrical and structural properties.

CNTs are a novel class of one-dimensional nanomaterials discovered by Iijima in 1991 [26]. The extraordinary mechanical properties and unique electrical properties of CNTs have stimulated extensive research activities worldwide, resulting in many special issues of journals [92, 93] and several books [94, 95]. CNTs are nanoscale cylinders of rolled up graphene sheets with an extensive range of variations, such as single-walled nanotubes (SWNTs) and multi-walled nanotubes (MWNTs) as shown in Figure 9.2 [96]. It can be seen that structures comprising one cylindrical tube are called SWCNTs. SWCNTs have a relatively smaller diameter, as low as 0.4 nm, and could be metallic or semiconducting in

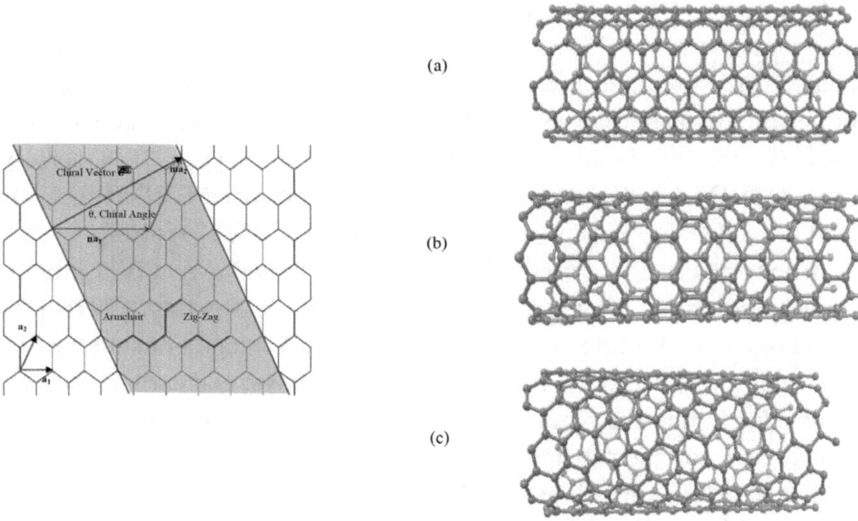

FIGURE 9.7 Schematic honeycomb structure of graphene sheet. Single-walled carbon nanotubes can be formed by folding the sheet along lattice vectors. The two basis vectors a_1 and a_2 are shown. Schematic illustration of the structures of carbon nanotubes: (a) armchair, (b) zigzag, and (c) chiral SWNTs.

nature, depending on their structure. MWCNTs can be considered as concentric SWCNTs that have increasing diameter and are coaxially disposed (Figure 9.2a). The number of walls present can vary from two (double wall nanotubes) to several tens, so that the external diameter can reach to 100 nm. The concentric walls are regularly spaced by 0.34 nm similar to the intergraphene distance in turbostratic graphite materials.

The methods for synthesizing CNTs include arc discharge, laser ablation, and chemical vapor deposition (CVD) [97]. Although both the arc discharge and laser ablation methods are control over the spatial arrangement of the produced nanostructures. Furthermore, some complex purification procedures are also required to remove amorphous carbon particles and entangled catalysts in order to obtain useful material. Among these three methods, only CVD allows controlled synthesis of CNTs efficient in producing high-quality nanotube material in large quantities.

The advantage of the CVD method is allowing a scaled-up production to the industrial level and producing CNTs in a predictive fashion with controlled length, positions, and orientations on the substrates. In the synthesis, controlling the CNT structure and its growth can be achieved through changing experimental parameters, such as process temperature, gas composition, pressure, and flow rates as well as catalytic materials. There are various different approaches used for CVD methods, including general thermal CVD [98, 99], floating catalyst CVD (FC−CVD) [100, 101], aerosol-assisted CVD (AA−CVD) [102, 103], plasma-enhanced CVD (PE−CVD) [104] as well as DC plasma-enhanced CVD (DC−PEVCD) [105].

Among these CVD methods, the FC−CVD and AA−CVD are two important ones in growing high-density and high-quality CNTs [86]. Compared to the traditional CVD method, the advantage of using the FC−CVD and AA−CVD methods is allowing the catalyst and carbon source to be introduced into the reaction chamber simultaneously to produce well-aligned CNTs on the substrates [101].

The AA−CVD involves pyrolysis of a mixed liquid aerosol, which consists of both a liquid hydrocarbon and a catalyst precursor [106]. In general, ferrocene, cobaltocene, cobalt nitrate, nickelocene, and iron pentacarbonyl are used as both catalysts and carbon sources. Additionally, acetylene, benzene, toluene, xylene, mesitylene, and tetrahydrofuran are also used as solvents for additional carbon sources. For example, Liu et al. [100] synthesized some aligned multiwalled carbon nanotubes with high purity on SiO_2 substrate by the AA-CVD method. In their method, an aerosol consisting of a liquid hydrocarbon source was decomposed over a catalyst for CNT growth. The aerosol droplets were produced by ultrasonication and transported by argon gas.

9.6.3.3 Carbon Nanofibers (CNFs)

The unique properties of carbon nanofibers (CNFs) have been explored in a number of applications, including selective absorption [107], energy storage [108], and polymer reinforcement [109] as well as catalyst supports [86, 110–111]. CNFs have lengths on the order of micrometers while their diameter varies between some tens of nanometers up to several hundreds of nanometers. The mechanical strength and electric properties of CNFs are similar to that of CNTs while their size and graphite ordering can be well controlled [112]. The primary distinguishing characteristic of CNFs from CNTs is the stacking of graphene sheets of varying shapes, producing more edge sites on the outer wall of CNFs than CNTs [113]. In comparison with conventional supports, the use of CNFs as catalyst support could increase the catalyst activity in DMFCs [114].

CNFs can be divided into the platelet CNFs, the fish-bone CNFs and the tubular CNFs, according to the different arrangement of graphene layers. Regarding their applications in catalyst supporting, the microstructure controllable CNFs have been used as support for noble metal catalysts [86, 115]. The CNF nanostructures consist of graphite sheets oriented in a direction dictated by the growth process. The fishbone CNFs can reveal their graphene layers stacking obliquely with respect to the fiber axis; the platelet CNFs consist of graphene layers oriented perpendicularly to the growth axis, and the ribbon CNFs display their graphene layers parallel to the growth axis.

Regarding CNF fabrication, the CVD method was normally used for the synthesizing CNFs [116–117], in which a carbon source was decomposed catalytically over a metal catalyst, such as Ni, Co, Fe, or their alloy. The catalytic decomposition of methane on Ni catalysts is one of the most extensively studied systems [111, 118]. For example, Sebastian et al. [119] synthesized CNFs by thermal CVD in a quartz tube electric furnace with methane as a carbon source and Ni:Cu:Al_2O_3 as a catalyst. In their experiment, a fixed bed reactor was used, in which CNFs were grown for 10 h. In general, different reaction conditions, varying temperatures (550°C–750°C) and gas space velocity (CH_4, air products) were used to obtain CNFs with different properties.

9.6.4 *CHALLENGES TO NANOCATALYSTS IN DMFCS*

There are several challenges to the development of CNT- and CNF–supported catalysts in term of fabrication and synthesis. Thus, this section highlighted the most significant limitation controlled by particle size, distribution, fabrication, and crystallinity.

9.6.4.1 Control of Particle Size

The size of Pt–Ru particles is a major challenge in the synthesis of CNT- and CNF–supported catalysts. Because the support can strongly affect catalyst utilization and activity, the synthesis of Pt nanoparticles supported by CNT and CNF is of fundamental importance. To control the particle size, uniform platinum nanotubes have been synthesized by directly mixing Ag nanowires and H_2PtCl_6 in a saturated solution of NaI at room temperature. The crystal structure of the resulting Pt nanotube was investigated in detail by field emission scanning electron microscopy, transmission electron microscopy, and x-ray diffraction [120]. Furthermore, catalyst particle size can also be controlled through CVD methods. There are various methods for CVD based on the different functions and chemicals used.

9.6.4.2 Distribution of Catalyst Nanoparticles

Due to the slow oxidation kinetics of methanol oxidation, catalyst loadings as high as 50 wt. % are required for DMFCs. Despite the high catalyst loading in DMFC cells, CO poisoning is prevalent. Thus, a considerable amount of research has been conducted in recent years to develop a method that distributes Pt/Ru evenly throughout the structure. Fujigaya et al. [121] described the fabrication of a CNT hybrid wrapped by pyridine-containing polybenzimidazole (PyPBI). Based on this method, highly homogeneous and remarkably efficient Pt loading onto the surface of CNTs through a coordination reaction between Pt and PyPBI has been achieved. They found that the wrapped PyPBI could serve as glue for immobilizing Pt nanoparticles onto the surface of CNTs without any strong oxidation process for the CNTs.

9.6.4.3 Fabrication of DMFC Electrode

In comparison with carbon black, CNT-supported electrocatalysts commonly exist in unusual shapes and have bulky specific volumes. As a result, CNT-supported electrocatalysts are difficult to fabricate into fuel cell electrodes by conventional means, such as painting, brushing, spraying, screen-printing, etc. Additionally, high catalyst loadings required for DMFCs are also difficult to achieve in CNT-supported electrocatalysts [122]. Common obstacles include thick catalyst layers, loose structures in coated layers, poor cell performance, and high ionomer content, leading to a high electrode resistance. These problems are exacerbated when catalysts with low amounts of metal and high amounts of CNT are used in the electrode catalyst layer.

9.7 CONCLUSIONS AND PROSPECTS

Electrocatalyst supports play a vital role in ascertaining the performance, durability, and cost of DMFC systems. Myriad nanostructured materials, including carbon nanotubes and nanofibers, have been exhaustively researched over the past few decades to improve existing, and also develop novel, PEMFC/DMFC catalyst supports. This article has reviewed the recent developments and investigations reported on various catalyst supports. Many developments and improvements can be seen in the structure, poisoning tolerance, and stability of various nanostructured carbonaceous supports over the recent years.

In conclusion, the uses of nanomaterials in fuel cells may significantly improve the electrocatalytic performance for high energy density and high power density while reducing the manufacturing cost. The prominent electrocatalytic behavior of the nanomaterials is contributed mainly from their unique physical–chemical properties such as sizes, shapes, pore structures/distribution, surface defects, and chemical properties. The great challenges to synthesize and further use various nanostructured catalysts, such as CNT- and CNF-supported catalysts are not only from chemistry but also from nanoengineering approaches. In general, the core-shell nanostructures could provide an economic and effective way to prepare precious metal catalysts for remarkably reducing the usage of the noble metals while the unique nanostructures, such as nanotubes and nanofibers are believed to provide high specific active surface area, superior conductivity, and better mass transport as well as high intrinsic catalytic activity. An amalgamation of these novel electrocatalyst supports and improved catalyst loading techniques could bring about revolutionary changes in the quest for high-performance, long-lasting DMFCs. However, more detailed investigations (MEA studies, continuous cycling, and accelerated degradation tests) are still required to understand the behavior of these materials under "real" fuel cell conditions.

APPENDIX

A. Diagram of fuel cells invention history

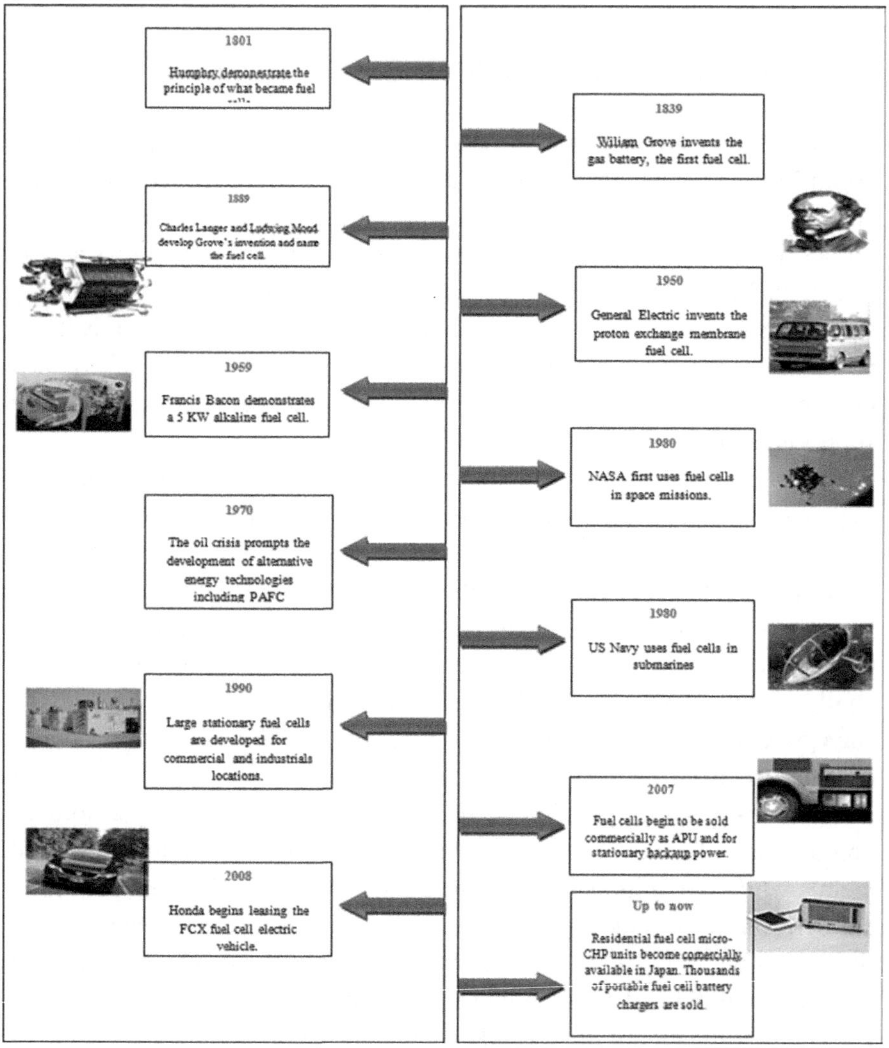

B. (a) Schematic representation of the structure of a carbon nanofiber; (b) enlarged section

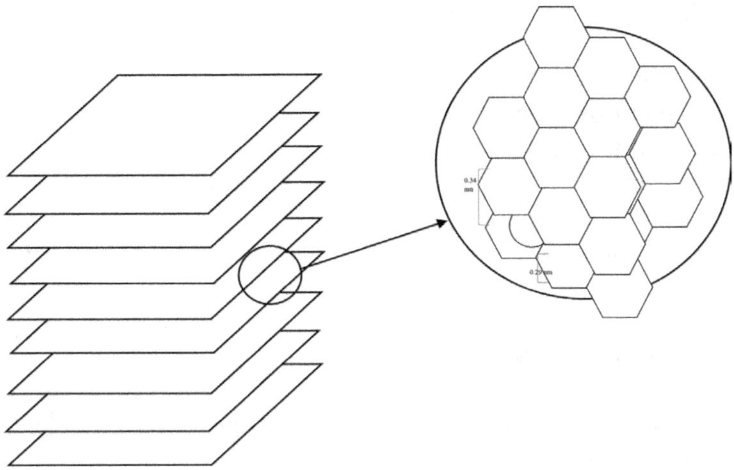

C. Schematic operation of (a) Solid Oxide fuel cells (SOFC) and (b) Proton Exchange Membrane fuel cells (PEM)

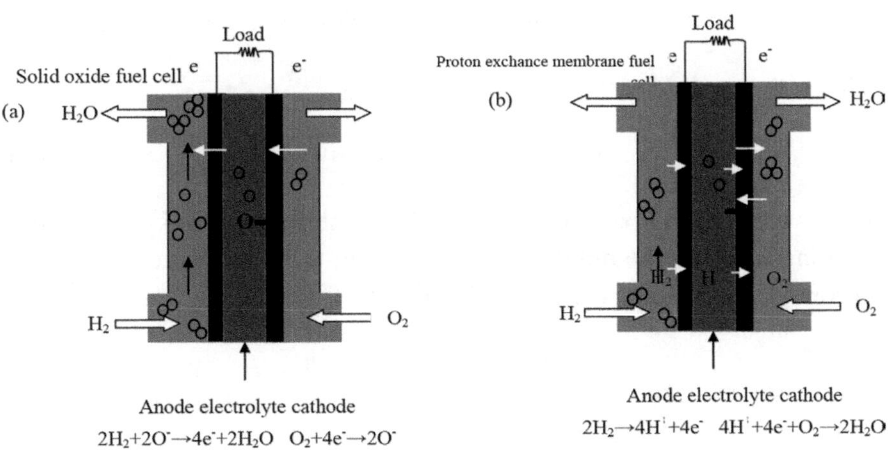

D. Useful Fuel Cell Equations

In this appendix, many useful equations are derived which are related to theses subjects:

- Oxygen usage rate
- Air inlet flow rate
- Air exit flow rate
- Hydrogen usage and the energy content of hydrogen
- Rate of water production
- Heat production.

D1. Equations of Oxygen and Air Usage

From the basic operation of the fuel cell, we know that four electrons are transferred for each mole of oxygen. Therefore,

$$\text{Charge} = 4F \times \text{amount of } O_2$$

Dividing by time, and rearranging:

$$O_2 \text{ usage} = I/4F \text{ moles s}^{-1}$$

This is for a single cell. For a stack of n cells

$$\text{O2 usage} = In/4F \text{ moles s}^{-1} \quad \text{(D.1)}$$

However, it would be more useful to have the formula in kg s^{-1}, without needing to know the number of cells, and in terms of power, rather than current. If the voltage of each cell in the stack is V_c, then

$$\text{Power, } P_e = V_c \times I \times n$$

So,

$$I = \frac{P_e}{V_c \times n}$$

replacing this into equation D.1 gives

$$O_2\,usage = \frac{P_e}{4.V_c.F} moless^{-1} \tag{D.2}$$

Changing from moles s^{-1} to kg s^{-1}

$$O_2\,usage = \frac{32\times10^{-3}.p_e}{4V_cF} kgs^{-1}$$

$$= 8.29\times10^{-8}\times\frac{P_e}{V_c} kgs^{-1} \tag{D.3}$$

This formula permits the oxygen usage of any fuel cell system of given power to be calculated. If V_c is not given, it can be calculated from the efficiency, and if that is not given, the figure of 0.65V can be used for a good approximation.

However, the oxygen used will normally be derived from air, so we need to adapt equation D.2 to air usage. The molar proportion of air that is oxygen is 0.21, and the molar mass of air is 28.97×10^{-3} kg $mole^{-1}$. So, equation B.2 becomes

$$Air\,usage = \frac{28.97\times10^{-3}.p_e}{0.21\times4\times V_c\times F} kgs^{-1}$$

$$= 3.57\times10^{-7}\times\frac{P_e}{V_c} kgs^{-1}$$

However, if the air were used at this rate, then as it left the cell it would be completely empty of any oxygen – it would all have been used. This is not useful, and in practice the airflow is well above stoichiometry, typically twice as much. If the stoichiometry is λ, then the equation for air usage becomes

$$Air\,usage = 3.57\times10^{-7}\times\lambda\times\frac{P_e}{V_c} kgs^{-1} \tag{D.4}$$

The kilogram per second is not, in fact, a very commonly used unit of mass flow. The following conversions to 'volume at standard conditions related' mass flow units will be found useful. The mass flow rate from equation B.4 should be multiplied by

- 3050 to give flow rate in standard $m^3\ h^{-1}$);
- 1795 to give flow rate in SCFM (or in standard $ft^3\ min^{-1}$);
- 5.1 × 104 to give flow rate in slm (standard $Lmin^{-1}$);
- 847 to give flow rate in sls (standard $L\ s^{-1}$).

D2. Equations of Air Exit Flow Rate

It is sometimes important to distinguish between the *inlet* flow rate of the air, which is given by equation B.4 above, and the *outlet* flow rate. This is particularly important when calculating the humidity, which is an important issue in certain types of fuel cells, especially proton exchange membrane (PEM) fuel cells. The difference is caused by the consumption of oxygen. There will usually be more water vapor in the exit air, but only 'dry air' at this stage is considered. Water production is given in Section D5. Clearly Exit airflow rate = Air inlet flow rate − oxygen usage. Using Equations D.3 and D.4 this becomes

$$Exit\,air\ flowrate = 3.57\times10^{-7}\times\lambda\times\frac{P_e}{V_c} - 8.29\times10^{-8}\times\frac{P_e}{V_c}kgs^{-1}$$

$$= (3.57\times10^{-7}\times\lambda - 8.29\times10^{-8})\times\frac{P_e}{V_c}kgs^{-1} \quad \text{(D.5)}$$

D3. Equations of Hydrogen Usage

The rate of usage of hydrogen is derived in a way similar to oxygen, except that there are two electrons from each mole of hydrogen. Equations D.1 and D.2 thus become

$$H_2\,usage = \frac{In}{2F}moless^{-1}$$

$$H_2\,usage = \frac{P_e}{2V_cF}moless^{-1} \quad \text{(D.6), (D.7)}$$

The molar mass of hydrogen is 2.02 × 10^{-3} kg $mole^{-1}$, so this becomes

$$H_2\,usage = \frac{2.02\times10^{-3}.p_e}{2\times V_c\times F}kgs^{-1}$$

$$=1.05\times10^{-8}\times\frac{P_e}{V_c}kgs^{-1} \qquad \text{(D.8)}$$

at stoichiometric operation. Obviously, this formula only applies to a hydrogen-fed fuel cell. In the case of a hydrogen/carbon monoxide mixture derived from a reformed hydrocarbon, things will be different, depending on the proportion of carbon monoxide present.

The result can be transformed to a volume rate using the density of hydrogen, which is 0.084 kgm^{-3} at normal temperature and pressure (NTP).

In addition to the rate of usage of hydrogen, it is often also useful to know the electrical energy that could be produced from a given mass or volume of hydrogen. The list presented in Table D.1 gives the energy in kilowatt-hour, rather than joules, as this is the measure usually applied for electrical power systems. In addition to the 'raw' energy per kilogram and standard liter, the 'effective' energy, taking into account the efficiency of the cell, is given. This is shown in terms of *Vc*, the mean voltage of each cell. If an equation with the efficiency is needed, then use the formula derived before,

$$efficiency = \frac{V_C}{1.48}$$

TABLE D.1 Raw and Effective Energy Content of Hydrogen Fuel

From	Energy content
Specific enthanlpy (HHV)	1.43×10^{8}J kg^{-1}
Specific enthanlpy (HHV)	39.7 $kWhkg^{-1}$
Effective specific electrical energy	$26.8\times V_c$ $kWhkg^{-1}$
Energy density at STP (HHV)	3.20$kWhm^{-3}$= 3.20 Wh SL^{-1}
Energy density at STP (HHV)	3.29$kWhm^{-3}$= 3.29 Wh SL^{-1}

*To obtain the lower heating value, multiply the HHV Figures given by 0.486.

D4. Equations of Water Production

In a hydrogen-fed fuel cell, water is produced at the rate of one mole for every two electrons. So, we again adapt equation D.2 to obtain

$$Water\ production = \frac{p_e}{2 \times V_c \times F} moless^{-1} \quad \text{(D.9)}$$

The molecular mass of water is 18.02×10^{-3} kg mole^{-1}, so this becomes

$$Water\ production = 9.34 \times 10^{-8} \times \frac{p_e}{V_c} kgs^{-1} \quad \text{(D.10)}$$

In the hydrogen-fed fuel cell, the rate of water production more or less has to be stoichiometric. However, if the fuel is a mixture of carbon monoxide with hydrogen, then the water production would be less – in proportion to the amount of carbon monoxide present in the mixture. If the fuel was a hydrocarbon that was internally reformed, then some of the product water would be used in the reformation process. For example, if methane is internally reformed, then half the product water is used in the reformation process, thus halving the rate of production.

It is sometimes useful to give an example figure to clarify a formula such as this. Let us take as an example a 1kW fuel cell operating for 1 h, at a cell voltage of 0.7 V. This corresponds to an efficiency of 47%. Substituting this into equation D.9 gives

$$The\ rate\ of\ water\ production = 9.34 \times 10^{-8} \times \frac{1000}{0.7}$$
$$= 1.33 \times 10^{-4} kgs^{-1}$$

So the mass of water produced in 1 h is

$$= 1.33 \times 10^{-4} \times 60 \times 60 = 0.48 kg$$

Since the density of water is 1.0 g cm^{-3}, this corresponds to 480 cm^3, which is almost exactly 1 pint. So, as a rough guide, 1 kWh of fuel cell generated electricity produces about 1 pint or 0.5 L of water.

D5. Equations of Heat Produced

As we discussed before, heat is created when a fuel cell operates. It was noted before that if all the enthalpy of reaction of a hydrogen fuel cell was converted into electrical energy then the output voltage would be 1.48V if the water product was in liquid form or 1.25V if the water product was in vapor form.

It clearly indicates that the difference between the actual cell voltage and this voltage represents the energy that is not converted into electricity – that is, the energy that is converted into heat instead. The cases in which water finally ends in liquid form are so few and far between that they are not worth considering. So only the vapor case is considered.

However, just the cooling effect of water evaporation was taken to account. It also means that energy is leaving the fuel cell in three forms: as electricity, as ordinary 'sensible' heat, and as the latent heat of water vapor. For a stack of n cells at current I, the heat generated is thus

$$Heating\ rate = nI(1.25 - V_c)W$$

In terms of electrical power, this becomes

$$Heating\ rate = P_e(\frac{1.25}{V_c} - 1)W \quad (D.11)$$

KEYWORDS

- **activated carbon**
- **electrochemical energy**
- **engineering approach**
- **fuel cells**
- **hydrogen storage**
- **nanofibers**
- **nanotubes**

REFERENCES

1. Jiang, S., Shen, P., Nanostructured and Advanced Materials for Fuel Cells. 2013, New York: CRC Press. 370.
2. Schimmel, H., et al., Hydrogen Adsorption in Carbon Nanostructures: Comparison of Nanotubes, Fibers, and Coals. Chemistry – A European Journal, 2003, 9, 4764–4770.
3. Thavasi, V., Singh, G., Ramakrishna, S., Electrospun nanofibers in energy and environmental applications. Energy and Environmental Science, 2008, 1, 205–221.
4. Cook, B., An Introduction to Fuel Cells and Hydrogen Technology. 2001, Vancouver.
5. Banerjee, S., Murad, S., Puri, I., Hydrogen Storage in Carbon Nanostructures: Possibilities and Challenges for Fundamental Molecular Simulations. Proceedings of the IEEE, 2006, 94, 1806–1814.
6. Züttel, A., et al., Hydrogen storage in carbon nanostructures. International Journal of Hydrogen Energy 2002, 27, 203–210.
7. Figueroa-Torres, M., et al., Hydrogen adsorption by nanostructured carbons synthesized by chemical activation. Microporous and Mesoporous Materials, 2007, 98, 89–93.
8. Im, J., et al., The study of controlling pore size on electrospun carbon nanofibers for hydrogen adsorption. Journal of Colloid and Interface Science, 2008, 318, 42–49.
9. Edwards, P., et al., Hydrogen and fuel cells: towards a sustainable energy future. Energy policy, 2008, 36(12), 4356–4362.
10. Crabtree, G., Dresselhaus, M., Buchanan, M., The hydrogen economy. Physics Today, 2004, 57(12), 39–44.
11. Mori, D., Hirose, K., Recent challenges of hydrogen storage technologies for fuel cell vehicles. International Journal of Hydrogen Energy, 2009, 34(10), 4569–4574.
12. Shinnar, R., The hydrogen economy, fuel cells, and electric cars. Technology in Society, 2003, 25(4), 455–476.
13. L.Vasiliev, L., et al., Hydrogen storage system based on novel carbon materials and heat pipe heat exchanger. International Journal of Thermal Sciences, 2007, 46, 914–925.
14. Oh, W., Park, J., Hydrogen Gas Generation of Metal-Activated Carbon for Fuel Cells. Journal of Chemical Industry and Engineering, 2007, 13, 578–584.
15. Sharon, M., et al., Synthesis of Carbon Nano-Fiber From Ethanol and It's Hydrogen Adsorption Capacity. 2004.
16. Ahn, C. C., et al., Hydrogen desorption and adsorption measurements on graphite nanofibers. Applied Physics Letters, 1998, 73.
17. Tibbetts, G. G., Meisner, G. P., C.Olk, H., Hydrogen storage capacity of carbon nanotubes, filaments, and vapor-grown fibers. Carbon, 2001, 39, 2291–2301.
18. Torres, M. Z. F., et al., Hydrogen adsorption by nanostructured carbons synthesized by chemical activation. Microporous and Mesoporous Materials, 2007, 98, 89–93.
19. Chahine, R., Bénard, P., Assessment of Hydrogen Storage on Different Carbons. Metal Hydrides and Carbon for Hydrogen Storage 2001.
20. R. Strobela et al., Hydrogen storage by carbon materials. Journal of Power Sources, 2006, 156, 781–801.
21. Chambers, A., et al., Hydrogen storage in graphite nanofibers. The journal of physical chemistry B, 1998, 102(22), 4253–4256.

22. Kroto, H., Fischer, J., Cox, D., The fullerenes. 2012, New York: Newnes. 358.
23. Park, C., et al., Further studies of the interaction of hydrogen with graphite nanofibers. The journal of physical chemistry B, 1999, 103(48), 10572–10581.
24. Yang, R., Hydrogen storage by alkali-doped carbon nanotubes–revisited. Carbon, 2000, 38(4), 623–626.
25. Atkinson, K., et al., Carbon nanostructures: An efficient hydrogen storage medium for fuel cells. Fuel Cells Bulletin, 2001, 4(38), 9–12.
26. Iijima, S., Helical microtubules of graphitic carbon. Nature, 1991, 354(6348), 56–58.
27. Ajayan, P., Zhou, O., Applications of carbon nanotubes, in Carbon nanotubes. 2001, Springer. 391–425.
28. Dillon, A., et al., Storage of hydrogen in single-walled carbon nanotubes. Nature, 1997, 386(6623), 377–379.
29. Hirscher, M., et al., Hydrogen storage in sonicated carbon materials. Applied Physics A, 2001, 72(2), 129–132.
30. Darkrim, F., Malbrunot, P., Tartaglia, G., Review of hydrogen storage by adsorption in carbon nanotubes. International Journal of Hydrogen Energy, 2002, 27(2), 193–202.
31. Browning, D. J., et al., Studies into the Storage of Hydrogen in Carbon Nanofibers: Proposal of a Possible Reaction Mechanism. Nano letters, 2002, 2, 201–205.
32. Rzepka, M., et al., Hydrogen Storage Capacity of Catalytically Grown Carbon Nanofibers. J. Phys. Chem., 2005, 109, 14979–14989.
33. James, M. B., et al., Activation of carbon nanofibers for hydrogen storage. Carbon, 2006, 44, 44.
34. Im, J., et al., The study of controlling pore size on electrospun carbon nanofibers for hydrogen adsorption. Journal of Colloid and Interface Science, 2008, 318, 42–49.
35. S.-Yoon, H., et al., KOH activation of carbon nanofibers. Carbon, 2004, 42, 1723–1729.
36. Byung-Joo, K., Young-Seak, L., Soo-Jin, P., Preparation of platinum-decorated porous graphite nanofibers, and their hydrogen storage behaviors. Journal of Colloid and Interface Science, 2008, 318, 530–533.
37. Lee, Y. S., J.Im, S., Preparation of Functionalized Nanofibers and Their Applications, in Nanofibers, Kumar, A., Editor. 2010.
38. Byung-Joo, K., Young-Seak, L., Soo-Jin, P., A study on the hydrogen storage capacity of Ni-plated porous carbon nanofibers. International Journal of Hydrogen Energy, 2008, 3-3, 4-1(1–2)–4-1(1–5).
39. Blackman, J., et al., Activation of carbon nanofibers for hydrogen storage. Carbon, 2006, 44, 44.
40. Jarvi, T.D., et al., Interaction of Hydrogen with Nanoporous Carbon Materials. Carbon, 1998, 33, 962–967.
41. Browning, D., et al., Studies into the Storage of Hydrogen in Carbon Nanofibers: Proposal of a Possible Reaction Mechanism. Nano letters, 2002, 2, 201–205.
42. Kim, D., et al., Electrospun Polyacrylonitrile-Based Carbon Nanofibers and Their Hydrogen Storages. Macromolecular Research, 2005, 13, 521–528.
43. Hacker, V., et al., Carbon nanofiber-based active layers for fuel cell cathodes – preparation and characterization. Electrochemistry Communications, 2005, 7, 377–382.
44. Guha, A., et al., Synthesis of novel Platinum/Carbon nanofiber electrodes for Polymer Electrolyte Membrane (PEM) fuel cells. Solid State Electrochem, 2001, 5, 131.

45. Perchthaler, M., Wallnöfer, E., Hacker, V., Performance of Carbon Nano-Fiber supported Membrane Electrode Assemblies. 2002, Graz University of Technology: Graz.
46. Wu, J., et al., A review of PEM fuel cell durability: degradation mechanisms and mitigation strategies. Journal of Power Sources, 2008, 184(1), 104–119.
47. Borup, R., et al., Scientific aspects of polymer electrolyte fuel cell durability and degradation. Chemical reviews, 2007, 107(10), 3904–3951.
48. Shao, Y., et al., Proton exchange membrane fuel cell from low temperature to high temperature: material challenges. Journal of Power Sources, 2007, 167(2), 235–242.
49. Zhang, J., et al., High temperature PEM fuel cells. Journal of Power Sources, 2006, 160(2), 872–891.
50. Cleghorn, S., et al., A polymer electrolyte fuel cell life test: 3 years of continuous operation. Journal of Power Sources, 2006, 158(1), 446–454.
51. Luo, Z., et al., Degradation behavior of membrane–electrode-assembly materials in 10-cell PEMFC stack. International Journal of Hydrogen Energy, 2006, 31(13), 1831–1837.
52. Cheng, X., et al., Catalyst microstructure examination of PEMFC membrane electrode assemblies vs. time. Journal of the Electrochemical Society, 2004, 151(1), p. A48–A52.
53. Coelho, N., et al., Carbon Nanofibers: a Versatile Catalytic Support. Materials Research, 2008, 11, 353–357.
54. Sebastián, D., et al., Carbon nanofibers as electrocatalyst support for fuel cells: Effect of hydrogen on their properties in CH4 decomposition. Journal of Power Sources, 2009, 192, 51–56.
55. Lee, K., et al., Progress in the synthesis of carbon nanotube- and nanofiber-supported Pt electrocatalysts for PEM fuel cell catalysis. Journal of Applied Electrochemistry, 2006, 507–522.
56. Li, W., et al., Platinum nanoparticles supported on stacked-cup carbon nanofibers as electrocatalysts for proton exchange membrane fuel cell. carbon, 2012, 48, 995–1003.
57. Chu, D., Jiang, R., Novel electrocatalysts for direct methanol fuel cells. Solid State Ionics, 2002, 148(3), 591–599.
58. Arico, A., et al., Optimization of operating parameters of a direct methanol fuel cell and physico-chemical investigation of catalyst–electrolyte interface. Electrochimica Acta, 1998, 43(24), 3719–3729.
59. Reddington, E., et al., Combinatorial electrochemistry: a highly parallel, optical screening method for discovery of better electrocatalysts. Science, 1998, 280(5370), 1735–1737.
60. Parsons, R., T. Vander Noot, The oxidation of small organic molecules: a survey of recent fuel cell related research. Journal of electroanalytical chemistry and interfacial electrochemistry, 1988, 257(1), 9–45.
61. Schmidt, T., et al., The oxygen reduction reaction on a Pt/carbon fuel cell catalyst in the presence of chloride anions. Journal of Electroanalytical Chemistry, 2001, 508(1), 41–47.
62. Matsumoto, T., et al., Reduction of Pt usage in fuel cell electrocatalysts with carbon nanotube electrodes. Chemical communications, 2004(7), 840–841.
63. Lamy, C., Belgsir, E., Leger, J., Electrocatalytic oxidation of aliphatic alcohols: application to the direct alcohol fuel cell (DAFC). Journal of Applied Electrochemistry, 2001, 31(7), 799–809.

64. Vinodgopal, K., et al., Fullerene-based carbon nanostructures for methanol oxidation. Nano Letters, 2004, 4(3), 415–418.
65. Liu, L., et al., Carbon supported and unsupported Pt–Ru anodes for liquid feed direct methanol fuel cells. Electrochimica Acta, 1998, 43(24), 3657–3663.
66. Lizcano-Valbuena, W., D. de Azevedo, Gonzalez, E., Supported metal nanoparticles as electrocatalysts for low-temperature fuel cells. Electrochimica Acta, 2004, 49(8), 1289–1295.
67. Anderson, M., Stroud, R., Rolison, D., Enhancing the activity of fuel-cell reactions by designing three-dimensional nanostructured architectures: catalyst-modified carbon-silica composite aerogels. Nano Letters, 2002, 2(3), 235–240.
68. Gasteiger, H., et al., Carbon monoxide electrooxidation on well-characterized platinum-ruthenium alloys. The Journal of Physical Chemistry, 1994, 98(2), 617–625.
69. Gojković, S., Vidaković, T., Đurović, D., Kinetic study of methanol oxidation on carbon-supported PtRu electrocatalyst. Electrochimica Acta, 2003, 48(24), 3607–3614.
70. Hamnett, A., Mechanism and electrocatalysis in the direct methanol fuel cell. Catalysis Today, 1997, 38(4), 445–457.
71. Wasmus, S., Küver, A., Methanol oxidation and direct methanol fuel cells: a selective review. Journal of Electroanalytical Chemistry, 1999, 461(1), 14–31.
72. Sun, X., et al., Composite electrodes made of Pt nanoparticles deposited on carbon nanotubes grown on fuel cell backings. Chemical Physics Letters, 2003, 379(1), 99–104.
73. Auer, E., et al., Carbons as supports for industrial precious metal catalysts. Applied Catalysis A: General, 1998, 173(2), 259–271.
74. Shao, Y., et al., Comparative investigation of the resistance to electrochemical oxidation of carbon black and carbon nanotubes in aqueous sulfuric acid solution. Electrochimica Acta, 2006, 51(26), 5853–5857.
75. Che, G., et al., Metal-nanocluster-filled carbon nanotubes: catalytic properties and possible applications in electrochemical energy storage and production. Langmuir, 1999, 15(3), 750–758.
76. Wu, G., Chen, Y., Xu, B., Remarkable support effect of SWNTs in Pt catalyst for methanol electrooxidation. Electrochemistry communications, 2005, 7(12), 1237–1243.
77. Saha, M., et al., Enhancement of PEMFC performance by using carbon nanotubes supported Pt/Co alloy catalysts. Asia-Pacific Journal of Chemical Engineering, 2009, 4(1), 12–16.
78. Yuan, Y., et al., Platinum decorated aligned carbon nanotubes: Electrocatalyst for improved performance of proton exchange membrane fuel cells. Journal of Power Sources, 2011, 196(15), 6160–6167.
79. Hernández-Fernández, P., et al., Functionalization of multi-walled carbon nanotubes and application as supports for electrocatalysts in proton-exchange membrane fuel cell. Applied Catalysis B: Environmental, 2010, 99(1), 343–352.
80. Golikand, A., Lohrasbi, E., Asgari, M., Enhancing the durability of multi-walled carbon nanotube supported by Pt and Pt–Pd nanoparticles in gas diffusion electrodes. International Journal of Hydrogen Energy, 2010, 35(17), 9233–9240.
81. Steigerwalt, E., Deluga, G., Lukehart, C., Pt-Ru/carbon fiber nanocomposites: Synthesis, characterization, and performance as anode catalysts of direct methanol fuel cells. A search for exceptional performance. The journal of physical chemistry B, 2002, 106(4), 760–766.

82. Jha, N., et al., Pt–Ru/multi-walled carbon nanotubes as electrocatalysts for direct methanol fuel cell. International Journal of Hydrogen Energy, 2008, 33(1), 427–433.
83. Thess, A., et al., Crystalline ropes of metallic carbon nanotubes. Science-AAAS-Weekly Paper Edition, 1996, 273(5274), 483–487.
84. Pantea, D., et al., Electrical conductivity of thermal carbon blacks: Influence of surface chemistry. Carbon, 2001, 39(8), 1147–1158.
85. Li, L., Xing, Y., Electrochemical durability of carbon nanotubes in noncatalyzed and catalyzed oxidations. Journal of the Electrochemical Society, 2006, 153(10), 1823–1828.
86. Rodriguez, N., Kim, M., Baker, R., Carbon nanofibers: a unique catalyst support medium. The Journal of Physical Chemistry, 1994, 98(50), 13108–13111.
87. Park, C., Baker, R., Catalytic behavior of graphite nanofiber supported nickel particles. 2. The influence of the nanofiber structure. The journal of physical chemistry B, 1998, 102(26), 5168–5177.
88. Bessel, C., et al., Graphite nanofibers as an electrode for fuel cell applications. The journal of physical chemistry B, 2001, 105(6), 1115–1118.
89. Yang, G., et al., Effective adhesion of Pt nanoparticles on thiolated multi-walled carbon nanotubes and their use for fabricating electrocatalysts. Carbon, 2007, 45(15), 3036–3041.
90. Okada, M., Konta, Y., Nakagawa, N., Carbon nano-fiber interlayer that provides high catalyst utilization in direct methanol fuel cell. Journal of Power Sources, 2008, 185(2), 711–716.
91. Joo, S., et al., Ordered mesoporous carbons (OMC) as supports of electrocatalysts for direct methanol fuel cells (DMFC): effect of carbon precursors of OMC on DMFC performances. Electrochimica Acta, 2006, 52(4), 1618–1626.
92. Terranova, M., Special issue on carbon nanotubes. Chemical Vapor Deposition, 2006, 12(6), 313–313.
93. Haddon, R., Carbon nanotubes. Accounts of Chemical Research, 2002, 35(12), 997–997.
94. Harris, P., Carbon nanotubes and related structures: new materials for the twenty-first century. 2001, Cambridge University Press. 438.
95. Saito, R., Dresselhaus, G., Dresselhaus, M., Physical properties of carbon nanotubes. Vol. 4. 1998, World Scientific. 350.
96. Dai, H., Carbon nanotubes: synthesis, integration, and properties. Accounts of Chemical Research, 2002, 35(12), 1035–1044.
97. Durrer, L., et al., SWNT growth by CVD on Ferritin-based iron catalyst nanoparticles towards CNT sensors. Sensors and Actuators B: Chemical, 2008, 132(2), 485–490.
98. Wong, Y., et al., Carbon nanotubes field emission devices grown by thermal CVD with palladium as catalysts. Diamond and related materials, 2004, 13(11), 2105–2112.
99. Dikonimos Makris, T., et al., CNT growth on alumina supported nickel catalyst by thermal CVD. Diamond and related materials, 2005, 14(3), 815–819.
100. Liu, H., et al., Aligned synthesis of multi-walled carbon nanotubes with high purity by aerosol assisted chemical vapor deposition: Effect of water vapor. Applied Surface Science, 2010, 256(14), 4692–4696.
101. Ci, L., et al., Preparation of carbon nanotubules by the floating catalyst method. Journal of materials science letters, 1999, 18(10), 797–799.

102. Barreiro, A., et al., Control of the single-wall carbon nanotube mean diameter in sulfur promoted aerosol-assisted chemical vapor deposition. Carbon, 2007, 45(1), 55–61.
103. Perez, H., et al., Evidence for high performances of low Pt loading electrodes based on capped platinum electrocatalyst and carbon nanotubes in fuel cell devices. Electrochimica Acta, 2010, 55(7), 2358–2362.
104. Li, Y., et al., Preferential growth of semiconducting single-walled carbon nanotubes by a plasma enhanced CVD method. Nano Letters, 2004, 4(2), 317–321.
105. Duy, D., et al., Growth of carbon nanotubes on stainless steel substrates by DC-PECVD. Applied Surface Science, 2009, 256(4), 1065–1068.
106. Saha, M., et al., 3-D composite electrodes for high performance PEM fuel cells composed of Pt supported on nitrogen-doped carbon nanotubes grown on carbon paper. Electrochemistry communications, 2009, 11(2), 438–441.
107. Park, C., et al., Use of carbon nanofibers in the removal of organic solvents from water. Langmuir, 2000, 16(21), 8050–8056.
108. Fan, Y., et al., Hydrogen uptake in vapor-grown carbon nanofibers. Carbon, 1999, 37(10), 1649–1652.
109. Wang, X., Wang, S., Chung, D., Sensing damage in carbon fiber and its polymer-matrix and carbon-matrix composites by electrical resistance measurement. Journal of materials science, 1999, 34(11), 2703–2713.
110. Rodriguez, N., A review of catalytically grown carbon nanofibers. Journal of Materials Research, 1993, 8(12), 3233–3250.
111. De Jong, K., Geus, J., Carbon nanofibers: catalytic synthesis and applications. Catalysis Reviews, 2000, 42(4), 481–510.
112. Vamvakaki, V., Tsagaraki, K., Chaniotakis, N., Carbon nanofiber-based glucose biosensor. Analytical chemistry, 2006, 78(15), 5538–5542.
113. Hao, C., et al., Biocompatible conductive architecture of carbon nanofiber-doped chitosan prepared with controllable electrodeposition for cytosensing. Analytical chemistry, 2007, 79(12), 4442–4447.
114. Mirabile Gattia, D., et al., Study of different nanostructured carbon supports for fuel cell catalysts. Journal of Power Sources, 2009, 194(1), 243–251.
115. Ismagilov, Z., et al., Development of active catalysts for low Pt loading cathodes of PEMFC by surface tailoring of nanocarbon materials. Catalysis Today, 2005, 102, 58–66.
116. Rodriguez, N., Kim, M., Baker, R., Promotional effect of carbon monoxide on the decomposition of ethylene over an iron catalyst. Journal of Catalysis, 1993, 144(1), 93–108.
117. Helveg, S., et al., Atomic-scale imaging of carbon nanofiber growth. Nature, 2004, 427(6973), 426–429.
118. Ermakova, M., Ermakov, D., Kuvshinov, G., Effective catalysts for direct cracking of methane to produce hydrogen and filamentous carbon: Part I. Nickel catalysts. Applied Catalysis A: General, 2000, 201(1), 61–70.
119. Sebastián, D., et al., Influence of carbon nanofiber properties as electrocatalyst support on the electrochemical performance for PEM fuel cells. International Journal of Hydrogen Energy, 2010, 35(18), 9934–9942.

120. Bi, Y., Lu, G., Control growth of uniform platinum nanotubes and their catalytic properties for methanol electrooxidation. Electrochemistry communications, 2009, 11(1), 45–49.
121. Fujigaya, T., Okamoto, M., Nakashima, N., Design of an assembly of pyridine-containing polybenzimidazole, carbon nanotubes and Pt nanoparticles for a fuel cell electrocatalyst with a high electrochemically active surface area. Carbon, 2009, 47(14), 3227–3232.
122. Zhu, Q., et al., Controlled synthesis of mesoporous carbon modified by tungsten carbides as an improved electrocatalyst support for the oxygen reduction reaction. Journal of Power Sources, 2009, 193(2), 495–500.

CHAPTER 10

TRENDS IN POLYMER CHEMISTRY

G. E. ZAIKOV

Russian Academy of Sciences, Moscow, Russia

CONTENTS

10.1 INTRODUCTION

In chemistry and in chemical technology often one uses such notions like: macromolecules, polymers, synthetic materials, plastics and elastomers. They have similar means but they are not synonyms.

Macromolecules are substances with a very large molecular mass, of linear structure or forming a special network. Sometimes it is difficult to give their chemical structures because the particular elements forming them exhibit an unrepeatable structure.

Polymers are macromolecules formed from repeatable units, called "mers." The number of mers in macromolecule determines the degree of polymerization.

Thermoplastic Polymers are utility materials, made from polymers, combined with different additives, such as fillers, plasticizers, stabilizers, dyes, pigments, modifiers and others. They can be processed in the melt.

Plastic Materials (Plastics) are synthetic materials made of polymers and additives.

Elastomers are synthetic materials, characterized by a large ability for deformations and elongation (up to 1200%), keeping at the same time intact the elastic properties. It means that after stopping the deformation force they recover their initial form.

A characteristic property of polymer structure is the presence of exceptionally large size constituent macromolecules, composed from a large number of atoms. Typical length of polymers oscillates between 100 nm (1 nm is a millionth part of millimeter) and 100–000 nm (100 μm). However, the diameter of individual linear polymer chain doesn't surpass the diameter of constituent molecules, for example, about 1 nm. Therefore, the structure of polymers can be observed only by the very high-resolution microscopes (atomic force microscope (AFM), scanning tunneling microscope (STM) and analyzed with help of the electron microscope (EM).

The length of particular macromolecules is determined by the number of constituent atoms, so their molecular mass. Because the particular macromolecules differ in the length of polymer chain for their description one introduces the notion of average molecular mass ($\overline{M}$). It appears that very often polymers prepared from the same monomer, characterized by the same average molecular mass exhibit different properties. Therefore, the notion of polydispersity of polymers was introduced. It describes the molecular mass distribution.

There exist several experimental techniques allowing a practical determination of the polydispersity degree for a polymer. Between them the most important are:

- method of fractional precipitation and solubilization;
- chromatographic technique exploiting the adsorption ability of macromolecules;
- fractioning in a centrifuge;
- measure of the diffusion rate;
- light scattering measurements;
- comparison of molecular masses determined by different techniques.

The polydispersity is usually expressed graphically using the molecular weight distribution curve. For this purpose, the experimentally found values for each fraction of polymer are spotted on a chart in the reference frame: degree of polymerization P – weight part of mass fraction dM/dP. Through the obtained in this way stepped chart an integration line is passed. This curve represents the weight parts of polymer fractions of with polymerization degree P (number of mers in the molecule) from P to (P + dP).

Taking the integral curve as the abscess one can plot in the coordinate system P, dM/dP the differential distribution curve. This curve exhibits a sharply marked maximum and allows a more illustrative observation of the degree of polydispersity of polymer than it is the case with the integral curve. Figure 10.1 shows, as examples, the differential distribution curves of molecular weights of cellulose nitrate and of polystyrene with different degrees of polydispersity.

The polydispersity of a polymer can be controlled, to a certain degree, by choosing in an adequate way the polymerization conditions. Polymers are obtained through:

- polymerization itself, for example, chemical binding of monomer molecules into one macromolecule without liberation of side products;

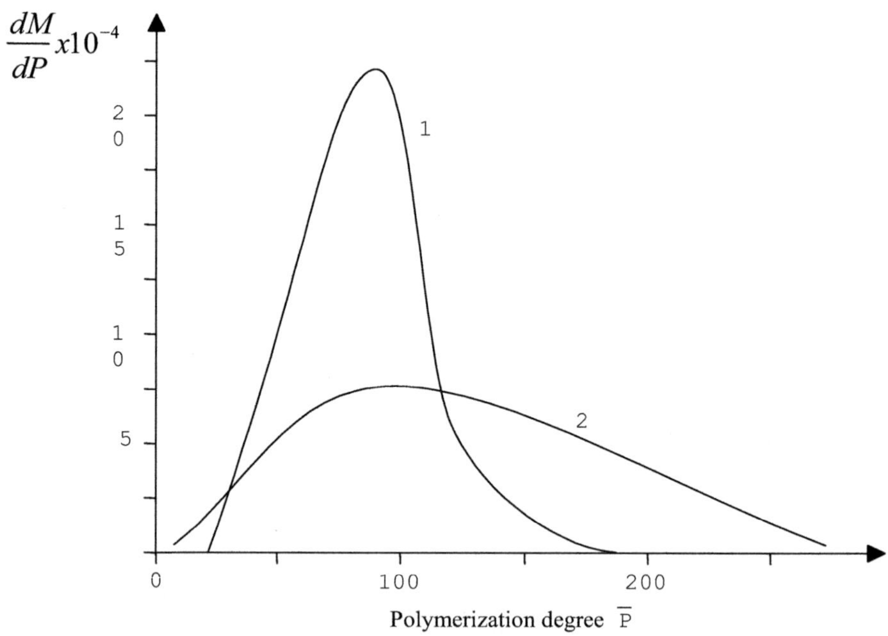

FIGURE 10.1 Differential distribution curves of molecular weight M for: (1) cellulose nitrate with an average degree of polymerization 860; (2) polystyrene with an average degree of polymerization 800.

- polycondensation (condensation polymerization), consisting on binding of a large number of molecules, containing reactive groups, into one macromolecule with liberation of small molecule side products (e.g., H_2O, HCl, NH_3, etc.);
- synthesis of macromolecules by living species (e.g., bacteria);
- chemical modification of natural or synthetic polymers.

The type and the properties of obtained polymer in a large measure depend on the monomer used. As a result of polymerization of bifunctional monomers, with one double bond or two reactive groups in molecule, one obtains linear polymers (chain like), which can melt and are soluble in solvents of similar polarity.

In the case of polymerization of multifunctional monomers (with two double bonds or at least three functional groups) one obtains in the first reaction step linear polymers, containing reactive groups, able to interact with neighboring chains and form cross linkages between polymer chains. As result one obtains a spatially cross-linked polymer, forming a very large 3D macromolecule (a block polymer can be just one macromolecule). Such polymers are usually insoluble and do not melt.

Polymers play a very important role in a lot of vital processes. They find also a large and increasing number of applications in, practically, each branch of industry as well as in everyday life. At present the polymer fabrication in the world is so important that its size for a given country may be used as an indicator of the degree of development and modernity of its chemical industry.

Macromolecules, composed of thousands of atoms, are the basic components of the living world. Polymer macromolecules are part of a large number of materials used by man since the prehistory, such as leather, natural fibers (linen, cotton, wool, silk), wood and rubber. Their number has increased significantly in recent years. A large number of new polymers and composites were obtained. The common characteristic of chemical structure of polymers is the macromolecule, which usually is a long chain, made of hundreds or thousands of mers, connected together by chemical bonds. The polymer chain ability to take different geometrical forms provides properties unattainable in the case of substances made of smaller molecules.

The long polymer chains may form regular series, forming crystalline phases, characterized by a high hardness and mechanical strength. Due to the perfect ordering of molecules materials with strength over passing that of steel were obtained.

The development of polymer chemistry was particularly important in the last, twentieth century. New types of polymers, with different properties, were obtained. They found application as construction materials, foils, synthetic fibers, synthetic rubbers, adhesives, ion exchangers, products to change the soil structure, coagulating agents, in medicines, electric conductors and semiconductors, photoconductors, etc. The diversity and versatility of polymers, provided by the possible infinite modification of their molecular and macroscopic structure guarantees a further development and increase of applications of these materials.

The twentieth century has known an exceptional development of polymer chemistry. New types of polymers, characterized by new properties, were synthesized. They found numerous applications, such as construction materials, foils, synthetic fibers, rubbers, adhesives, paints, ionic exchangers, agents amending the soil structure, coagulants, medicine and medicine components, packaging materials, semiconductors and others. The diversity of polymer properties, possible due to an infinite variability of the molecule and macromolecule structures, is a best guarantee of further growth of their applications.

Nowhere a large diversity of polymers is observed as in biological materials. Among them one distinguishes proteins and polycarbohydrates, which are part of all living organisms: vegetal and animal. Another important biopolymers are enzymes (globular proteins), being catalysts of processes connected with the life, polynucleotides (DNA and RNA), responsible for genetic information during the whole development of living organisms. The biopolymers have to fulfill well-defined functions. The sequence of amino acids in the chains of proteins is well established and unchangeable, ideally repeated in all molecules of a given protein. Such repeatability was still not obtained by using the modern, available, polymer synthesis methods. The performances of nature in this field demand the polymer scientists to use more and more elaborated synthesis methods. The polymer science develops in an exceptionally dynamic way and the new discoveries will be of importance in different scientific and technical areas.

Polymers are basic elements of synthetic materials and nowadays they find applications in all areas of life and of industrial activity.

10.2 METHODS OF POLYMERIZATION OF VINYL COMPOUNDS

10.2.1 FREE RADICAL POLYMERIZATION OF VINYL COMPOUNDS

10.2.1.1 Effect of Monomers on the Course of Polymerization

Vinyl monomers are derivatives of ethylene, in which one or more hydrogen atoms have been replaced by other substituents. A characteristic feature of the construction of connections of this type is the electron pair-sharing (e), and thus the presence of a double carbon–carbon bond. The sp^2 orbitals present in monomer molecule make that each carbon atom is in the middle of a triangle, in which the vertices are: hydrogen atom of substituent and the second carbon atom.

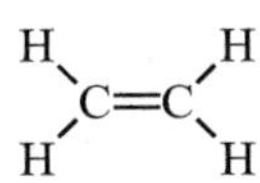

Each of the angles between the bonds in such a system is equal to 120°. By creating sp^2 orbitals, each carbon atom uses only two of its three p orbitals. The remaining p orbital is composed of two equal loops, one of which lies above and the second below the plane defined by the three sp^2 orbitals and is filled by one electron. By a combination of the two p orbitals of carbon a new π bond is created, consisting of an electron cloud located above the plane and defined by the atoms and another one lying below this plane. Due to the lateral orbital overlap, the binding energy of the double π bond between carbon atoms is greater than that a single δ bond.

The binding occurs only when the p-orbitals may overlap. It means that it is possible only when all six atoms responsible for the establishment of the bond (2C atoms and 4 H atoms) are lying in the same plane. Therefore, an ethylene molecule is flat.

Thus the double C = C bond consists of a strong δ bond and a weak π bond. The total binding energy in this case amounts to 682 kJ/mol and is larger than the binding energy characteristic of a single carbon–carbon bond in the ethane molecule, which is of 368 kJ/mol. Therefore, the distance between carbon atoms in the molecule of an unsaturated compound is smaller than the distance between carbon atoms in the ethane molecule. It means that the double carbon–carbon bond is shorter than the corresponding single bond.

The free radical polymerization process consists on breaking the double bond between carbon atoms in the monomer molecule, followed by a chain reaction of its growing. From the above considerations it follows that the polymerization process is an exothermic reaction. The monomer molecules can polymerize only if the process is accompanied by a decrease of free energy ($\Delta G < 0$):

$$\Delta G = \Delta H - T\Delta S$$

where: ΔG – change of the free energy of system; ΔH – change of the enthalpy of the system, equal to the heat of reaction with the opposite sign; ΔS – change of the entropy of the system.

The polymerization process runs always with a decrease of the entropy of system. The value of $T\Delta S$ at 298 K is of 31.38–41.84 kJ/mol. For $\Delta G < 0$ the following condition must be fulfilled:

$$\Delta H < 31.38 \text{ kJ/mol.}$$

The double bond energy in vinyl monomers is of 609 kJ/ mol. In the expense of that two single carbon–carbon bonds with energy of 2 × 351.5 kJ/mol = 703 kJ/mol are formed. The difference of these thermal effects is equal to the theoretical value of the polymerization heat Q:

$$Q = 703–609 = 94 \text{ kJ/mol}$$

In practice it turns out that there are significant differences between the various polymerization heats of different monomers. This is caused by both: the loss of coupling energy in the conversion of monomer molecules into the polymer molecule and the energy loss associated with the formation of stresses in the polymer chain as a result of mutual interaction of substituents (steric effect). The polymerization heats of selected monomers are listed in Table 10.1.

The steric effect in some monomers is so large that their polymerization is often nearly impossible. Examples of such monomers containing exceptionally large substituents are the 1,2-dibenzil (1,2-diphenylethane) and the maleic acid.

The mechanism of polymerization is influenced also by the size and the type of the polarization of double bond in the monomer molecule, which depends on its structure. Depending on the type of functional groups adjacent to the double bond, monomers can be divided into three types:

- Polymerizing by free radical mechanism (containing substituents affecting slightly π electrons);
- Polymerizing according to the cationic mechanism (containing substituents dislocating π electrons); and

TABLE 10.1 Heat of Polymerization for Unsaturated Monomers

No.	Monomer	Chemical formula	Heat of polymerization* (kJ/mol)
1	Ethylene	$H_2C{=}CH_2$	93
2	Propylene	$H_2C{=}CHCH_3$	85.7
3	Isobutylene	$H_2C{=}CH(CH_3)_2$	51.4
4	1,3-butadiene	$H_2C{=}CH\text{-}CH{=}CH_2$	72.3
5	Styrene	$H_2C{=}CHC_6H_5$	68.97
6	Vinyl chloride	$H_2C{=}CHCl$	95.7
7	Vinylydene chloride	$H_2C{=}CCl_2$	75.2
8	Vinyl acetate	$H_2C{=}CHCOOCH_3$	89
9	Acrylonitrile	$H_2C{=}CHCN$	72.3
10	Methyl acrylate	$H_2C{=}CHCOOCH_3$	78.5
11	Methyl methacrylate	$H_2C{=}CCH_3COOCH_3$	56.4
12	Acid methacrylate	$H_2C{=}CCH_3COOH$	66
13	Tetrafluoroethylene	$F_2C{=}CF_2$	155.4
14	Isoprene	$H_2C{=}CCH_3\text{-}CH{=}\ CH_2$	74.4

*Heat of polymerization is defined as the energy necessary to change the enthalpy of a liquid monomer into an amorphous or a partially crystalline polymer.

- Polymerizing according to the anionic mechanism (having substituents attracting π electrons in double bond and facilitating formation of a carbanion).

Vinyl monomers polymerize readily through the free radical mechanism. An important factor determining their reactivity is the possibility of the resonance stabilization of formed free radicals. Larger is the value of energy corresponding to equilibrium between the resonance forms of free radicals coming from a given monomer, larger is the reactivity of this monomer in the polymerization reaction.

10.2.1.2 Course of Free Radical Polymerization

When testing the radical polymerization mechanism it was demonstrated that it takes place in three steps:

- initiation of polymerization;
- polymer chain growth (propagation);
- ending of chain (termination).

10.2.1.2.1 Initiation of Polymerization

Initiation of free radical polymerization is a process in which the creation of a free radical on the carbon atom of monomer molecules takes place.

Factors enabling the initiation of the free radical polymerization are:

- free radicals, resulting from the degradation of peroxide or azo-initiators, as well as during the redox reactions
- heat energy (thermal polymerization)
- UV irradiation (photopolymerization)
- X-rays or gamma rays (radiation polymerization)
- ultrasounds (ultrasonic polymerization).

Most frequently the polymerization process is initiated by free radicals obtained through the decomposition of hydroperoxides, alkyl peroxides, dialkyl peroxides, acyl peroxides, carboxylic ester peracids, salts of (tetraoxo)sulfuric acid, hydrogen peroxide, aliphatic azo-compounds and bifunctional azobenzoin initiators. The rate of decomposition of different initiators into free radicals depends on their structure and on temperature. A measure of the efficiency of the initiator in the polymerization process is the *half-decomposition period.*

Table 10.2 presents the characteristics and the *half-decomposition* periods of the most common free radical initiators.

A typical example of a initiator decomposition into free radicals is the chemical decomposition of dibenzoyl peroxide (called also benzoyl peroxide):

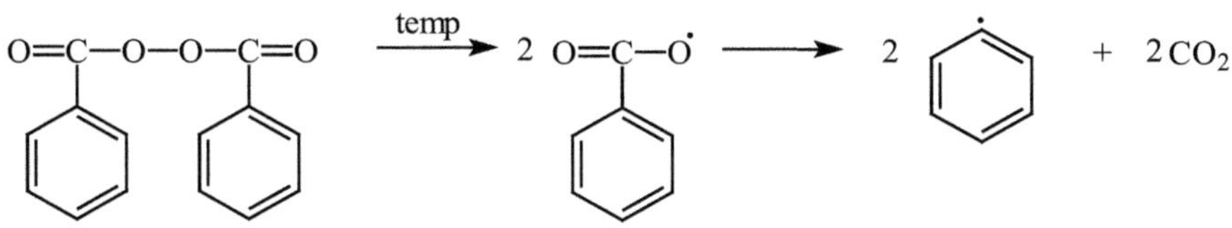

As shown by experimental studies in the polymerization process involving both benzoyl radicals and phenyl radicals their ratio in the reaction mixture is 1:1.

Initiators may disintegrate into free radicals with a simultaneous liberation of nitrogen. The most widely used initiator from this group of compounds is the dinitrile of azoisobutyric acid, commonly called azoisobutyronitrile (AIBN):

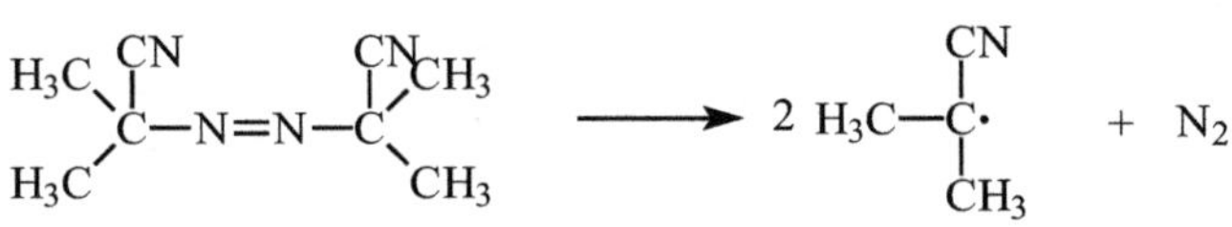

The emerging cyanopropyl radicals not only can initiate the polymerization reactions but may also recombine with the formation of *dinitril tetramethyl succinate* or a corresponding ketenimine.

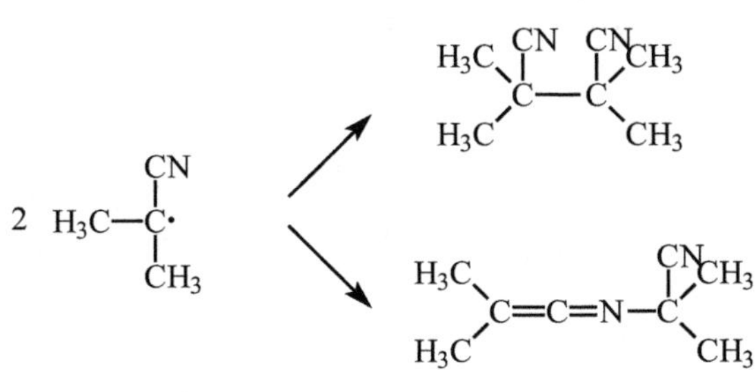

TABLE 10.2 Characteristics of Selected Radical Polymerization Initiators

No.	Initiator	Chemical formula	Molecular mass.	Melting temp. [°C]	Boiling temp. [°C(kPa)]	Density[kg/m3]	Refract. index n20D	Half-decomp. period [H/°C]
1	dibenzoyl peroxide	$C_6H_5COOO\text{-}COC_6H_5$	242.22	100 with Decom.	—	—	1.4056	2/85
2	acetyl-t-butyl peroxide	$CH_3COOO–C(CH_3)_2CH_3$	135.15	—	42.5 (1.46)	9455	1.5007	14/100
3	benzoyl-t-butyl peroxide	$C_6H_5COOO–C(CH_3)_2CH_3$	194.22	8	76.2	1043	1.3872	20/100
4	di-t-butyl peroxide	$(CH_3)_3COOC(CH_3)_3$	146.22	—	12.5 (2.66)	793	1.4013	20/120
5	t-butyl hydroperoxide	$(CH_3)_3$- -COOH	90.12	–4	5 (0.266)	896	1.524	20/115
6	k-cumene hydroperoxide	$(CH_3)_2C$- $-C_6H_5OOH$	152.18	—	53 (0.311)	1062	—	25/113
7	azoisobutyric acid dinitrile (AIBN)	$(CH_3)_2C(CN)\text{-}N{=}N\text{-}(CN)C(CH_3)_2$	164.2	—	103 with decomp.	—	—	1.5/80
8	potassium persulfate	$K_2S_2O_8$	270.32	100 with decomp.	—	2477	—	1.25/80
9	ammonium persulfate	$(NH_4)_2S_2O_8$	228.21	120 with decomp.	—	1982	—	1.25/80
10	hydrogen peroxide	H_2O_2	34.01	–0.41	150.2	1422	—	—

Free radicals are formed also in the redox reactions. A characteristic feature of these reactions is the possibility to obtain free radicals at low temperatures. They are particularly useful when initiating the low temperature and the emulsion polymerizations.

The most widely used redox systems, serving for the production of free radicals, are the mixtures of diluted in the reaction environment solutions of hydroperoxides or organic peroxides with transition metal ions:

$$HOOH + Fe^{2+} \longrightarrow Fe^{3+} + OH^- + {}^{\bullet}OH$$

$$ROOH + Co^{2+} \longrightarrow RO^{\bullet} + OH^- + Co^{3+}$$

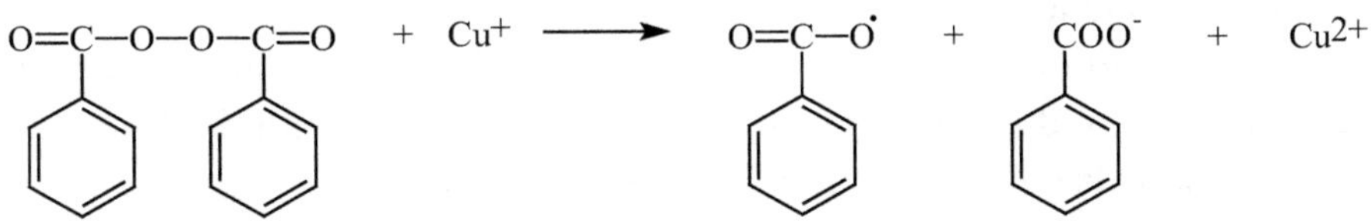

In the process of emulsion polymerization as initiating system one uses often a mixture of potassium persulfate and sodium thiosulfate:

$$S_2O_8^{2-} + S_2O_3^{2-} \longrightarrow SO_4^{-\bullet} + SO_4^{2-} + S_2O_3^-$$

$$SO_4^{-\bullet} + H_2O \longrightarrow HSO_4^- + {}^{\bullet}OH$$

Sometimes, as a reducing agent in the initiating system, the tertiary amines are also used:

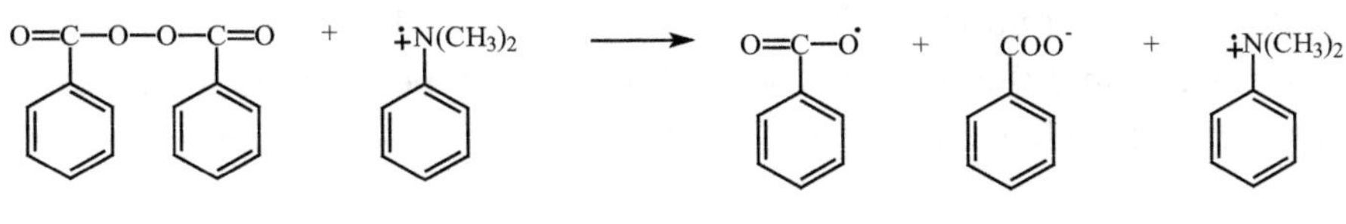

$$C_6H_5COO^- + C_6H_5\overset{+}{N}{}^{\bullet}(CH_3)_2 \longrightarrow C_6H_5COOH + {}^{\bullet}C_6H_4N(CH_3)_2$$

$$2\,{}^{\bullet}C_6H_4N(CH_3)_2 \longrightarrow (CH_3)_2NC_6H_4{-}C_6H_4N(CH_3)_2$$

An interesting redox system serving to initiate polymerization reaction in the aquatic environment is also a solution of potassium permanganate and a reducing organic substance, such as, for example, oxalic acid, ascorbic acid, lactic acid, citric acid, thioglycollic acid, ethylene glycol or glycerol, which generates free radicals in acidic environment in the presence of monomer.

The mechanism of the polymerization reaction with such system is the following: manganese atom in potassium permanganate, which is in the +7 degree of oxidation is reduced as a result of the reaction with monomer molecule to the oxidation degree of +3, with a simultaneous formation of a by-product

$$Mn^{+7} + \text{monomer} \longrightarrow Mn^{+3} + \text{oxidation by-product}$$

The manganese ions in +3 oxidation states may be subject to a reversible disproportionation reaction or react first with the water molecules, and then with the reducing agent (Red) and monomer (M) with the formation of free radicals (M*), initiating the polymerization.

$$2\ Mn^{+3} \rightleftharpoons Mn^{+4} + Mn^{+2}$$

$$Mn^{+3} + H_2O \rightleftharpoons (MnOH)^{+2} + H^+$$

$$(MnOH)^{+2} + Red \rightleftharpoons (Mn(OH)Red)^{+2}$$

$$(MnOH)^{+2} + M \rightleftharpoons (Mn(OH)M)^{+2}$$

$$(Mn(OH)Red)^{+2} + (Mn(OH)M)^{+2} \rightleftharpoons M^{\bullet} + (MnOH)^{+2} + Mn^{+2} + H_2O + Red$$

Sometimes the free radical polymerization can be initiated thermally without a use of initiators. This process generally takes place in an anaerobic system, in different ways, depending on the type of monomer used. An example of such a reaction may be the styrene polymerization:

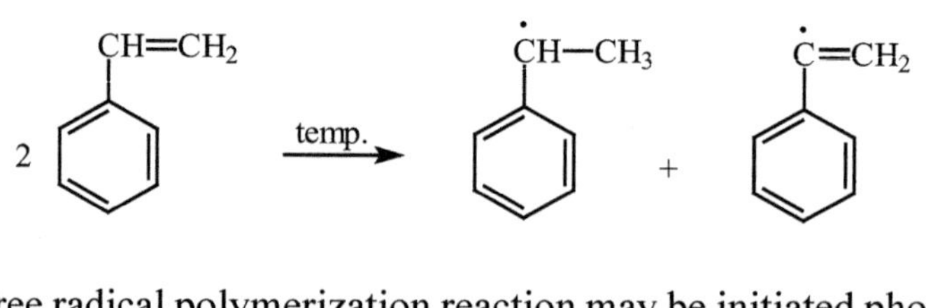

The free radical polymerization reaction may be initiated photochemically too. This process consists on the absorption by monomer molecule of a quantum of light energy resulting in its passage to an excited state. The photochemical initiation of polymerization takes place frequently in the presence of photosensitizers (optical sensitizers), which readily absorb the energy and play an important role in its transmission. The main advantage of the photochemical reaction is its total independence on temperature and the possibility of initiating the polymerization process at temperatures often lower than those used with other sources of free radicals generation.

The photochemical polymerization process can be easily controlled by using a light source with a narrow range of emission wavelengths and the possibility of introducing into the reaction environment other types of initiators, disintegrating into free radicals under the UV radiation.

Examples of such initiators are disulfides, benzoin and dibenzoil:

$$RSSR \xrightarrow{h\nu} 2\ RS^{\bullet}$$

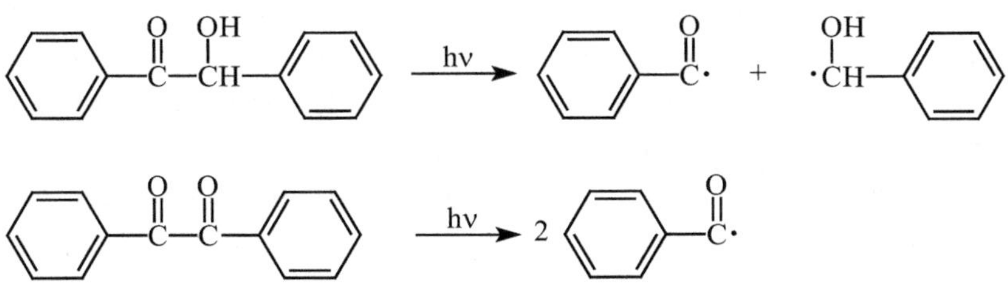

An important factor characterizing the stage of initiation of polymerization is the induction period. This is the time counted from the initiation of polymerization to the beginning of the growth of chains. It depends on the reaction temperature, structure and monomer concentration as well as on the presence of inhibitors.

Inhibitors are compounds, which inhibit the polymerization process during the initiation and growth of the chain. Typical inhibitors include hydroquinone, pyrocatechol and its derivatives, aromatic p-phenylenediamine amines, N-phenyl-2-naphthylamine, and also trinitrobenzene, picric acid, copper salts, and others. The reaction of free radical with a monomer molecule depends on the reactivity of double bonds of the latter.

For example, the free radical polymerization of ethylene runs extremely difficult and is necessary to be performed at high temperature (420–470 K) and at high pressure of the order of 100 MPa.

The presence of an electronegative substituent (halogen, nitrile or carboxyl group) or an electropositive (methyl or amine group) in the monomer molecule causes a polarization of double bonds, which manifests itself in the increase of the molecular dipole moment and the corresponding increase in the polymerization rate. From this rule, however, exceptions are because the butadiene molecules have no own dipole moment. However, they are easily polarized and thus readily polymerize.

10.2.1.2.2 Chain Growth Reaction

Second important step in the course of the polymerization process is the increase of the chain length, called propagation. This increase is associated with the attachment of new monomer molecules, initially to a free radical, created at the initial stage of the polymerization, and then to the continuously growing macroradical.

The chain growth rate is so high that practically it has no effect on the kinetics of polymerization. The polymer molecule composed from about 10–000 monomer molecules is formed in less than one second. One assumes that the activity of the growing polymer radical does not depend on the chain length. The activation energy of the polymer chain growth reaction is 16.7–41.9 kJ/mol.

While the speed of the polymerization initiation depends also on the monomer activity, the rate of increase of the chain length depends on both the monomer activity and the activity of the growing polymer radical.

It was shown that the polarized monomer molecule reacts easier with the free radical. However, free radicals, arising from the polarized monomer molecules, are always less active than the radicals of unpolarized monomer molecules. This is due to the fact that the activity of free radical depends on the ability of free electrons to act upon the π bonding of the monomer molecule. The shift of electron due to the coupling effect reduces the probability of the course of this reaction, and hence the radical activity. The coupling effect affects more the radical activity than that of monomer. Therefore, the chain growth reaction rate in the radical polymerization depends primarily on the activity of free radicals. The type of substituent in the monomer molecule has an impact not only on the rate of the macroradical growth, but also on the configuration of the resulting polymer macromolecule.

The addition of a monomer molecule, containing a substituent X, to the growing macroradical may occur in the "head to tail" (Scheme A) or "head to head" configuration (Scheme B).

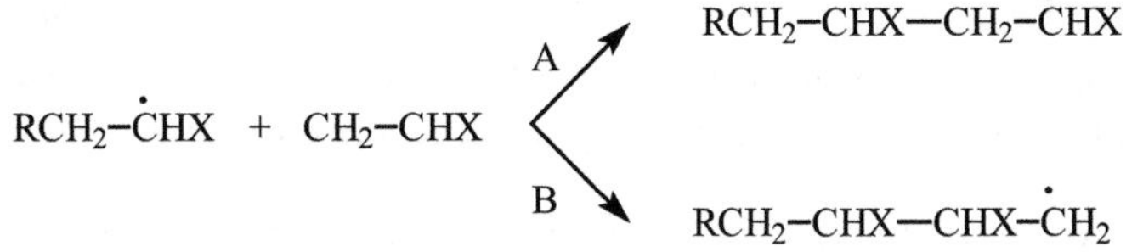

As a result of the reaction of two polarized groups: the growing radical and monomer molecule the addition takes place in agreement with the charge density distribution. The coupling effect provides the stability to the growing polymer radical and therefore more likely is obtaining the structure of "head to tail" type. In principle, more probable is also attachment of a monomer molecule in which the most stable radicals are formed. However, obtaining of polymers with a regular arrangement of mers by the free radical polymerization method is difficult. This is explained by small differences in the activation energy of monomer molecules addition in different positions.

A large amount of "head to head" connections was found in some polymers such as poly(vinylidene fluoride) [-CH_2CF_2-] and poly(vinyl fluoride) [-CH_2CHF-] by using the high-resolution nuclear magnetic resonance spectroscopy technique.

10.2.1.2.3 Ways to End the Chain Growth

The completion of the polymer chain growth during the free radical polymerization may occur in the following situations:

- through the reaction of the formed macroradical with another radical (the so-called recombination), resulting from the initiator disintegration or from the secondary reactions.

$$RM_n^{\bullet} + R_1^{\bullet} \longrightarrow RM_nR_1$$

$$RM_n^{\bullet} + H^{\bullet} \longrightarrow RM_nH$$

- through the recombination of two macroradicals

$$RM_n^{\bullet} + RM_m^{\bullet} \longrightarrow R(M)_{n+m}R$$

where: M = CH_2CHX; RM^*_n and RM^*_m – growing macroradicals.

- through the disproportionation reaction by the transfer of hydrogen atom from one of the growing chain to another one.

$$R{-}(CH_2CHX)_n{-}CH_2\dot{C}HX + R{-}(CH_2CHX)_m{-}CH_2\dot{C}HX \longrightarrow$$

$$\longrightarrow R{-}(CH_2CHX)_n{-}CH{=}CHX + R{-}(CH_2CHX)_m{-}CH_2CH_2X$$

- through the transfer reaction of kinetic chain activity as a result of a collision with inactive molecule (monomer, solvent, polymer or otherwise), contained in the reaction environment. It is followed by the completion of the macroradical chain growth with a simultaneous emergence of a new radical.

$$R{-}(CH_2CHX)_n{-}CH_2\dot{C}HX + R_1H \longrightarrow R{-}(CH_2CHX)_n{-}CH_2CH_2X + R_1^{\bullet}$$

Completion of the chain growth as a result of recombination or disproportionation of reactants stops the growth of the polymer macromolecule and ends the kinetic chain free radical reaction. In that case no new free radicals, able to initiate the growth of a new macromolecule, are formed.

During the chain transfer reaction, new free radicals emerge, which, depending on their structure may show different activity.

The transfer of the chain to another monomer molecule is followed by the increase of next macroradical. In that case the polymerization reaction rate does not change. The transfer of the chain on a solvent molecule, able to create relatively persistent free radicals leads to the formation of oligomeric products terminated by large substituents, coming from the decomposition of solvent molecule. These substituents determine the properties of the resulting compounds, called telomeres. This issue will be discussed larger in the chapter devoted to the polymerization in solution.

The transfer of chain to the polymer molecule gives rise to the creation of a new macroradical in which the unpaired electron is not at the end of the chain, but in its middle. The formed in this way new chain is a branch of the already existing one. This type of reaction is called the grafting reaction.

$$RM_n^{\bullet} + RM_kMM_lH \longrightarrow RM_nH + RM_k\dot{M}M_lH$$

$$RM_k\dot{M}M_lH + mM \longrightarrow RM_k\underset{\underset{\dot{M}_m}{|}}{M}M_lH$$

Very often one introduces into the medium of polymerization reaction some compounds, which serve as a chain transfer agents. They are known in the polymer chemistry under the name of moderators or molecular weight regulators. As examples one can cite: chlorinated hydrocarbons, dodecyl mercaptan and thioglycollic acid. These compounds are characterized by the presence in molecule of a mobile atom, able to detach away on impact with the growing macroradical.

If we denote by AB the moderator molecule than its action can be demonstrated as follows:

$$RM_n^{\bullet} + AB \longrightarrow RM_nA + B^{\bullet}$$

If the resulting radical B^* has a similar activity to that of the completed macroradical, than the action of AB moderator is limited to reducing the molecular mass of the product and have no significant influence on the kinetics of polymerization.

However, if B^* is less active than the completed macroradical, than the rate of polymerization is reduced. In that case the AB compound plays the role of retarder in this process.

A special case of delaying the polymerization process is its inhibition. The radical B^* originating f from the inhibitive substance, that is from the inhibitor, is completely unreactive and causes a gradual termination (ending) of the kinetic chain growth reaction. The time needed to wear the total amount of inhibitor is called the period of inhibition.

10.2.1.2.4 Influence of Oxygen on the Course of the Polymerization Process

Oxygen, present in the reaction environment (usually air), plays an important role in the process of polymerization. In some cases, such as during the high temperature and high-pressure ethylene polymerization it plays the role of initiator. In other cases, oxygen, depending on the reaction conditions accelerates or delays the process of polymerization. In extreme cases oxygen acts as an inhibitor.

These phenomena are explained by the chemical reactions between oxygen and free radicals present in the reaction environment. In that case the peroxide radicals are formed, as shown below:

$$R^{\cdot} + O_2 \longrightarrow R{-}O{-}O^{\cdot}$$

The reactivity of these peroxide radicals determines the role oxygen plays in the given type of polymerization. Often, especially in the process of emulsion polymerization, one uses an anaerobic atmosphere by blowing the free space of reaction device with free of oxygen nitrogen.

10.2.1.3 Kinetics of Radical Polymerization

The rate of free radical polymerization depends on the speed of the initiation, growth and chain termination reactions.

The initialization process starts with the initiator decomposition into free radicals, which reacting with monomer molecules initiate the chain polymerization reaction:

$$I \xrightarrow{k_d} 2R^{\bullet}$$

$$R^{\bullet} + M \xrightarrow{k_1} RM^{\bullet}$$

where: I – initiator molecule; M – monomer molecule; R^*, M^* – free radicals; k_d – rate constant of the radicals formation; k_i – initiation rate constant ($dm^3mol^{-1}s^{-1}$).

Not all free radicals initiate effectively the polymerization process as part of them is deactivated by the recombination process:

$$2R^{\bullet} \rightarrow R - R$$

The initiator decomposition reaction is of first order. The rate of formation of new free radicals V is given by the formula:

$$V_i = \frac{d[R^{\bullet}]}{dt} = -\frac{d[I]}{dt} = 2k_d[I]$$

where: [R*] – radical initiator concentration (mol/dm^3); [I] – initiator concentration (mol/dm^3).

Since not the all formed in this way free radicals initiate the polymerization process effectively a corrective factor f, called the capacity to initiate the reaction, was introduced. It defines the fractional part of free radicals reacting with monomer molecules and can be written in the following way:

$$V_i = 2fk_d[I] = k_i[I]$$

As mentioned previously, the chain reaction growth consists on the addition of consecutive monomer molecules to the nascent macroradical. Its speed V_w is given by the formula:

$$V_w = -\frac{d[M]}{dt} = k_w[M_n^\bullet][M]$$

where: k_w – the rate constant of chain growth (dm^{3}mol^{-1}s$^-$1); [M] – monomer concentration (mol/dm^3); [M*] – concentration of macroradicals (mol/dm3).

In considering this it is assumed that:

- the reactivity of radicals does not depend on their chain length;
- the average length of the macroradical of the resulting polymer is large, (it allows to neglect the minor consumption of monomer in the initiation reaction and assume that its total amount is involved in the chain growth reaction);
- the number of radicals present in the system during the polymerization process is constant (it means that the rate of formation of radicals is equal to V_i and the decay rate to V_z).

The rate of the termination reaction Vz, due to the recombination of two macroradicals, is given by the following equation:

$$V_z = -\frac{d[M_n^\bullet]}{dt} = k_z[M_n^\bullet]^2$$

Given that:

$$V_i = V_z$$

one gets

$$k_i[I] = k_z[M_n^\bullet]^2 \qquad [M_n^\bullet] = \sqrt{\frac{k_i}{k_z}\cdot[I]}$$

It is assumed that the overall polymerization speed V is equal to the speed of chain growth V_w. By substitution of [Mn] in formula for V_w one obtains the following equation for V:

$$V = V_w = k_w\sqrt{\frac{k_i}{k_z}\cdot[I]}\cdot[M]$$

where: k_w – the rate constant for the polymerization reaction.

This means that the speed for free radical polymerization is directly proportional to the square root of the initiator concentration and the reaction is of first order with respect to the monomer concentration.

In the initial period of polymerization, with a low degree of conversion, the monomer concentration is constant (with its large excess) and therefore the polymerization rate depends solely on the square root of the initiator concentration

$$V = const\cdot\sqrt{[I]}$$

The constants and the activation energy values for radical polymerization of selected monomers are given in Table 10.3.

In a completely different way run the polymerization at its final stage, when the degree of conversion of substrates is large. In this case a self-acceleration takes place and a departure from first order law is observed. This phenomenon is called the gel or Norrish – Tromsdorff effect. It is mainly due to the increased viscosity of the polymerization medium and the diminished probability of meeting macroradicals and their recombination. In that case one observes also a decrease of thermal conductivity of the system. It results in the temperature increase, thus in the acceleration of reaction. The gel effect leads also to the increase of molecular mass of polymer, because the inhibition of the termination process does not affect the growth rate of the chain.

TABLE 10.3 Rate Constants and Activation Energies for Radical Polymerization of Selected Monomers

No.	Monomer	Temp. [K]	Rate constant		Activation energy		Kinetic polymer chain transfer constant Cm
			growth *kw*	Termination *kz*	growth	Termination.	
1	ethylene	356	470	1050	8.2	—	5
2	butadiene	283	8.4	—	2.6	—	—
3	styrene	273	6.91	1.83	6.5	2.8	0.118
4	styrene	333	176	72	—	—	6
5	vinyl chloride	298	6200	1100	3.7	4.2	6.25
6	vinylidene chloride	288	2.3	0.023	25	40	—
7	vinyl acetate	293	586	3040	7.32	5.24	0.94
8	acrylonitrile	298	14500	2000	4.1	5.4	0.105
9	butyl acrylate	308	13	0.018	2.1	0	—
10	methacrylonitrile	298	26	21	11.5	5	2.08
11	4-vinylpyridine	298	12	3	—	—	—

*After Physicochemical Guide (Poradnik fizykochemiczny;in Polish), WNT, Warszawa 1974, p. 383.

The up to now known kinetic equations enable us to determine the average degree of polymerization P, which is expressed by the ratio of chain growth rate to the completion rate:

$$\overline{P} = \frac{V_w}{V_z} = \frac{k_w[M^\bullet][M]}{k_z[M^\bullet]^2} = \frac{k_w[M]}{k_z[M^\bullet]} = \frac{k_w[M]}{\sqrt{k_z k_i} \cdot \sqrt{[I]}}$$

At the initial stage of polymerization, assuming existence of an excess and constant monomer concentration, the average degree of polymerization, and thus the average molecular mass of the polymer is inversely proportional to the square root of the initiator concentration:

$$\overline{P} = \frac{const}{\sqrt{[I]}}$$

If the polymerization process is carried out in the presence of the molecular mass regulator (moderator), then the average degree of polymerization depends on the regulator concentration and on the degree of conversion s:

$$\overline{P} = \frac{[M]}{[S]} \cdot \frac{s}{1-(1-s)^2}$$

where: s – degree of conversion at a given time; [M] – initial concentration of monomer; [S] – initial concentration of regulator; c – transfer constant, equal to the ratio of rate constant of regulator consumption to rate constant of chain growth.

10.2.1.4 Influence of Parameters on the Course of Radical Polymerization

The most important parameter influencing the course of polymerization is temperature. The overall activation energy of the polymerization process is given by:

$$E = \frac{1}{2}E_i - (E_w - \frac{1}{2}E_z)$$

where: *E* – polymerization energy; *Ei* – energy of the reaction initiation; *Ew* – energy of the chain growth; *Ez* – chain termination energy.

In the course of polymerization initiation of benzoyl peroxide $E_i = 125.4$ kJ/mol. Since for most of monomers $E_w = 29.1$ kJ/mol, thus $E_z = 12–20$ kJ/mol then it follows that the polymerization activation energy $E > 0$. For this reason the polymerization reaction rate clearly increases with temperature. Thus the degree of polymerization and the same the average molecular mass of the resulting polymer decrease.

In the case of photochemically and radiation initiated polymerization $E_i = 0$. Therefore, the polymerization energy $E < 0$ and the degree of polymerization increases with increasing temperature.

The activation energy of side branching reaction is larger than the activation energy for their growth. Therefore, an increase of temperature during the polymerization favors the formation of branched compounds.

Temperature affects also the structure of the resulting polymer. It contributes to the reduction of its stereoregularity (spatial arrangement of substituents in the polymer molecule) and causes an increase of polymer polydispersity (individual macromolecules differ significantly from each other in molecular mass). At higher temperatures another side reactions are possible, such as, for example, the degradation (breakdown of the polymer chain into smaller fragments) and cross-linking (formation of lateral bonds between the polymer chains).

Another important parameter influencing the course of polymerization is the pressure. A moderate increase of pressure shows no apparent effect on the polymerization reaction rate. Pressures above 100 MPa increase the rate constant of chain growth and by the same the speed of polymerization reaction. The increase of pressure results also in the increase of average molecular weight of the formed polymer and improves the regularity of its spatial structure.

During the polymerization in solution an important role plays also the monomer concentration. With the increasing monomer concentration the rate of polymerization and the polymer molecular mass increase. The influence of interactions between molecules of monomer and solvent are discussed in the chapter devoted to the polymerization in solution.

10.2.1.5 Study of the Polymerization Kinetics

The most widely used way to study the kinetics of polymerization is the dilatometric method. It allows to carry out the measurements on a continuous basis and provides precise and reproducible results.

Its principle is based on the measuring the difference between the specific volume of monomer and specific volume polymer. As a consequence of the running polymerization process is always the volume contraction of the studied system. This contraction is proportional to the changes in concentration of monomer or polymer in the reaction mixture. Measuring such changes and expressing them in function of time allows to determine the rate of polymerization. The densities and the contraction coefficients of selected monomers at different temperatures are listed in Table 10.4.

The implementation of the comprehensive studies of the kinetics of polymerization or copolymerization (polymerization of two different monomers) is a difficult task. According to the earlier provided information the kinetics of polymerization consists of initiation rates, growth, completion and a series of side reactions. It requires often an additional research using other methods.

The application of dilatometric method for kinetic studies of polymerization or copolymerization processes is usually based on the determination of the dependences of the speed of the process on the following parameters:

- concentration of initiator [I];
- concentration of monomer or comonomers (monomers in the copolymerization process) [M];
- value of absolute temperature (T);
- concentration of inhibitor;
- composition of comonomer mixture.

The quantities measured in the dilatometric method are the changes in the height of the liquid column (h) during the reaction time (t). The degree of transformation (W), which is a measure of the decrease of meniscus of the reaction mixture in the capillary of dilatometer, is obtained from the following formula:

$$W = \frac{\Delta V}{V_m L^T} x100$$

TABLE 10.4 Densities and Contraction Coefficients for Selected Monomers

No.	Monomer	Temperature [K]	Density [kg/m³]	Contraction at 1% conversion [%]
1	styrene	283	915	0.1327
2	styrene	293	906	0.1405
3	styrene	303	897	0.147
4	styrene	323	878	0.163
5	styrene	333	869	0.1704
6	styrene	343	860	0.1778
7	methyl methacrylate	293	941	0.2282
8	methyl methacrylate	303	929	0.233
9	methyl methacrylate	313	918	0.2379
10	methyl methacrylate	323	907	0.2427
11	methyl methacrylate	333	899	0.2468
12	methyl methacrylate	343	887	0.2511
13	methyl methacrylate	353	876	0.2552
14	vinyl chloride	303	892	0.372
15	vinyl chloride	313	872	0.4
16	vinyl chloride	323	852	0.433
17	vinyl chloride	333	831	0.47
18	acrylonitrile	293	806	0.323
19	acrylonitrile	303	795	0.332
20	acrylonitrile	313	784	0.341
21	acrylonitrile	323	773	0.35
22	acrylonitrile	333	762	0.36
23	acrylonitrile	343	751	0.369
24	acrylonitrile	353	740	0.378
25	methyl acrylate	293	960,7	0.2189

TABLE 10.4 Continued

No.	Monomer	Temperature [K]	Density [kg/m³]	Contraction at 1% conversion [%]
26	methyl acrylate	303	936,2	0.2286
27	methyl acrylate	323	912	0.239
28	vinyl acetate	298	925	0.2206
29	vinyl acetate	303	892	0.235
30	vinyl acetate	355	850	0.2355

After S. Połowiński: Techniques of measurement and research in physical chemistry of polymers (Techniki pomiarowo-badawcze w chemii fizycznej polimerów, in Polish), WPŁ, Łódź, Poland 1975.

where: W – degree of conversion) [%]; ΔV – volume change of the system; V_m – initial volume of monomer; L^T – contraction coefficient of the system at temperature T.

$$L^T = \frac{V_m - V_p}{V_m} = \frac{\frac{1}{d_m} - \frac{1}{d_p}}{\frac{1}{d_m}} = 1 - \frac{d_m}{d_p}$$

where: V_m, V_p – specific volume of the appropriate monomer (m) and polymer (p) at temperature T; d_m, d_p – density of monomer (m) and of polymer (p) at the measurement temperature T.

The initial speed of polymerization V is calculated from the formula:

$$V = \frac{dM}{dt} = \frac{\Delta V}{V_{zb}\Delta t} \cdot \frac{1}{MX} [mol/dm^3 s]$$

where: [M] – monomer concentration in mol/dm³; ΔV – change of the volume of monomer (solution) at time Δτ; V_{zb} – volume of the dilatometer reservoir, equal to the initial volume of monomer (block polymerization) or of solution (polymerization in a solvent); M – molecular mass of monomer; X – absolute change of the specific volume of the system, resulting from the full (100%) monomer conversion at temperature T.

10.2.1.6 Stereochemistry of Radical Polymerization

As a result of the polymerization of unsaturated monomers containing one or two substituents at the same carbon atom one obtains polymers with different spatial structure:

- isotactic, which is characterized by one type of basic configuration units (with chiral or pseudochiral atoms in the main chain) with the same type of sequence;
- syndiotactic, in which the molecules can be described as an alteration of the enantiomeric basic configuration units;
- atactic, in which the particles have a random distribution of equal number of possible basic configuration units.

The stereochemistry of monomer molecules addition to the macroradical depends on:

- mutual interaction between the end carbon atom of the growing macroradical and the approaching monomer molecule,
- configuration of the penultimate unit (mer) repeatable in the chain.

The terminal carbon atom with unpaired electron in the growing macroradical molecule exhibits probably a flat sp^2 hybridization. For the monomer CH=CXY (Y = H, X or R) polymerization to occur the molecule must approach the end carbon atom of the macroradical and set in the mirror position at which the same substituents are on the same side (scheme "a") or in nonmirror position at which the same substituents are on opposite side (scheme "b") as shown below:

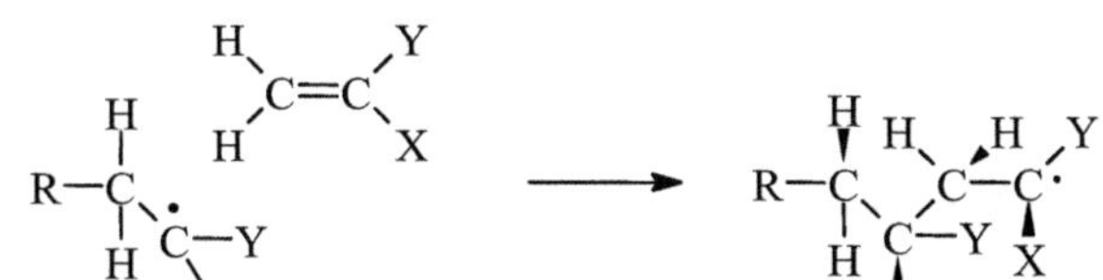

(a)

isotactic polymer

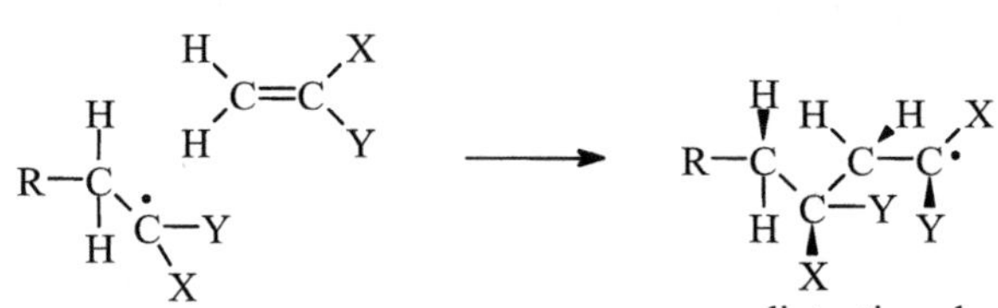

(b)

syndiotactic polymer

The stereochemistry of polymerization reaction is maintained when before the attachment of the next monomer molecule no free rotation around the macroradical takes place.

When the polymerization stereochemistry is preserved then the mirror approach (reaction "a") will always lead to the formation the isotactic polymer and the nonmirror approach to the formation of the syndiotactic polymer (reaction "b"), respectively.

If the mutual interaction between the reacting monomer molecule and the mer substituents of the penultimate carbon atom of macroradical is significant, the also the conformational factors may influence the way how the attachment is realized. It will depend on which of the two interactions: the steric or the electrostatic one will be the weakest one.

In the free radical polymerization process, conducted usually at elevated temperatures, these effects are insignificant and the reaction usually leads to the formation of atactic polymer only. However, in some cases, like, for example, in free radical polymerization of methyl methacrylate at temperature below 0°C one obtains a crystalline polymer with syndiotactic structure, as it was proven by the high resolution nuclear magnetic resonance spectroscopy. These results confirm the rule that according to which the degree of stereoregularity decreases with increasing temperature.

The stereochemistry of free radical polymerization depends on a set of steric and polar effects. A significantly larger stereoregularity is obtained in ionic polymerization, and in particular the coordination one.

The tacticity of obtained polymer is characterized by the ratio kws/kwi of reaction rate constants of chain growth with the formation of syndiotactic (kws) and isotactic (kwi) structures. Namely when:

(a)

$$\frac{k_{ws}}{k_{wi}} = 1$$

an atactic polymer is formed,

(b)

$$\frac{k_{ws}}{k_{wi}} = \infty$$

a syndiotactic polymer is formed,

(c)

$$\frac{k_{ws}}{k_{wi}} = 0$$

an isotactic polymer is formed.

10.2.1.7 Polymerization of Dienes

Polymerization of dienes is a special case of the polymerization of unsaturated compounds and its course depends on the structure of diene used.

Dienes with isolated double bonds can polymerize with the formation of cross-linked polymers, because the two double bonds present in the molecule can react independently of each other. However, in some cases in addition reactions both double bonds of monomer molecules can take part, leading to the formation of cyclic polymers. This reaction occurs particularly when it leads to the formation of rings with five or six-members. During the cyclization reaction side vinyl groups are formed, which polymerize with formation of cyclic polymers.

In the case of cyclopolymerization the most favorable configuration is *cis*, while in the normal polymerization of vinyl monomers the dominating configuration is the *trans* one.

Dienes with π electron conjugated double bonds, such as 1,3-butadiene and its derivatives undergo an addition process in both positions: 1.2 and 1.4. As a result of the polymerization process of 1.2 type a polymer containing side vinyl groups is formed, while 1.4 addition reaction leads to a polymer with double bond in the chain. In that case both configurations *cis* and *trans* are equally possible.

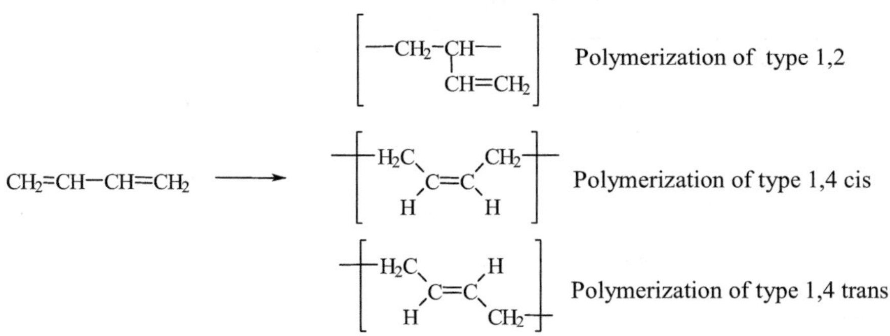

The nature of the formed polymer structure depends on the type of the diene monomer and of the initiator as well as on the polymerization reaction conditions. In general, lowering the reaction temperature leads preferentially to type 1.4, which is more desirable than 1.2 because of a greater flexibility of these polymers.

In the case of polymerization of substituted dienes such as 2-methyl-butadiene (isoprene), the structure of the resulting product may be more complicated because it is possible to create polymers with structure of type 1.2 and 3.4, as well as 1,4 *cis* and 1,4 *trans*. Moreover, structures "head to tail" and "head to head" type may be formed. The results of the analysis of synthetic polyisoprene have shown that, in general dominating is the "head to tail" arrangement and the 1,4 *trans* configuration. The natural rubber is 1,4 *cis*-polyisoprene, whereas gutta-percha and balata have 1,4 *trans* structure.

The parts of the structural elements of three most important diene monomers: butadiene, isoprene and chloroprene (2-chlorobutadiene) in the structure of different polymers obtained by free radical polymerization at different temperatures are given in Table 10.5.

In the case of chloroprene the content of structural groups of type 1.4 is much higher compared to the others. This may be due to the large difference in the electron density of two double bonds present in the monomer molecule.

TABLE 10.5 Structure of Diene Polymers Obtained by the Free Radical Polymerization Method

No.	Monomer	Polymerization temperature [K]	Content of the structure type [%]			
			1,4 *cis*	**1,4 *trans***	**1,2**	**3,4**
1	butadiene	253	6	77	17	—
		263	9	74	17	—
		278	15	68	17	—
		293	22	58	20	—
		373	28	51	21	—
		448	37	43	20	—
		506	43	39	18	—

TABLE 10.5 Continued

No.	Monomer	Polymerization temperature [K]	Content of the structure type [%]			
			1,4 *cis*	1,4 *trans*	1,2	3,4
2	isoprene	253	1	90	5	4
		268	7	82	5	5
		283	11	79	5	5
		323	18	72	5	5
		373	23	66	5	6
		423	17	72	5	6
		476	19	69	3	9
		530	12	77	2	9
3	chloroprene	227	5	94	1	0.3
		283	9	84	1	1
		319	10	85	2	1
		373	13	71	2.4	2.4

10.2.2 CATIONIC POLYMERIZATION

The cationic polymerization is the process of a monomer or a mixture of monomers conversion into the polymer by a cationic mechanism in the presence of catalysts.

10.2.2.1 Characteristic of Cationic Polymerization

Cationic polymerization of unsaturated compounds proceeds through the stage of carbanion cations, called also carbocations. Typical catalysts for this reaction are strong protic acids such as sulfuric acid, perchloric and trifluoroctane or the Lewis acids, which include halides of elements III, IV and V groups of the periodic table (Friedel-Crafts catalysts), such as: boron trifluoride, aluminum trichloride, tin tetrachloride and titanium tetrachloride. The activity of Friedel-Crafts catalysts increases significantly the presence of small quantities of cocatalysts, that is the compounds, which most often are the source of protons.

As cocatalysts one uses water, hydrochloric acids and chlorinated aliphatic hydrocarbons. The quantities of cocatalysts introduced into the reaction environment are very small, since the increase of their concentration facilitates the completion of the chain reaction, what leads to the formation of polymers of lower molecular mass. The use of water in excess leads to the catalyst deactivation as a result of its hydrolysis.

Monomers used in the cationic polymerization processes are characterized by a high electron density around the unsaturated carbon atom and show a tendency to form a relatively permanent carbanion cation. Among the others they include isobutylene, vinyl alkyl ethers, styrene, α-methylstyrene, isoprene, coumarone and indene.

As a result of the reaction of a Friedel-Crafts catalyst with a cocatalyst a complex compound is formed which dissociates into ions according to the scheme:

$$BF_3 + H_2O \longrightarrow [BF_3{*}H_2O] \rightleftharpoons H^+ + [BF_3OH]^-$$

$$SnCl_4 + HCl \longrightarrow [SnCl_4{*}HCl] \rightleftharpoons H^+ + [SnCl_5]^-$$

The resulting protons H^+ join the monomer molecule, giving it the structure of the carbanion cation. In the case of substituted alkyl monomers the proton attachment follows the Markovnikov rule. It means that it is attached to the unsubstituted methylene group:

$$R{-}\underset{\substack{|\\H}}{C}{=}CH_2 + H^+ \longrightarrow R{-}\underset{\substack{|\\H}}{\overset{\oplus}{C}}{-}CH_3$$

The resulting carbocation, similarly as the free radical, is very reactive because of the tendency of the endowed with a positive charge carbon atom to complete the octet of electrons.

In carbocation the carbon atom with a deficit of electrons binds three other atoms by means of its orbitals and therefore the system exhibits trigonal bonds (bonds are directed towards the vertices of an equilateral triangle). For this reason the center of active carbocation is flat. The carbon atom with a deficit of electrons and three atoms connected to it lie in the same plane. In accordance with the laws of electrostatics, the stability of a charged system increases with the increasing charge delocalization. The electron donating substituents located in the molecule (methyl groups) help to reduce the positive charge of the carbon atom with a deficit of electrons. The charge diffusion stabilizes the carbocation. It follows from this theorem that with the increasing number of alkyl groups in the molecule increases also the sustainability of the emerging from it carbocation. This can also explain the special flexibility of isobutylene for cationic polymerization.

During the cationic polymerization important is the kind of solvent used. A particularly positive role plays polar solvents such as, for example, nitrobenzene. In their presence the polymerization reaction proceeds many times faster than when using non-polar solvents. In the solvolysis process the solvent may facilitate the formation of carbocations through the so-called nucleophilic power.

The formation of tertiary cations occurs relatively easily and their reactivity to a small extent depends on the nucleophilic solvent. It depends mainly on its polarity. The formation of secondary cations requires a large nucleophilic assistance and their reactivity depends both on the nucleophilic power and on the polarity of solvent used. The nucleophilic assistance differs from S_{N2}-type attack that it does not lead to a product but only to the formation of carbocation, or a molecule with a carbocation character.

If the solvent used is characterized by a moderate polarity only than the bond between the cation and the anion can be mainly of electrostatic character. In this case the formed system is called the ion pair. The active center of the growing chain in the cationic polymerization process may occur as:

- free carbocation;
- carbocation molecule solvated by the solvent molecules;
- component of the ion pair of carbocation with the counterion derived from the catalyst;
- component of the solvated ion pair;
- pseudoion (covalent) active center.

The general mechanism of the cationic polymerization consists of three stages: initiation, chain growth and completion (termination).

As mentioned above, the most common way of initiating the cationic polymerization is the attachment of proton to the double carbon bond. There is a number of other ways to initiate this reaction. They include polymerization in the presence of diazonium salts, charge transfer complexes and catalyzed by metal salts in a heterogeneous system.

Very interesting is the recently developed cationic polymerization method, catalyzed with diazonium compounds. It runs at room temperature and the formed polymers exhibit a high molecular weight. An example of such a reaction is the polymerization of 4-methoxystyrene under the influence of 4–4-*nitrobenzenediazonium hexafluorophosphate,* in which the poly(4-methoxystyrene) with a molecular weight of more than 100–000 is obtained:

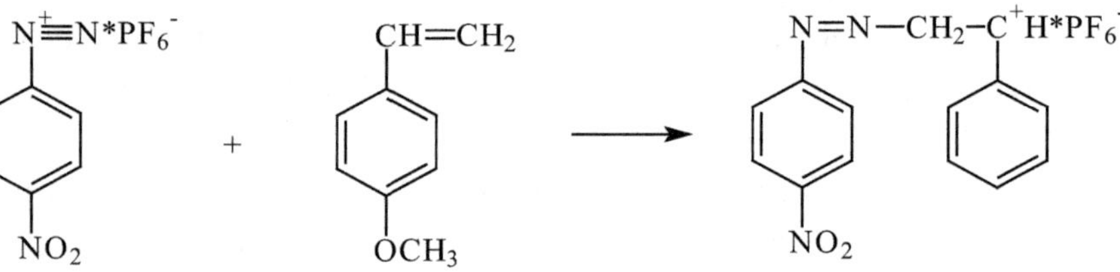

It should be noted that the diazonium compounds are unstable and decompose under heating or light illumination with the emission of nitrogen.

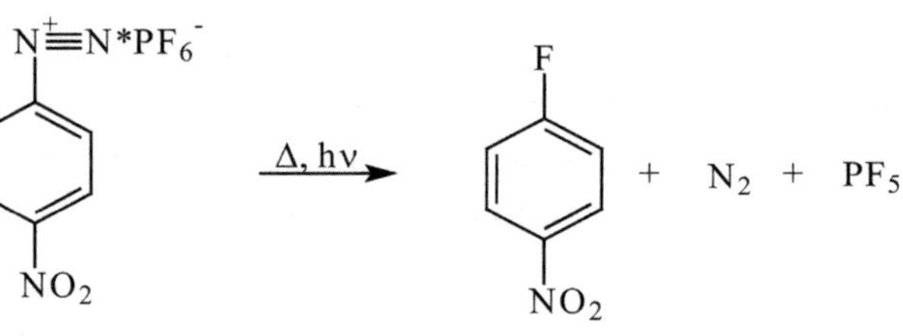

An example of the cationic polymerization initiated by a charge transfer complex is the reaction of Yamada, which can be illustrated by the polymerization of isobutylene with vanadium oxychloride complex with naphthalene in the heptane environment. The resulting complex decays with the formation of radical ions, which then react with isobutylene according to the following reactions:

$$VOCl_3 + C_{10}H_8 \longrightarrow \text{complex CT} \rightleftharpoons VOCl_3^{-} + C_{10}H_8^{+\cdot}$$

$$C_{10}H_8^{+\cdot} + H_2C{=}C(CH_3)_2 \longrightarrow C_{10}H_8 + \dot{C}H_2{-}C^{+}(CH_3)_2$$

$$2\dot{C}H_2{-}C^{+}(CH_3)_2 \longrightarrow (CH_3)_2C^{+}{-}CH_2{-}CH_2{-}C^{+}(CH_3)_2$$

To the new heterogeneous systems used to initiate the cationic polymerization belong perchlorates and trifluoromethyl sulphonates of magnesium, aluminum, cobalt, nickel and gallium. In contrast the lithium and silver salts do not show any catalytic activity. Initiating mechanism of this reaction relies on creation of an appropriate ion pair:

$$2\dot{C}H_2-C^+(CH_3)_2 \longrightarrow (CH_3)_2C^{+}-CH_2-CH_2-C^+(CH_3)_2 + nCH_2=C(CH_3)_2 \longrightarrow \text{polymer}$$

The formed carbocation reacts with the consecutive monomer molecules to create polymers. The cationic polymerization process requires high purity monomers and solvents, as impurities such as water, alcohols and acids play role of the chain transfer agent deactivating carbocations and decreasing the molecular weight of polymer.

The durability of the formed carbanion cation has a decisive influence on the growth rate of polymer chain. More lasting are the carbocations located at the chain ends larger is the reaction growth rate. So, this is an inverse relationship to that which takes place in the previously discussed process of free radical polymerization.

Presently this phenomenon is explained by the possibility of forming a π electron complex between carbocation located at the end of the chain and the new monomer molecule, rather than a covalent bond. It takes place during the attachment of a free radical to the monomer molecule. A large impact on the growth rate may have also the used solvent because of the differences in solvation of the growing ion. The use of a solvent of relatively low dielectric constant in the polymerization process causes that the ends of the chain appear mainly as a pair of ion with the counter-ion. Anion located near carbocation may also affect the growth rate, what makes the whole process more complex.

The formed during the polymerization carbanion macrocations may be subject of termination deactivity. The chain transfer can take place:

- in reaction with the monomer molecule:

$$\sim\sim CH_2-\underset{\displaystyle R}{\underset{|}{C}}{}^+HX^- + CH_2=\underset{\displaystyle R}{\underset{|}{C}}H \longrightarrow \sim\sim CH=\underset{\displaystyle R}{\underset{|}{C}}H + CH_3-\underset{\displaystyle R}{\underset{|}{C}}{}^+HX^-$$

- as a result of reaction with another polymer molecule with the formation of a more stable ion(grafting):

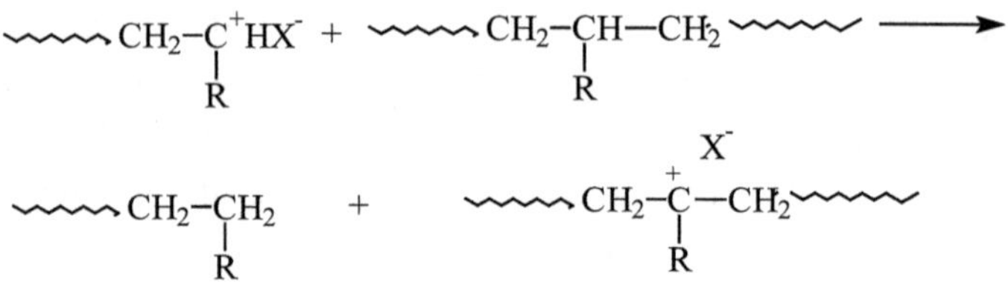

- as a result of electrophilic substitution reaction to the solvent molecule, for example, benzene:

- as a result of the ring alkylation:

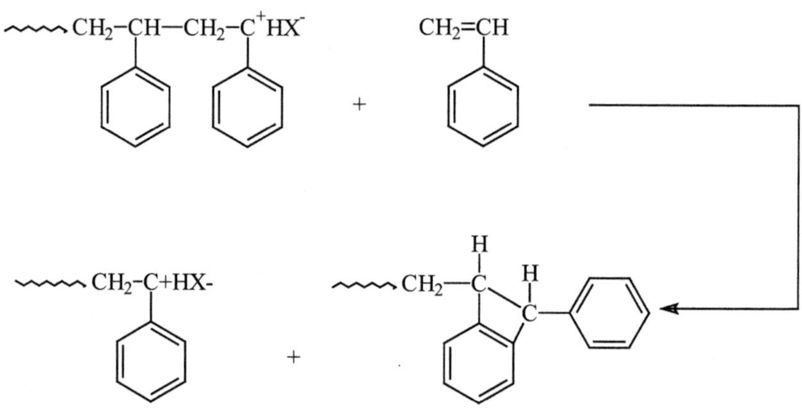

In the reaction consisting on the carbocation charge transfer to the carbon atom located in the main chain of another polymer molecule (grafting) a chain ramification takes place.

The completion of cationic polymerization occurs only when the catalyst is somehow deactivated during the process. Otherwise, the cationic polymerization shows several features of the so-called living polymerization process, what is of a great practical importance. In the living polymerization process the carbocation is not terminated and when adding next portion of the same or another polymer it is initiating further the polymerization reaction.

In several cases it is also possible to eliminate the transfer reaction. This is attributed to the covalent, pseudo ionic propagation reaction, running without involvement of reactive carbanion ions.

One of the examples illustrating the living cationic polymerization process is the polymerization of alkyl-vinyl ethers initiated by the mixture of hydrogen iodide with iodine. In this process the system is stabilized by a suitably strong interaction of carbocation with counterion:

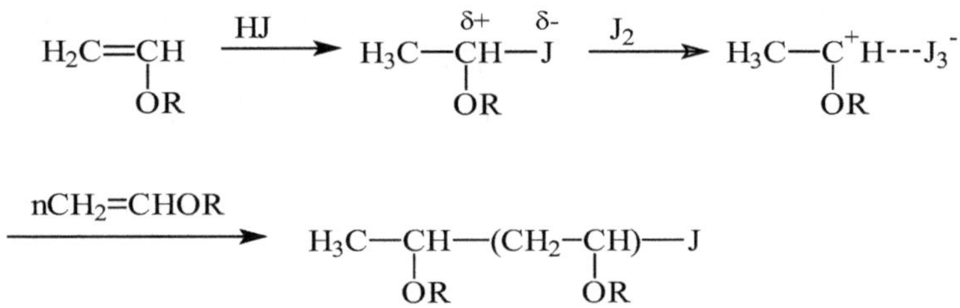

The concept of living polymerization will be detailed in Chapter 2.3.3. According to the mechanism of polymerization proposed by Higashimura its initiation consists on formation of an equimolar adduct of vinylether and hydrogen iodide. Then the binding of carbon – iodine in iodine adduct is activated with iodine, what in turn means a relaxation and polarization of the bond allowing attachment of further monomer molecules, up to their extinction in the reaction environment. The reaction, conducted in the presence of zinc iodide or other metal iodides as activators, proceeds with a conservation of the proportionality of molecular mass of the formed polymer to the degree of monomer conversion, even above the room temperature. The resulting polymer is characterized by a small dispersion of molecular masses (all resulting polymer molecules exhibit similar length chains).

The process of formation of living polymers allows the synthesis of block copolymers (in which the chain of one polymer is attached to the second polymer), because after all molecules of monomer A have reacted in the system there is no chain termination. Following introduction of monomer B to the system, capable of cationic polymerization, the reaction starts again with formation of a new block copolymer.

A-(A)n-A-B-(B)m-B

The formed block copolymer molecules are composed of linearly connected blocks either directly or by a constituent not being part of the blocks.

10.2.2.2 Kinetics of Cationic Polymerization

The kinetics of cationic polymerization determination was carried out in the presence of Friedel-Crafts catalysts and was specified for the cases in which the chain terminationtakes place. In the consideration the transfer reaction of chain activity to the monomer is ignored, because it runs with a low efficiency.

The rate of the catalyzing reaction is proportional to the concentrations of catalyst and monomer:

$$A + HB \rightleftharpoons AB^-H^+$$

$$AB^-H^+ + M \xrightarrow{k_i} HM^+AB^-$$

$$V_i = k_i[A][M]$$

where: A – cationic catalyst, such as, for example, BF_3, $AlCl_3$, $SnCl_4$, $SbCl_5$, etc.; HB – cocatalyst, for example, H_2O, CCl_3COOH and other; M – monomer; k_i – rate constant of initiation reaction; [A] – catalyst concentration; [M] – monomer concentration.

The reactions of the growth and of the chain termination can be described by the following equations:

$$HM_n^+AB^- + M \xrightarrow{k_W} HM_{n+1}^+AB^-$$

$$HM_n^+AB^- \xrightarrow{k_Z} M_n + HAB$$

In the steady-state of catalysis the reaction rate (V_i) is equal to that of the chain termination (V_z):

$$V_i = V_z$$
$$V_z = k_z[HM^+]$$
$$k_i[A][M] = k_z[HM_n^+]$$
$$[HM_n^+] = \frac{k_i}{k_z}[A][M]$$

The overall polymerization rate (V_p) is equal to that of the chain growth (V_w):

$$V_p = V_w = k_w[HM_n^+][M]$$

where: k_w and k_z – rate constants for chain growth and termination, respectively.

Substituting the previously calculated value of the macrocation concentration $[HM^+_n]$ to the above formula one obtains

$$V_p = \frac{k_w k_i}{k_z}[A][M]^2 = k[A][M]^2$$

The activation energy of cationic polymerization is always smaller than that of the free radical polymerization, and is of 50.2–62.7 kJ/mol.

A characteristic feature of the cationic polymerization is its high rate at low temperatures. A classical example of such a process, which was applied on the industrial scale, is the polymerization of isobutylene in the presence of boron trifluoride, which runs at temperatures below 200 K. Polymers with very high molecular mass are then formed within a few seconds. At temperatures above 290 K formed are only oligomers of isobutylene with a low molecular mass.

The high speed of cationic polymerization process makes difficult the study of the kinetics of this reaction.

10.2.3 ANIONIC POLYMERIZATION

10.2.3.1 Characteristics of Anionic Polymerization Process

The reaction of the attachment of nucleophilic agent to the unsaturated double bond of monomer leads to the formation of carbanion monomer capable of reacting with the next monomer molecules with the creation of polymer macromolecule. This type of reaction, proceeding through the stage with carbanion anion is called by the anionic polymerization.

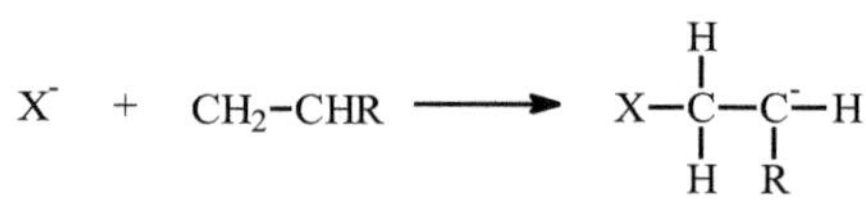

Using this method one can polymerize only the monomers, which contain substituents such as nitro group, nitrile, carboxyl, ester, vinyl and phenyl, capable of stabilizing the carbanion anion through the resonance or inductive effect.

Typical examples of monomers, which can polymerize in anionic way are: 2-nitropropene, acrylonitrile, methacrylonitrile, esters of acrylic and methacrylic acid, butadiene, styrene and izopropene.

A decisive influence on the effectiveness of the anionic polymerization has the chemical structure of used monomer. In the extreme case, very reactive monomers can be polymerized using very weak bases as nucleophilic agents.

Examples of such highly reactive monomers are: vinylidene cyanides, already polimerizing under influence of so weak nucleophilic agent as water and 2-nitropropene, whose polymerization is catalyzed by potassium bicarbonate:

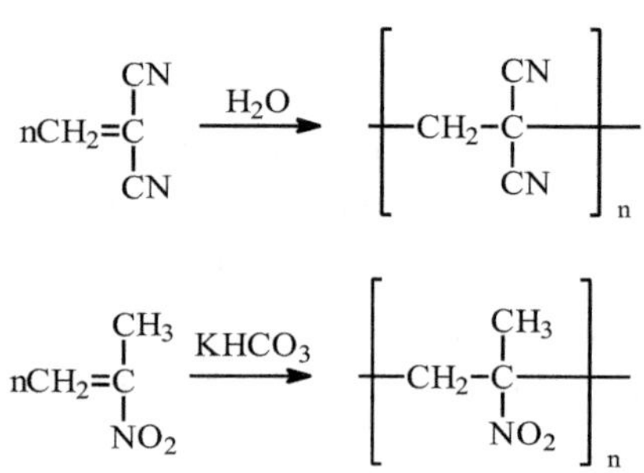

Unfortunately such cases are not common. Most frequently one has to use for this purpose compounds showing a strong nucleophilic effect.

Generally catalysts used for the anionic polymerization can be divided into two basic types:

- reacting through the connection of a negative ion;
- reacting by an electron transfer.

A few examples of catalysts reacting through attachment of a negative ion are: amides, litoorganic compounds, Grignard compounds and other metalloorganics.

The second group of anionic polymerization catalysts includes alkali metals and their additive complexes with some aromatic compounds.

The mechanism of anionic polymerization may change depending on the type of nucleophilic catalyst used, on reaction environment and on temperature.

The active site at the end of the growing chain can occur as:

- free anion

$$\sim\sim\sim M^{-}$$

- anion solvated by the solvent molecules S

$$\sim\sim\sim M^{-}\bigstar yS$$

- component of the ion pair

$$\sim\sim\sim M^{-}A^{+}$$

- component of the solvated ion pair

$$\sim\sim\sim M^{-}A^{+}\bigstar yS$$

Between different forms of the active carbanion center an equilibrium state is established:

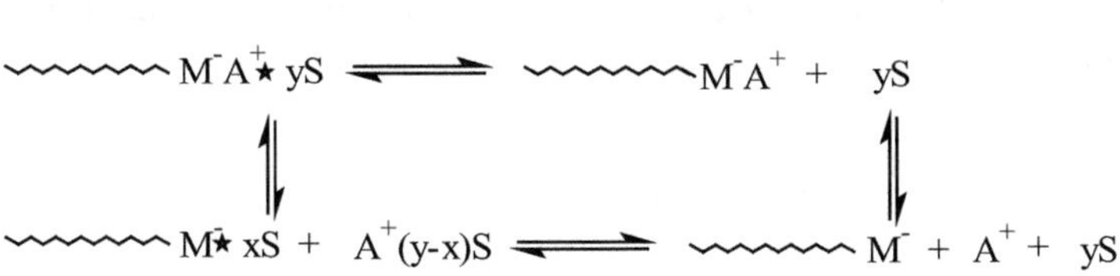

The active center, occurring in the form of a free ion, is more reactive in anionic polymerization than the active center being a component of an ionic pair. This is due to the polarization of monomer molecule approaching the active site.

Free ion produces an electric field stronger than a pair of ions and thus facilitates the polarization.

10.2.3.2 Mechanisms of Anionic Polymerization

The anionic polymerization, similarly as the cationic polymerization, takes place in three stages:

- initiation:

$$AB \rightleftharpoons A^- + B^+$$

$$A^- + M \xrightarrow{k_i} AM^-$$

- chain growth (propagation)

$$AM^- + nM \xrightarrow{k_w} AM^-_n$$

- chain termination:

$$AM^-_n + H^+ \xrightarrow{k_z} AM_nH$$

where: A, B – catalyst; M – monomer; k_i, k_w, k_z – rate constants of initiation, growth and chain termination, respectively.

A typical example of anionic polymerization, running with participation of free carbanions, is the polymerization of styrene initiated by potassium amide in liquid ammonia. The mechanism of this reaction presents as follows:

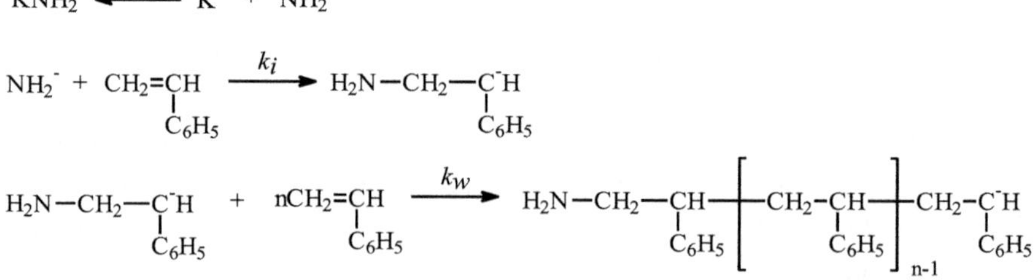

Completion of this reaction occurs by chain transfer to the solvent, which is ammonia:

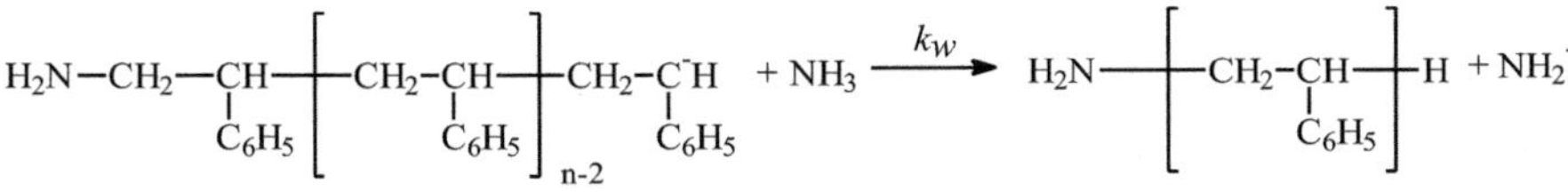

The resulting chain of polystyrene is ended on one side with the amino group. The anion NH_2^- formed from the solvent molecule, reconstruct with the K^+ cation the potassium amide.

Basing on the above equations one can write the following formulas for the rates of: initiation (V_i), growth (V_w) and chain termination (V_z) reactions:

$$V_i = k_i[NH_2^-][M]$$
$$V = V_w = k_w[M_n^-][M]$$

Taking account of stationary conditions, according to which $V_i = V_z$, one gets

$$k_i[NH_2^-][M] = k_z[M_n^-][NH_3]$$
$$[M_n^-] = \frac{k_i}{k_z}\frac{[NH_2^-][M]}{[NH_3]}$$

By substituting the calculated quantities to the equation defining the reaction rate V one gets the final result:

$$V = V_w = \frac{k_w k_i}{k_z}\frac{[NH_2^-][M]^2}{[NH_3]}$$

It follows from the above considerations that the rate of the anionic polymerization of styrene in the presence of potassium amide is proportional to the concentration of catalyst and the square of monomer concentration.

The degree of polymerization of substrates in this reaction is directly proportional to the monomer concentration and inversely proportional to the concentration of ammonia:

$$\bar{P} = \frac{V_w}{V_z} = \frac{k_w[M_n^-][M]}{k_z[M_n^-][NH_3]} = \frac{k_w[M]}{k_z[NH_3]}$$

An example of the anionic polymerization with a participation of ion pairs is the polymerization of styrene in the presence of butyllithium:

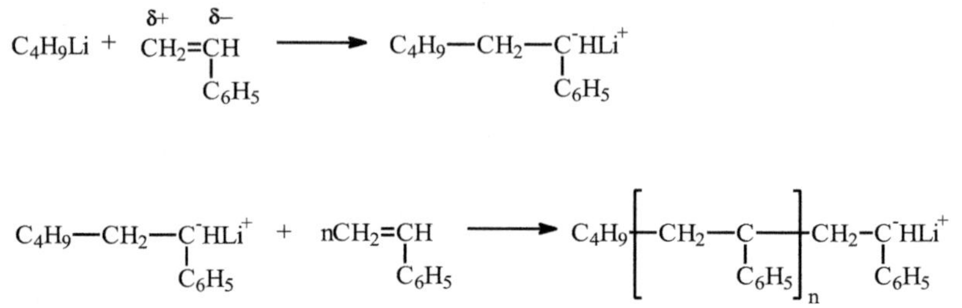

When such a reaction is carried out at room temperature, in a mixture of benzene with tetrahydrofuran, then it involves the participation of both unsolvated and solvated tetrahydrofuran ionic pairs. For this reason the kinetic scheme of such a process is more complicated, since one has to take into account three different rate constants of chain growth (depending on solvation) as well as the equilibrium constants between the various forms of active centers.

In a quite different way proceeds polymerization with an electron transfer. In that case, during the reaction of monomer with a metal from the first group of the periodic table of elements, an anionic radical is formed, which dimerizes immediately with formation of a bianion:

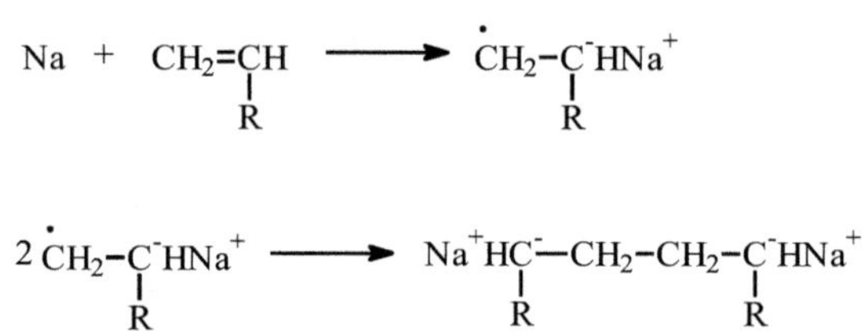

The evidence demonstrating such course of the reaction is the formation of corresponding dicarboxylic acids after an addition of carbon dioxide to the equimolar mixture of catalyst and monomer.

In some cases a mixed free-radical – anionic polymerization process is observed.

In a similar way proceeds the process of styrene polymerization, assisted by sodium naphthalene, formed easily in the reaction of sodium with naphthalene:

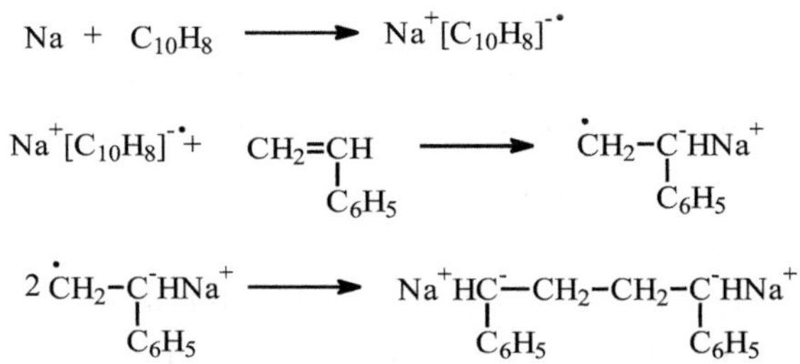

There exist also another methods of anionic polymerization, catalyzed by pyridine derivatives or by phosphines. An example of the latter reaction is the polymerization of maleic anhydride in the presence of triphenylphosphine, which is not homopolymerizing in the presence of radical initiators. The present reaction proceeds through the stage of ylide by using a free π electron pair.

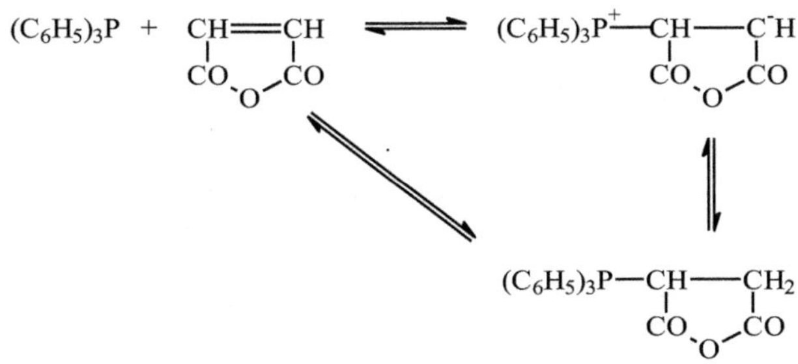

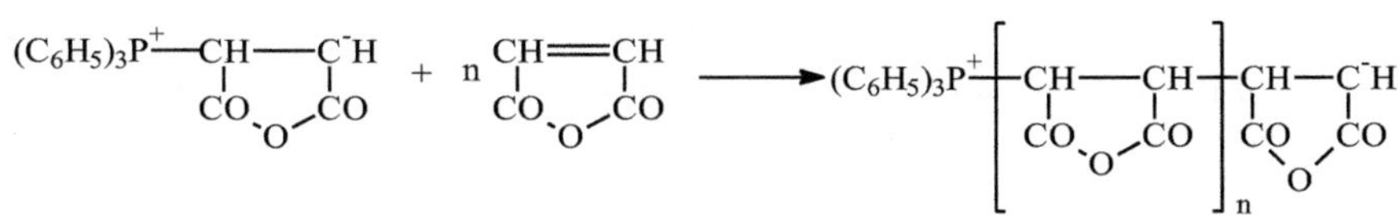

The activation energy of this reaction depends on the type of solvent used and amounts to 26.4 kJ/mol in acetic anhydride and to 39.8 kJ/mol in dimethylformamide.

The termination or the chain transfer reactions in anionic polymerization runs through:

- elimination of proton with formation of an unsaturated bond at the chain end;
- transfer of proton from the solvent molecule to monomer or to polymer;

- isomerization to an inactive ion;
- an irreversible reaction of the active carbanion center at the end of the chain with a monomer or a solvent molecule.

In a clean reaction environment the process of the termination or the chain transfer is rare. The formed polymer contains at the end of the chain a carbanion, which can be, even after a prolonged storage, an active center, starting the polymerization process after introduction of a new portion of monomer.

10.2.3.3 "Living" Polymerization

The concept of "living" polymerization was introduced for the first time by M. Szwarc in 1956. According to his definition the "living" polymerization takes place if:

$$ki > kw \text{ and } kz = 0$$

where: k_i, k_w, k_z – rate constants of initiation, growth and termination of polymer chain, respectively.

If these conditions are satisfied, the concentration of active centers is constant and equal to the initial concentration of nucleophilic agent (catalyst), the polymerization rate is constant. The average molecular weight (M) increases linearly with the increasing degree of conversion. The distribution of molecular weights is narrow and close to the Poisson distribution

$$\frac{\bar{M}_w}{\bar{M}_n} < 1.1$$

Introduction of the concept of living polymerization and its practical use was possible owing to the research done on the course of anionic polymerization of styrene and dienes. During the styrene polymerization, using sodium naphthalene as catalyst; its green color changes to red of styrene anion.

The red color is maintained even at 100 percent of conversion, and the addition of another portion of styrene or other polymerizing anionic monomer induces a further polymerization. In the living polymerization all macromolecules present in the system are terminated by active centers, which retain their activity during the polymerization, even after depletion of monomer. This creates a practical possibility of obtaining block copolymers by a successive introduction of different monomers into the system:

$$-[M_1]_{n-1} - \bar{M} + (m+1)M_2 \rightarrow -(M_1)_n - (M_2)_m - \bar{M}_2$$

The possibilities of obtaining block copolymers from "living" polymers are presented in Table 10.6.

Some of them are used on industrial scale. The systems, which accomplish all the strict requirements of the "living" polymerization, include only the anionic polymerization of vinyl and cyclic monomers.

TABLE 10.6 Block Copolymers Obtained from "Living" Polymers

No	"Living" polymer	Second monomer
1	Polystyrene	isoprene
2	Polystyrene	α-methylstyrene
3	Polystyrene	1-vinyl naphthalene
4	Polystyrene	acrylonitrile
5	Polystyrene	methacrylonitryle
6	Polystyrene	methyl methacrylate
7	Polystyrene	methyl methacrylate
8	Polystyrene	2-vinylopyridine
9	Polystyrene	4-vinylopyridine
10	Polystyrene	ethylene oxide
11	Polystyrene	dimethylosiloxane
12	Polyisoprene	styrene
13	Polyisoprene	ethylene oxide
14	Polı(α-methylstyrene)	styrene
15	Poly(4-vinylopyridine)	methyl methacrylate
16	Poly(2-vinylopyridine)	styrene
17	Poly(4-vinylopyridine)	4-bromostyrene
18	Poly(methyl methacrylate)	butyl methacrylate
19	Poly(methyl methacrylate)	isopropyl acrylate
20	Poly(methyl methacrylate)	acrylonitrile

Carbanions derived from such monomers like styrene and butadiene, are not subject to any other reactions outside the propagation. During the anionic polymerization of methacrylates a chain transfer on the esther-COOR group of monomer or polymer can take place. Therefore, a formation of living polymers in this system is difficult.

It turned out that in many cases one uses the term of "living" polymerization for processes that only partially fall within the definition given by Szwarc. Therefore, some authors aware of this, introduce the concept of the "pseudo-living" or "quasi-living" polymerization. These concepts apply when the termination and transfer of chain rate constants are equal to zero and the condition: *ki* > *kw* is not satisfied, or the propagation reaction is reversible, or a reversible chain transfer to polymer takes place.

Living polymers can also react with the appropriate chemical reagents (XY) to give polymers containing terminal functional groups, such as macromers:

$$-[M]n - M^{-} + XY \rightarrow -(M)n - MY + X^{-}$$

10.2.3.4 Star-Shaped Polymers

In last years the obtained by anionic polymerization living polymers were used for the synthesis of a new class of polymers called star-shaped polymers.

The synthesis of such polymers is represented schematically in the following examples:

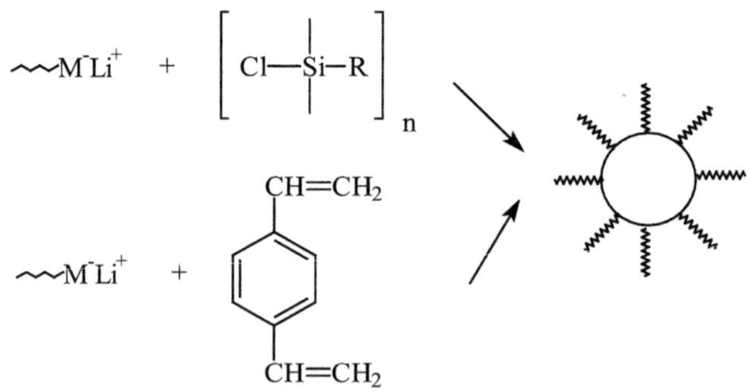

The first type synthesis allows a better control of the arm lengths, which exhibit similar average molecular weight.

In the case of the application of p-divinylbenzene for this purpose the core macromolecule is a cross-linked polymer. Despite that the presence of long, linear arms, derived from the "living" polymer makes the macromolecule soluble. It turned out that not all active centers in the cross-linked core of poly divinylbenzene are accessible in the same way. For this reason the initiation of the reaction proceeds slowly and leads to the star-shaped polymer with arms of different lengths. An important advantage of this method is the possibility to obtain the star block copolymers. An example of such a compound is a copolymer containing up to 30 arms. Each of them is a block copolymer of styrene (M = 2700–30,000) and ethylene oxide (M = 3000–6000). In addition to this information it should be added that recently the Japanese researchers elaborated a method of synthesis of star polymers based on the Grignard compounds:

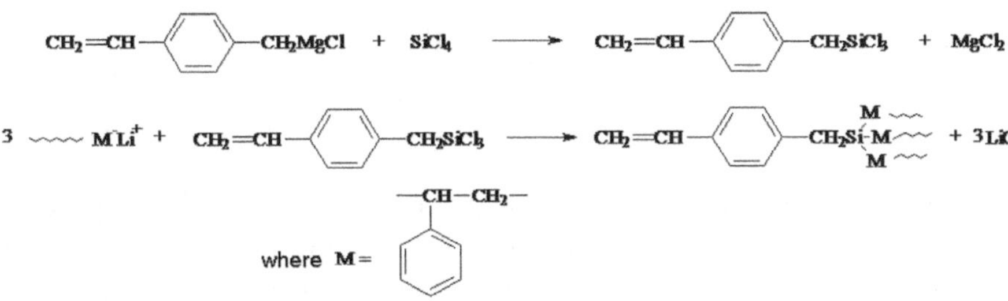

The presence of free vinyl groups in these products was used to obtain grafted copolymers. Grafted copolymer is a grafted polymer formed of more than one type of monomer.

In 1988, J. Roovers with coworkers have synthesized star-shaped polymers with a high packing. For that they have used the hydrosilylated poly(1,2-butadiene) with a low molecular weight:

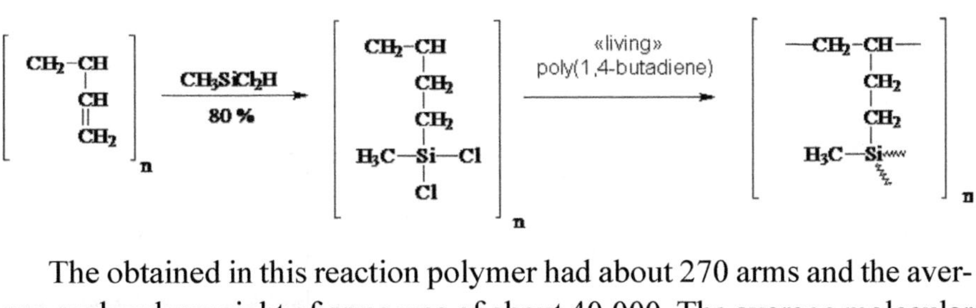

The obtained in this reaction polymer had about 270 arms and the average molecular weight of arms was of about 40,000. The average molecular weight of the entire macromolecule was of 11,000,000.

Strictly speaking the obtained reaction product is a grafted copolymer. However, due to the small size of the core (average molecular weight of used poly(1,2-butadiene is of about 9000) in comparison with the whole macromolecule (M = 11,000,000) and the large number of chains the synthesized macromolecule s a model of a star-shaped molecule.

The above-described star-shaped polymers are soluble and can serve as models to verify the theory describing conformations of macromolecules in solutions with a high degree of branching.

As mentioned earlier, the anionic polymerization of methacrylates is accompanied by the chain transfer reaction. A more accurate understanding of this process has enabled the development of anionic polymerization with a group transfer group (GTP – Group Transfer Polymerization), which can be illustrated by the following example:

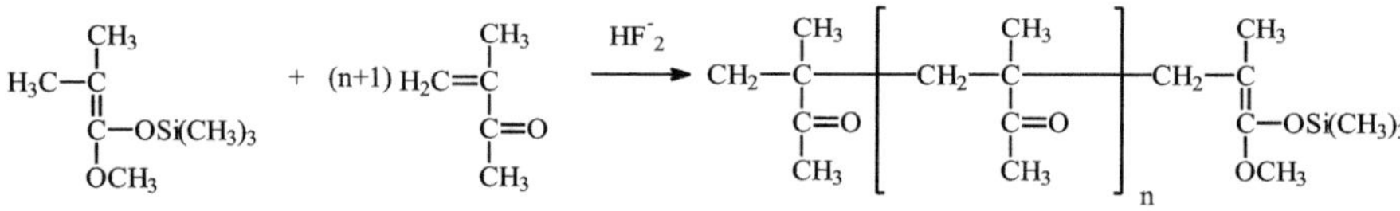

Basing on the modeling studies, which have shown that in the process of propagation the initial dissociation of O-Si bonds and the formation of an ion active site are not required, the following mechanism of this reaction is assumed:

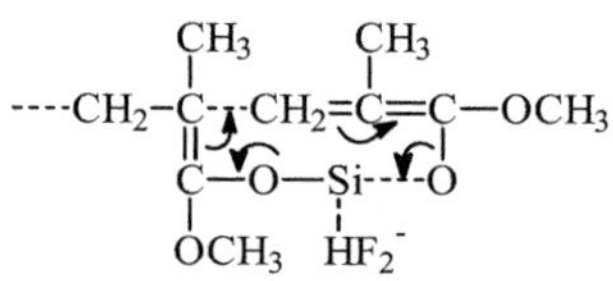

The polymerization process with the group transfer was used for the synthesis of ladder polymers:

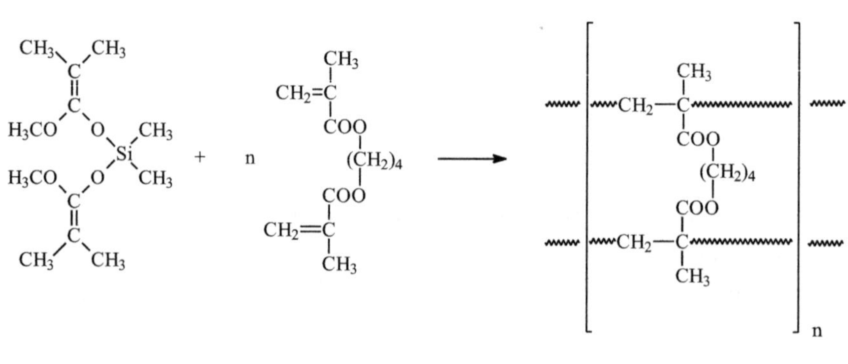

Introduction of ethylene or propylene to the dimethacrylate system gives rise to formation of a cross-linked core (P_n). The resulting product has structure of a star-shaped ladder polymer:

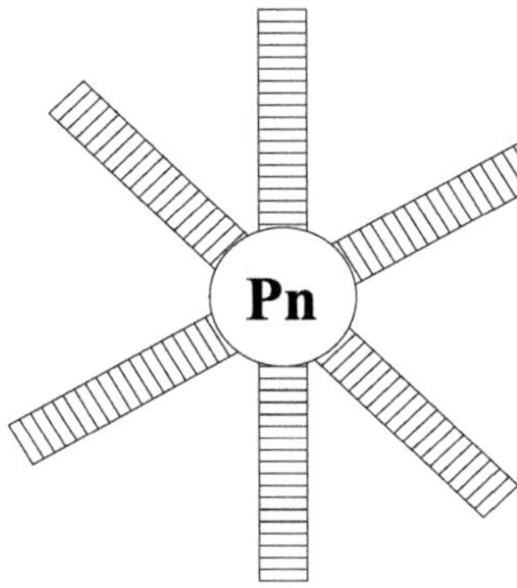

From the above considerations it follows that the possibilities of application of anionic polymerization are large with a real perspective of its use on the industrial scale.

10.2.3.5 Stereochemistry of Anionic Polymerization

The spatial structure of the products of anionic polymerization depends on the type of catalyst used and to a large extent on the degree of counterion association. An important role plays here the type of solvent used and the temperature of process conduction.

When using anionic catalysts in a homogenous environment at low temperatures the polar solvents favor the formation of syndiotactic structure, while the non-polar solvents privilege the formation of isotactic polymer. It is believed that the mechanism of polymerization in this case is connected with the formation of ion pairs.

Anion located at the end of the chain has a flat structure and is stabilized by resonance. The monomer molecule can come close to the active center from the front, giving the syndiotactic structure of polymer, or from behind, creating an isotactic structure. It depends on the degree of associative links of carbanion with the counterion and on the size of substituents in the polymer molecule.

In particular, of importance is the use of anionic polymerization to obtain polydienes. Depending on the type of used catalyst and on operating conditions one can obtain in this way polydienes of different structures, including polymers with overwhelming "cis-1,4" structure, similar to that of natural rubber. The influence of the used catalyst on the course of anionic polymerization of butadiene and isoprene and the structure of formed polymers are given in Table 10.7.

It is worthy to note that the first synthetic polybutadiene, produced on a large scale, was obtained by the anionic polymerization method in the presence of metallic sodium. From the names of butadiene and natrium (sodium) it took the trade name "Buna".

10.2.3.6 Effect of Solvent on the Course of Anionic Polymerization

The course of anionic polymerization in solution depends on the type of solvent used and particularly on its polarity. Polar solvents are easier

TABLE 10.7 Effect of Catalyst in Anionic Polymerization of Butadiene and Isoprene on the Structure of the Resulting Polymer Chain

No.	Catalyst	Monomer	Solvent	Chain microstructure [% mol]			
				cis-1,4	*trans*-1,4	1,2	3,4
1	C_2H_5Li	butadiene	n-hexane	43	50	7	—
2	C_2H_5Li	butadiene	tetrahydrofuran	0	9	91	—
3	Na	butadiene	—	10	25	65	—
4	$C_{10}H_8Na$	butadiene	tetrahydrofuran	0	9	91	—
5	K	butadiene	—	15	40	45	—
6	$C_{10}H_8K$	butadiene	tetrahydrofuran	0	17	83	—
7	Rb	butadiene	—	7	31	62	—
8	Rb	butadiene	tetrahydrofuran	0	25	75	—
9	Cs	butadiene	—	6	35	59	—
10	n-C_4H_9Li	isoprene	n-heptane	93	0	0	7
11	n-C_4H_9Li	isoprene	tetrahydrofuran	0	30	16	54
12	C_6H_5Na	isoprene	n-heptane	0	47	8	45
13	C_6H_5Na	isoprene	tetrahydrofuran	0	38	13	49
14	$C_6H_5CH_2K$	isoprene	n-heptane	0	52	10	38
15	$C_6H_5CH_2K$	isoprene	tetrahydrofuran	0	43	17	40
16	Rb	isoprene	—	5	47	10	38
17	Cs	isoprene	—	4	51	8	37

to solvate the catalyst. It was checked carefully in the case of anionic polymerization initiated by lithium or by lithorganic compounds. In addition, the larger value of dielectric constant of polar solvents facilitates dissociation of ionic bonds of carbon-lithium anion, present in the catalyst. The polarity of the used solvent affects the chain termination rate as well as the structure of the resulting polymer. Using of nonpolar solvents causes formation of oligomeric side products during the anionic polymerization. A very useful solvent to carry out the anionic polymerization is tetrahydrofuran. It was also demonstrated that the polydispersity of resulting polymer decreases with the increasing size of counterion.

The solvent polarity affects significantly the structure of polymer. Isoprene polymerized by anionic polymerization method in hydrocarbons forms 70–90% of cis-1,4- variety of polyisoprene, whereas in tetrahydrofuran it yields mainly the poly(3,4-isoprene). A similar effect was observed during the polymerization of 1,3-butadiene. During the polymerization in hydrocarbons mainly poly(1,4-butadiene) is obtained, whereas in tetrahydrofuran the main product is poly(1,2-butadiene). These drastic changes occur even when a small amount of polar solvent such as tetrahydrofuran is added to the hydrocarbon reaction environment in which the anionic polymerization runs. Other, ether or anionic co-solvents do not show such large influence on the reaction like does tetrahydrofuran. The ratio of the variety cis to the variety of trans polybutadiene obtained in hydrocarbon environment is usually of 2 to 3 and is maintained even at high content of vinyl groups originating from poly(1,2-butadiene).

10.2.4 COORDINATION POLYMERIZATION

10.2.4.1 Characteristics of Coordination Polymerization Process

The coordination polymerization is a new type of synthesis of macromolecular compounds to obtain polymers with a regular spatial structure. It was preceded by the basic research of K. Ziegler and his coworkers in the field of synthesis and application of organometallic compounds. The studies have led in 1953 to the discovery of a new complexing compound consisting of triethylaluminum and titanium trichloride.

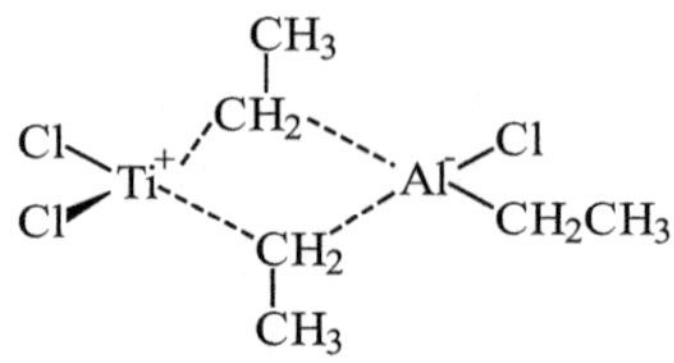

It turned out that Ziegler obtained a complex catalyzing polymerization of ethylene at normal pressure.

At the same time an Italian scholar G. Natta received for the first time the isotactic polypropylene by polymerization of propylene on the catalyst complex composed of titanium trichloride and triethylaluminum.

These discoveries have revolutionized the research on the polymerization process. The catalysts based on organometallic complexes are called Ziegler – Natta catalysts.

The importance of discovery is the best demonstrated by the launching in Ferrara (Italy) of industrial production of isotactic polypropylene already in 1957 by an Italian company Montecatini.

But it turned out that the various organometallic complexes vary in effectiveness and stereospecificity in catalytic activities in the polymerization process and in different behavior for different monomers. An example is the already discussed complex: titanium trichloride with aluminum triethyl. It is a very good catalyst for the synthesis of polyethylene, but is not suitable for polymerization of propylene due to the small stereospecificity. The quantity of produced isotactic polypropylene is then twice less than the one obtained using the Natta complex of titanium trichloride with aluminum triethyl.

In further studies it appeared that the activity of Ziegler – Natta catalysts depends not only on the chemical structure of the complex molecule but also on its crystal structure, and therefore on the way of its preparation.

For example, the titanium trichloride can be prepared by reducing titanium tetrachloride with hydrogen, metallic titanium or with aluminum. It can be activated additionally by using physical or chemical methods. The greatest activity, without additional treatments, shows δ trichloride of titanium. The characteristics of titanium trichloride as a catalyst for propylene polymerization are presented in Table 10.8.

TABLE 10.8 Characteristics of Titanium Trichloride as a Catalyst for Propylene Polymerization

No.	Designation of the form	Preparation	Stereospecificity (amount of isotactic polypropylene) [%]
1	α (alfa)	reduction of $TiCl_4$ by hydrogen or titanium at 673 K	80–90
2	β (beta)	reduction of $TiCl_4$ with hydrogen in an electric arc or decomposition of CH_3TiCl_3 in hydrocarbons	40–50
3	γ (gamma)	heating of beta form for 2–3 hours or decomposition of CH_3TiCl_3 without solvent	80–90
4	Δ (delta)	reduction of $TiCl_4$ with metallic aluminum (the product contains still $AlCl_3$)	>90

After F. Andreas, K. Grobe, Propylene chemistry (Chemia propylenu, in Polish), WNT, Warszawa 1974, p. 372.

Currently a large number of Ziegler-Natta catalysts are known. They are produced by the reaction of transition metal halides derivatives (Tr, Cr, Th, Zr, V, Nb, Ta, Mo, Co) with organic derivatives of metals from groups I, II and III (Li, Mg, Al), playing the role of reducing agents. The products of these reactions have the structure of complex compounds. The influence of the used metal halide on the catalytic activity of the complex compound with aluminum triethyl during the propylene polymerization is presented in Table 10.9.

TABLE 10.9 Influence of Metal Halides on the Catalytic Activity of the Complex with Aluminum Triethyl

No.	Metal halide	Chemical formula	Yield of isotactic polypropylene [%]
1	titanium dichloride	$TiCl_2$	80–90
2	α-titanium trichloride	$TiCl_3$	85–90
3	β-titanium trichloride	$TiCl_3$	40–50
4	titanium tribromide	$TiBr_3$	44
5	titanium triiodide	TiJ_3	10

TABLE 10.9 Continued

No.	Metal halide	Chemical formula	Yield of isotactic polypropylene [%]
6	zirconium dichloride	$ZrCl_2$	55
7	vanadium trichloride	VCl_3	70–75
8	chromium chloride	$CrCl_3$	36
9	titanium tetrachloride	$TiCl_4$	45–50
10	titanium tetrabromide	$TiBr_4$	40–42
11	titanium tetraiodide	TiJ_4	46
12	titanium tetraalkoxy	$Ti(OR)_4$	traces
13	zirconium tetrachloride	$ZrCl_4$	52
14	vanadium tetrachloride	VCl_4	48
15	vanadium chloride	$VOCl_3$	32

The strong interest of industry for Ziegler – Natta catalysts led to the discovery of new types of bonds of diverse, often much more active and applicable for different purposes forms. Currently catalytic systems operating in a homogenous as well as a heterogeneous environment system are known. They allow to make the coordination polymerization in solution or in a fluidized phase.

Several recent studies have shown that the activity of the classical Ziegler-Natta catalyst can be increased by introducing into the system additional chemical compounds, most frequently the magnesium chloride. Sometimes similar effect is obtained by the addition of ethyl benzoate. By the fact it was shown that the time of course of the polymerization reaction has no apparent effect on the average rate of isotacticity and on the molecular weight.

The influence of the catalyst, operating in a heterogeneous system, on the rate of isotacticity and efficiency of the reaction is given in Table 10.10.

Some new types of Ziegler-Natta catalysts such as the complex of tetrabenzyltitan with methyl aluminosiloxane allow obtaining the syndiotactic polymers with a good yield.

A characteristic feature of the coordination polymerization, which distinguishes it from other processes of this type, is the fact that each attached molecule is at the beginning coordinated by a metal atom of the catalyst, for example, the atom of titanium. In this way it gets a well-determined

TABLE 10.10 Effect of Catalyst on the Course of the Coordination Polymerization of Styrene and Propylene at Temperature of 313 K

No.	Catalyst	Cocatalyst	Styrene polymerization		Propylene polymerization	
			Activity [mol/(Ti*h)]	Isotatic rate [%]	Activity [mol/(Ti*h)]	Isotatic rate [%]
1	$Ti(OC4H9)4/Mg(OH)Cl$	$Al(CH3)3$	0.4[a]	~100	100[b]	12
2	$TiCl3/MgCl2$	$Al(C2H5)3$	25[c]	96.7	14,000[d]	0.9
3	$TiCl3$ (Solvay type)	$Cp2Ti(CH3)2$	30[c]	99.2	70[d]	99.2
4	$TiCl4/COOC2H5/MgCl2$	$Al(C2H5)3$	190[c]	93.4	8360[d]	59.7

Explanations: C_p – cyklopentadiene;
[a]40 cm^3 of heptane and 10 cm3 of styrene, 24 H;
[b]30 cm^3 of heptane and propylene at a pressure of 2 MPa, 24 H;
[c]40 cm^3 of heptane and 10 cm3 of styrene, 0.5 H;
[d]100 cm^3 of heptane and propylene at a pressure of 0.1 MPa, 2 H;
after K. Soga et al., Macromol. Chem. Rapid Commun. 11, 229 (1990).

orientation with respect to the growing chain. It facilitates also the formation of polymers with stereoregular structure. The addition of the monomer molecules to the growing chain of polymer takes place at its base and not at the final fragment of the chain, as it was the case in the classical free radical or ionic polymerization. The growth of chain of such a polymer can be compared to the hair growth.

10.2.4.2 The Mechanism of Coordination Polymerization

The mechanism of the coordination polymerization may vary depending on the used coordinate catalyst. If for the synthesis of a catalytic complex a metal halide is used, in which the metal atom is at a higher degree of oxidation, then the coordinate – radical polymerization takes place. This is possible due to the possibility of free radical formation by the reaction of reduction.

$$TiCl_4 + AlR_3 \longrightarrow RTiCl_3 + AlR_2Cl$$

$$RTiCl_3 \longrightarrow R^{\bullet} + TiCl_3$$

$$R^{\bullet} + nM \longrightarrow \text{polymer}$$

The formed free radicals "R •" initiate the process of polymerization. Its efficiency depends on the type of monomer used. Polymers obtained by the coordinate – radical polymerization exhibit a less regular structure than those obtained by the coordinate – anionic polymerization. This is due to the presence of free radicals in the system initiating the polymerization of monomers without participation of active centers of the coordinate catalyst.

If in the formation of the Ziegler-Natta catalyst complex is involved a component, which is an electron acceptor than the acidity of such a system is sufficient to initiate the coordinate – caionic polymerization. In the reaction of such kind involved are monomers capable of resonance stabilization of carbon ion cation, such as, for example, vinyl ethers. The reaction is probably initiated by the proton produced by reaction of an acid fragment of catalyst with water, which in turn joins the monomer molecule, coordinated by the metal atom.

Due to the nature of the initiation the resulting polymer does not contain alkyl groups, originating from the catalyst.

The polymerization on Ziegler-Natta catalysts most frequently proceeds along the coordinate – anionic mechanism in active sites located on the surface of catalyst. It is believed that the organometallic component of catalyst activates a specific place of the polymer chain, alkylating the transition metal atom located on the surface. The literature describes a number of possible courses of such a reaction. Currently as the most possible for this reaction are considered two mechanisms:

- bimetallic mechanism;
- monometallic mechanism.

According to the bimetallic mechanism of monomer polymerization on the catalytic titanium-aluminum system, it is assumed that at the first stage a complex between the π bond of monomer and the titanium atom is formed, which is connected with the aluminum atom by alkyl groups. It is followed by polarization of the transition metal – alkyl group bond and the formation in the transition state of a hexacomponent ring, incorporating the monomer molecule.

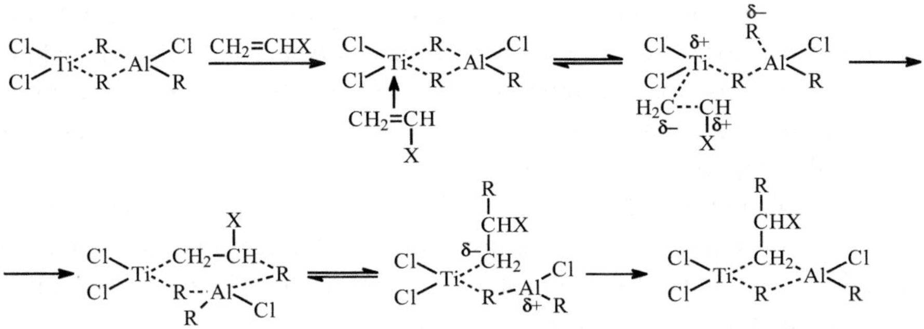

The reaction proceeds further in analogic way with the next monomer molecules, up to the formation of polymer.

The monometallic mechanism, developed by P. Cossee, presupposes the existence of an active center, which is the transition metal atom with a coordination number of 5. This center is activated by replacement of the alkyl group by a metaloorganic cocatalyst (e.g., trialkylaluminum).

The titanium or other transition metal atom coordinates the monomer molecule, which then joins. At the same time the location of vacancy in octahedron moves. The polymer chain moves back, so that the free space (vacancy) takes its starting position. This last step is necessary to preserve the stereoregularity of polymer.

The reaction of coordination polymerization by the monometallic mechanism can be expressed as follows:

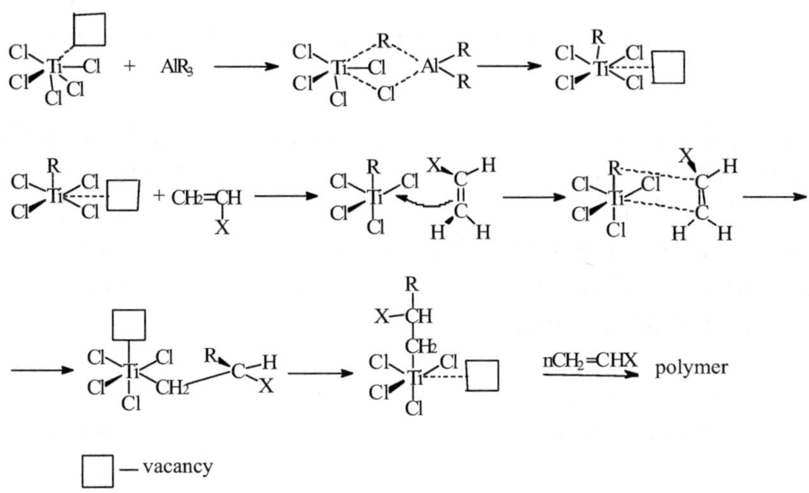

For both mechanisms of the coordination polymerization the polymer chain grows away from the surface of the catalyst as a result of attaching the next, pre-coordinated monomer molecules. The alkyl group, coming from trialkylaluminum or another metaloorganic cocatalyst becomes to be the final group of the polymer chain.

In the Ziegler – Natta polymerization takes also place the transfer reactions of the chain activity as demonstrated by the fact that the molecular weight reaches a relatively constant value in a short time after the initiation of reaction. These reactions include:

- transfer to the monomer molecule;

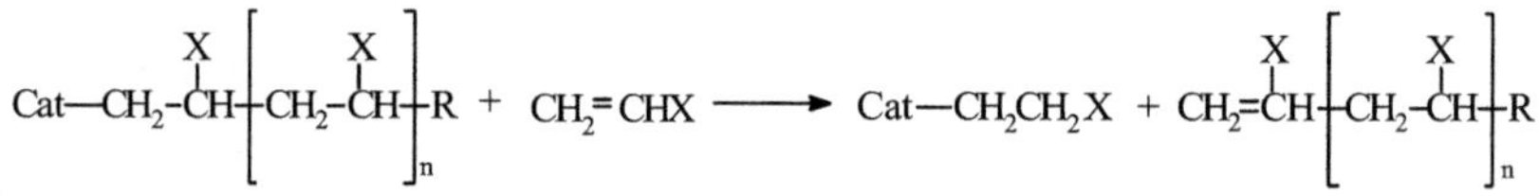

- transfer to the alkyl group of organometallic compound;

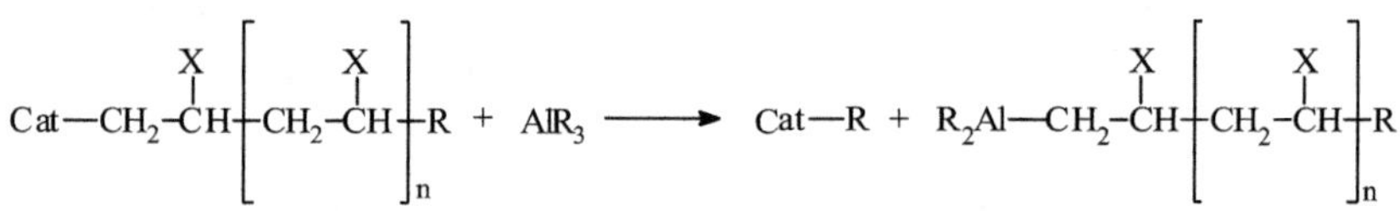

- transfer of a proton, associated with the termination of the chain growth;

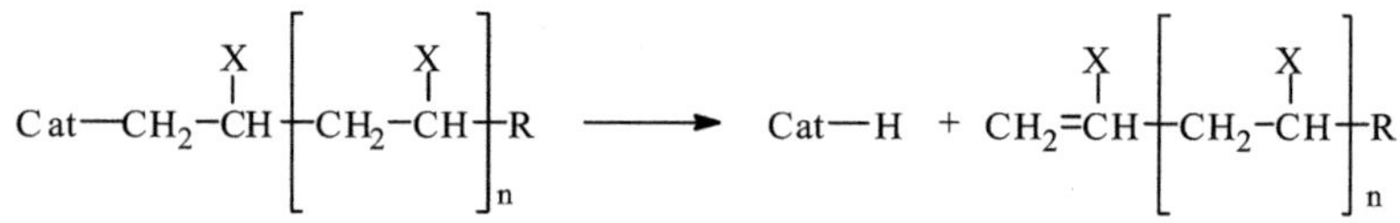

The presence of methyl groups and unsaturated bonds in polymer, confirmed by the infrared spectroscopy demonstrates the ability of such a course of reaction.

Introduction to the system of compounds active hydrogen atoms causes the termination of the chain growth. It proves the coordinate – anionic mechanism of this reaction.

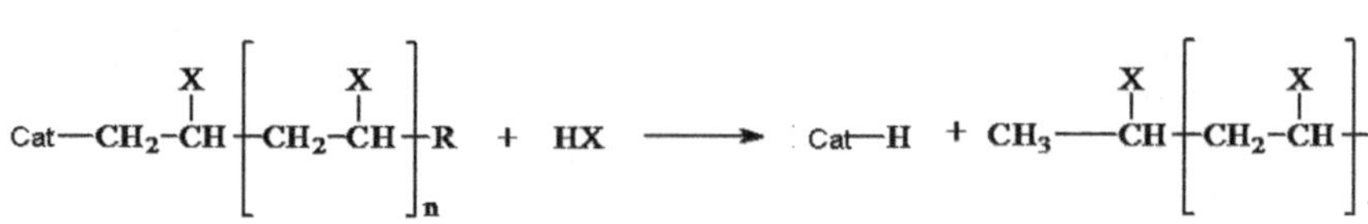

Another evidence confirming the coordinate – anionic mechanism of reaction is the relatively long life of growing chains. It makes possible obtaining of block copolymers by adding a second monomer after the reaction of the first. The process of polymerization in the presence of Ziegler-Natta catalysts is not affected by the typical radical transfer reactions. It means that the process is not a free radical one.

10.2.4.3 Coordination Polymerization of Dienes

As a result of polymerization of 1,3-butadiene and its derivatives one obtains synthetic rubbers (elastomers) whose properties depend on the structure of the formed products. The stereospecific Ziegler-Natta catalysts offer new opportunities for the synthesis of rubbers with a defined structure. It turned out that by choosing an appropriate catalyst one can be obtain the following polymer structures: "*cis*-1.4," "*trans*-1.4," isotactic "1.2," and syndiotactic "1.2," all in a relatively pure form. The influence of the catalyst structure on the stereospecific polymerization of butadiene is shown in Table 10.11.

TABLE 10.11 Effect of Catalyst on the Stereospecific Polymerization of Butadiene

No.	Catalyst	Yield [%]	Polymer structure	References
1	R3Al-VCl4	97–98	trans-1,4	J. Polym. Sci. 48, 219 (1960)
2	R3Al-VCl3	99	trans-1,4	J. Polym. Sci. 48, 219 (1960)
3	R3Al-VOCl3	97–98	trans-1,4	J. Polym. Sci. 48, 219 (1960)
4	di(π-cyclooctadiene) nickel–HJ	~100	trans-1,4	J. Polym. Sci. B. 5, 785 (1967)
5	R3Al-TiJ4	93–94	cis-1,4	J. Polym. Sci. 48, 219 (1960)
6	R3Al-CoCl2•pyridine	90–97	cis-1,4	Dokł. Akad. SSSR. 135, 847 (1960)
7	R2AlCl-CoCl2	96–97	cis-1,4	J. Polym. Sci. 48, 219 (1960)
8	R3Al-Ti(OC6H9)4	90–100	1,2	J. Polym. Sci. 38, 45 (1959)
9	Al(C2H5)3-V(OCOCH3)3	~90	syndiotactic 1,2	J. Polym. Sci. 48, 219 (1960)
10	Al(C2H5)3-Cr(C6H5CN)3 Al/Cr = 2	~100	syndiotactic 1,2	Macromol. Chem. 77, 114 (1964)
11	Al(C2H5)3-Cr(C6H5CN)3 Al/Cr = 10	~100	isotactic 1,2	Macromol. Chem. 77, 126 (1964)

The structure of the emerging polyene depends on the structure and the crystal form of the catalyst, the valence of transition metal and the configuration of the monomer molecule (cis or trans).

Depending on the used catalyst one or both diene double bonds are coordinated. Coordination of one of two bonds only leads to the polymerization such as "1.2." The coordination of both bonds provides polymer with "1.4" structure. According to another theory the coordination of π-allyl structure takes place and the polymer structure is determined by the movement direction of the approaching monomer molecule. If the monomer molecule is approaching the center of an active catalyst from the CH_{2-}titanium bond side the formed polymer has the "1.4" structure type. When the monomer molecule is approaching from the CH_{2-}titanium bond then the polymerization is of "1.2" type.

A simplified diagram is as follows:

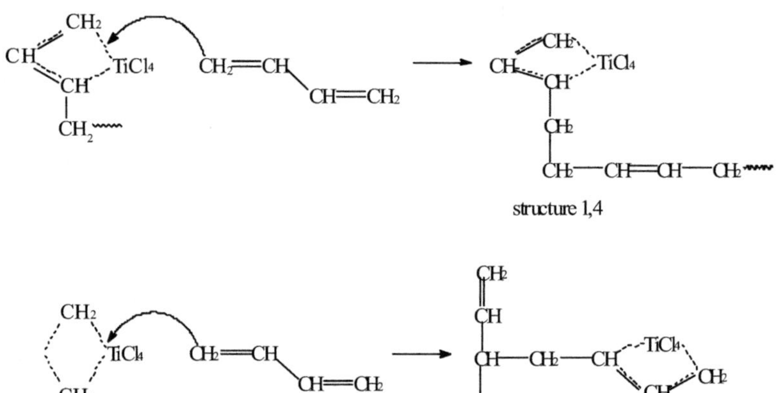

structure 1,4

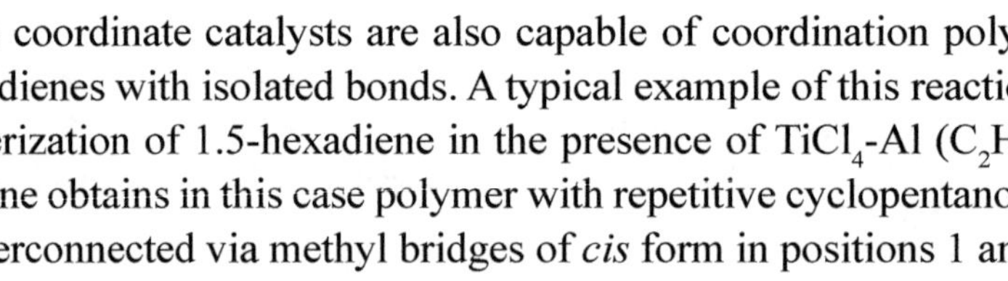

The coordinate catalysts are also capable of coordination polymerization of dienes with isolated bonds. A typical example of this reaction is the polymerization of 1.5-hexadiene in the presence of $TiCl_4$-Al (C_2H_5) complex. One obtains in this case polymer with repetitive cyclopentanone rings and interconnected via methyl bridges of *cis* form in positions 1 and 3:

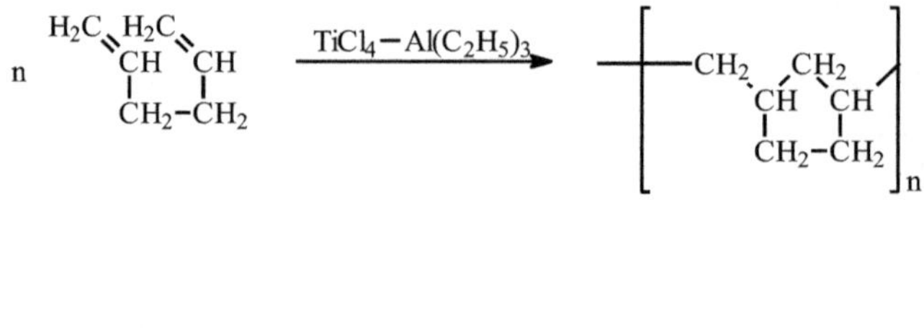

If soluble catalysts are used then the "1.2" polymerization can be carried out. Similarly, from the 2,6-dimethyl-1,6-heptadiene is formed a polymer with cyclohexane rings:

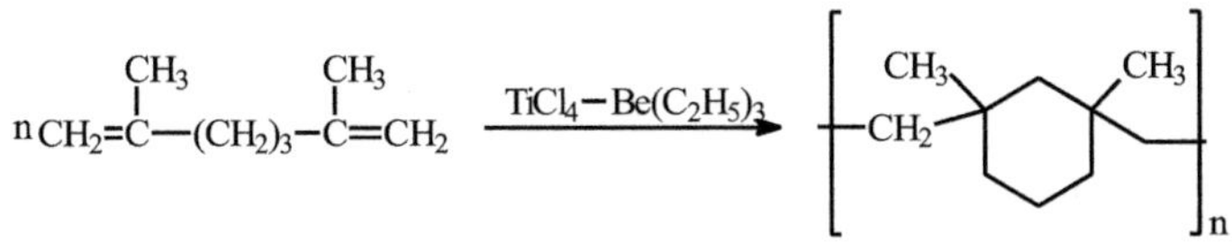

Another example is the polymerization of 1,5-cyclooctadiene leading to the formation of a polymer with a bicyclic repeat unit:

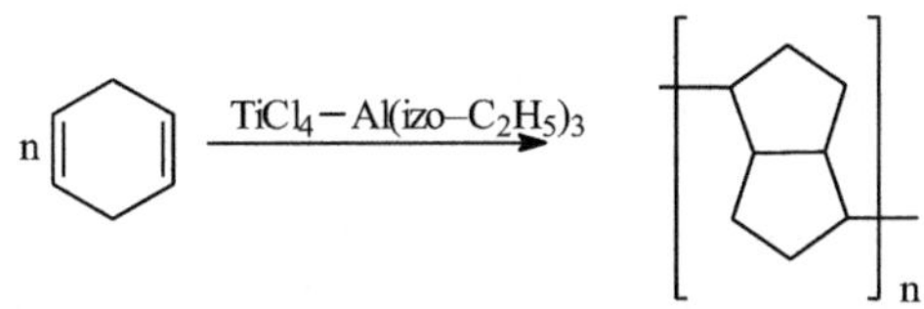

The coordinate catalysts enable the synthesis on an industrial scale of isoprene rubbers, which in terms of composition, structure and physico-chemical properties are similar to natural rubber. Also with these catalysts one can obtain the cis-butadiene rubbers of a regular structure, known for valuable utilization performances.

The difference in the structure explains several properties of these polymers. Atactic polymers are formed of poorly packed molecules and do not form crystallites. Thus they exhibit a lower static strength. The syndiotactic and isotactic polymers crystallize relatively easily and have a greater static strength and density than the atactic. The formation of varieties of syndiotactic polymers is more likely than of the isotactic ones. Syndiotactic molecule is characterized by a spiral conformation, because it provides the regularity of the recurrence of the mers, containing the tertiary carbon atoms.

The data presented in Table 10.12 confirm these conclusions.

A series of catalytic systems was also developed, in which the mechanism of action is similar to Ziegler-Natta catalysts. These are the oxides of nickel, cobalt, vanadium and molybdenum deposited on the surface of aluminum and of chromium oxide on silica gel. Such catalysts contain different promoters in the form of metal alkyl. They can be also used for the synthesis of polydienes.

TABLE 10.12 Crystallizable Stereoisomers of Polybutadiene-1.4-Diene

Stereoisomer	Melting temperature [K]	Glass transition temperature [K]	Crystal repeat unit [nm]	Density [g/cm³]
1,2-isotactic	393	263	65.0	0.96
1,2- syndio isotactic	427	–	51.4	0.96
trans-1,4	408	190	49.0	1.01
cis-1,4	263–274	163	–	–

After B.A. Dogadkin: Elastomer Chemistry (Chemia elastomerów, in Polish), WNT, Warszawa 1976, p.88.

10.2.4.4 Metallocene Catalysts

The name of metallocenes is derived from ferrocene, which is a complex compound of iron with dicyclopentadiene. The metallocene complexes are new types of Ziegler-Natta catalysts. The general formula of these catalysts is identified as Cp_2MeCl_2, where Cp denotes the cyclopentadiene ring or its derivatives, Me is metal atom, which the most often is zirconium (Zr), hafnium (Hf), titanium (Ti), scandium (Sc), thorium (Th) or another rare earth element.

Some examples of chemical structures of metallocene catalysts are shown below:

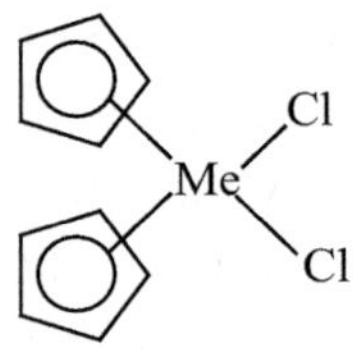

General formula

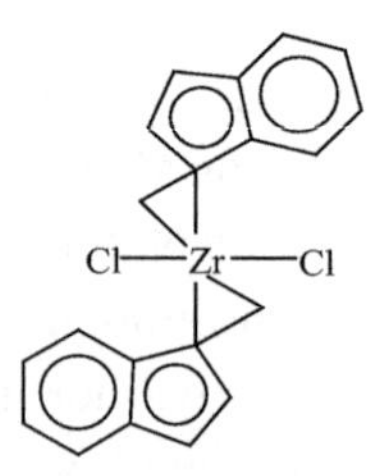

diindeno dichloro zirconium

The activity of these catalysts increases after the addition of a cocatalyst, which is an oligomeric methylaluminoxane:

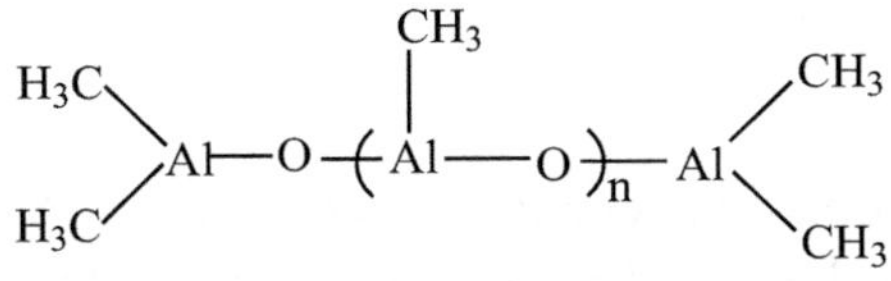

where n = 2–20.

Often this compound is presented in a simpler form:

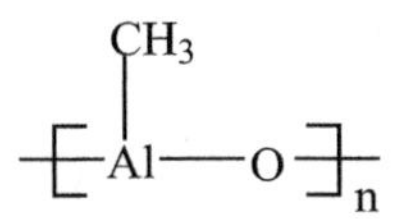

During the reaction the metallocene catalyst molecule with methylaluminoxane a displacement of the chlorine atom to the molecule of cocatalyst and of methyl group to the metal atom takes place with the formation of an active center $[Cp_2Me^+ CH_3]^{\bullet}$.

The reaction proceeds according to the scheme:

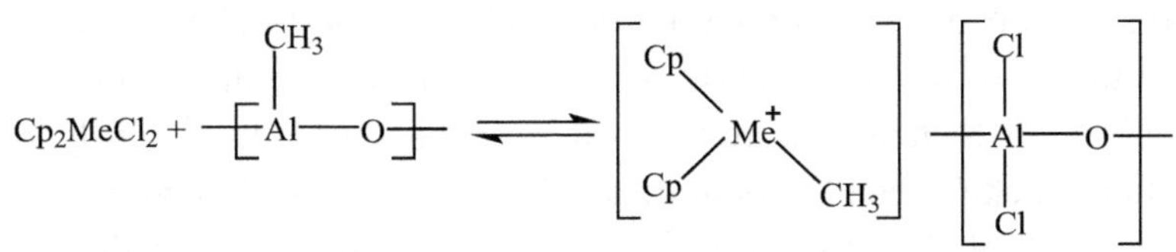

To the formed catalytic center is attached the monomer molecule through π electrons of double C=C bond, being properly oriented at the same time. As a result of successive attachments of monomer molecules one gets an increase of the chain, resulting in the formation of an isotactic polymer.

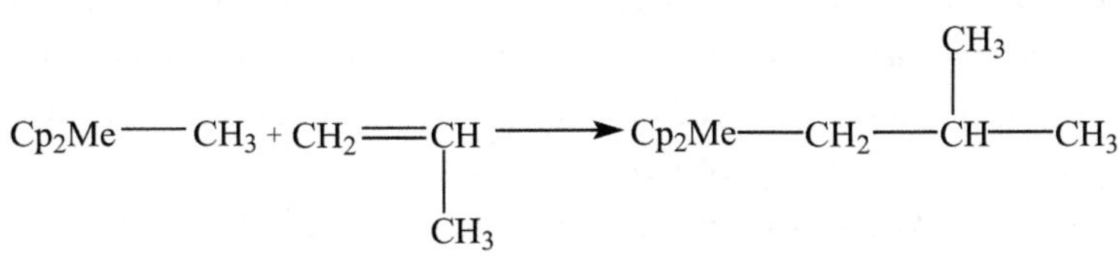

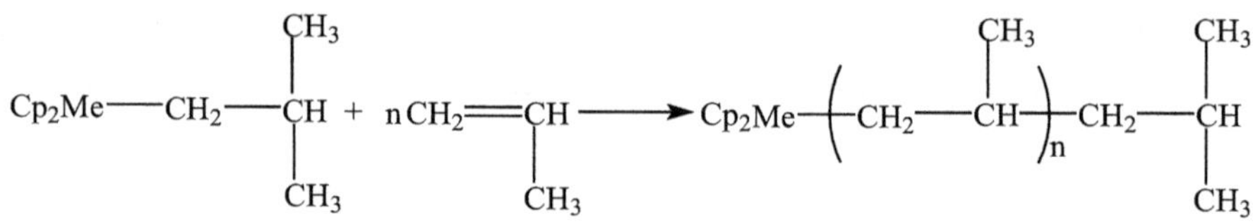

The chain termination reaction proceeds through β-elimination with formation of a double bond at the chain end or by the reaction with a hydrogen molecule.

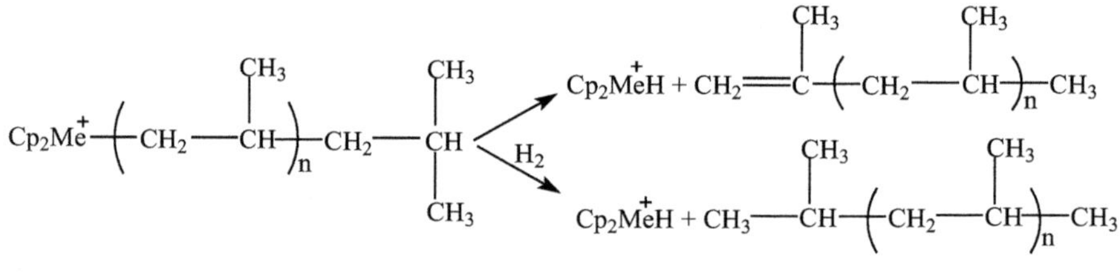

10.2.4.5 Coordination Catalysts Based on Metal Oxides

Oxides of some transition metals, deposited on well-shredded material such as diatomaceous earth, clay, alumina, charcoal or silicon dioxide cause polymerization of ethylene and of some other vinyl monomers. The mechanism of this process is probably very similar to the mechanism of Ziegler-Natta catalysis. The initiation of the polymerization reaction takes place in the active centers of catalyst in which are placed metal atoms such as chromium, molybdenum, vanadium, nickel, cobalt, niobium, tantalum, tungsten and titanium.

Catalysts of this type are obtained by two methods. In the first, the medium is saturated with metal ions, and then is heated in the presence of air to a high temperature in order to obtain a metal oxide. The second method consists on the deposition on support of a metal oxide with a higher degree of oxidation, such as, for example, anhydride chromic acid (CrO_3) and its subsequent reduction by hydrogen or by carbon monoxide.

The type of the reductive agent used and the conditions under which this process is realized influence significantly the subsequent activity of the catalyst. Through the reduction of deposited chromium trioxide by hydrogen one obtains a green colored catalyst (the color comes from the formed Cr_2O_3), whereas in the case of reduction with the carbon monoxide the catalyst takes a blue color. In the case of transmission of excess carbon monoxide over the catalyst it changes its color to violet and becomes inactive. This is due to the absorption of carbon monoxide on the catalyst surface and the blocking of active centers.

The catalysts based on metal oxides are susceptible to poisoning by compounds such as water, oxygen and acetylene.

The catalyst activation affects not only its activity but influences also the molecular weight of the emerging later polymer. For example, the complex CrO_3 with diatomaceous earth, activated at the temperature of 500°C (773 K), causes formation of polyethylene with an average molecular weight 3.5 times greater than the product obtained on the catalyst from the same starting material activated at the temperature of 850°C (1123 K). The average molecular weight of the produced polymer depends also on the content of metal oxide such as, for example, chromium oxide on the support.

When the chromium concentrations is reduced from 0.75% to 0.001% the average molecular weight of formed polymer increases by 40%. This is due to the increase in the number of ethylene molecules per one active center of the catalyst.

The monomer polymerization on catalysts always runs in three stages:

- adsorption of monomer (initiation);
- reaction on the surface (chain growth); and
- desorption (termination of the chain).

The course of polymerization of ethylene on chromium catalyst (Cat – Cr) can be presented schematically as, for example:

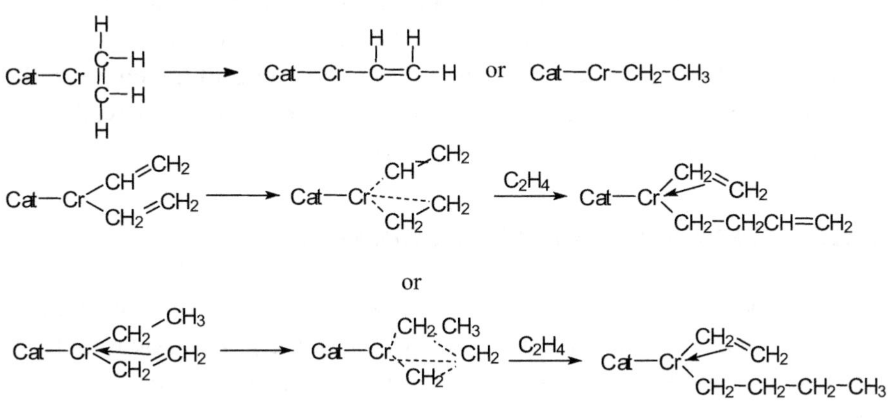

The chain termination reaction is presented as follows:

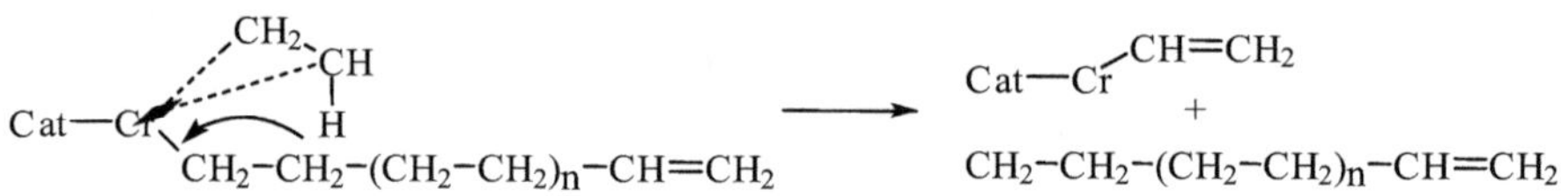

or

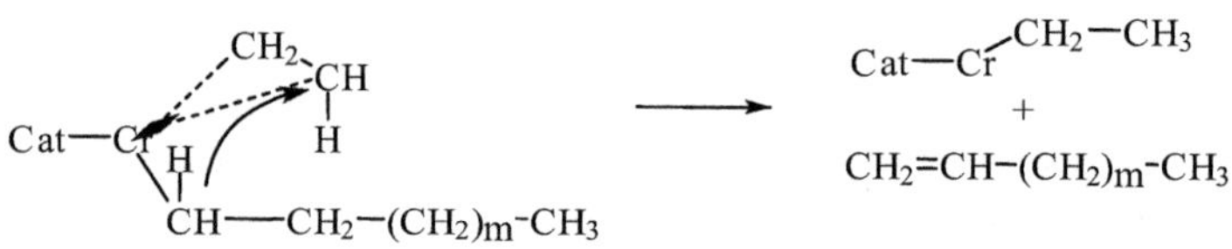

From the above considerations it follows that the center of active catalyst, based on chromium oxide, not only participates in the growth of the polymer chain, but affects also the orientation of monomer molecules. The catalyst operates in a stereospecific way and therefore facilitates the formation of tactic polymers.

One of possible ways for manufacturing polyethylene on an industrial scale is the Philips method, introduced in 1957. The catalyst used for the reaction is the chromium trioxide (CrO_3) disposed on a support composed of a mixture of silica and alumina. In this method a 5–7 percent solution of ethylene in selected hydrocarbons is prepared. Then it is heated to 420 K under a pressure of 3.5 MPa. Under these conditions ethylene is polymerized virtually in a single cycle. The solvents used, which may be aliphatic or aromatic hydrocarbons, allow to control the molecular mass of produced polyethylene as well as the reaction rate. It was found at the same time that the rate of polymerization increases with increasing molecular weight solvent.

The structure and the properties of polyethylene obtained with the catalyst based on chromium trioxide are similar to those of polyethylene synthesized using the Ziegler-Natta catalyst.

10.2.5 ELECTROCHEMICAL POLYMERIZATION

The electrochemical initiation of polymerization involves the formation of radicals, ions, or ion radicals as a result of the reduction or oxidation reactions on a monomer, or a specially introduced electrochemical initiator. These reactions occur directly on the cathode or anode surface.

In organic chemistry well known is the electrosynthesis reaction of Kolbe, in which the electrolysis of sodium salts of carboxylic acids, carried out under appropriate conditions, leads to the formation of free radicals at the anode. These free radicals undergo decarboxylation and linking with the formation of saturated hydrocarbons:

$$R{-}COONa \rightleftharpoons R{-}COO^{-} + Na^{+}$$

on anode:

$$R{-}COO^{-} - e \longrightarrow R{-}COO^{\bullet} \longrightarrow R^{\bullet} + CO_2$$

When conducting the reaction in the presence of monomer, the formed free radicals can initiate the polymerization reaction. The method is used for polymerization of styrene, butadiene, isoprene, vinyl acetate and methyl methacrylate. Similar effects can be achieved by the use of other active radicals such as trifluoromethyl $\bullet CF_3$, or fluorine •F. These radicals are formed at the anode as a result of the electrolysis of trifluoroacetic or hydrofluoric acid. Application of these systems allows performing the free radical polymerization of tetrafluoroethylene and trifluorochloroethylene at the anode, at normal pressure and at a temperature below 273 K. The polymerization of these compounds, when initiated by peroxides, takes place at high pressures and at high temperatures. The effect of temperature and of current density on the process of electrochemical polymerization is illustrated by the results obtained during the polymerization of N-vinylcarbazole in the solution of methylene chloride at a concentration of around 0.1 mol/dm^3 and in the presence of tetrabutylammonium tetrafluoroborate (32 mmol/dm^3). The data presented in Table 10.13, shows clearly that lowering the reaction temperature affects in a decisive extent the molecular weight of the formed polymer.

The type of electrode material used influences strongly the molecular weight of polymer formed during the electrochemical polymerization. This is well illustrated by the example of polymerization of methyl methacrylate at the presence of nitric acid. The obtained results are presented in Table 10.14.

10.2.6 RADIATION-INDUCED POLYMERIZATION

The radiation-induced polymerization is a process initiated by the ionizing radiation. There are several known methods of its initiation by α, β, γ rays as well as with neutron beams or with Roentgen radiation. However, the most frequently it is initiated by γ rays or by a beam of fast electrons.

TABLE 10.13 Electroinitiated Polymerization of N-Vinylcarbazole

No.	Current density [mA/cm^2]	Reaction temperature [K]	Conversion degree [%]	Molecular mass*
1	1	298	98	18,500
2	2	298	92	26,400
3	5	298	90	40,300
4	7	298	88	37,000
5	10	298	88	37,000
6	10	273	90	401,000
7	10	223	96	602,000
8	10	203	94	757,000

After E.B. Mano, B.A.L. Calafate: J. Polym. Sci. Polym. Chem. Ed. 21,829 (1983).
*Molecular mass determined by the Viscosimetric method in benzene and at 298 K.

TABLE 10.14 Effect of the Anode Material on the Molecular Weight of Poly(methyl methacrylate) Formed During the Electrochemical Polymerization Methyl Methacrylate Performed with the Current Intensity of 5 mA and in the Presence of Nitric Acid During 80 Minutes

No.	Material of anode	Efficiency [%]	Molecular mass [Mv]
1	graphite	81.8	275–000
2	platinizated graphite	42.5	295–000
3	gold	11.1	420–000
4	platinum	4.24	680–000
5	aluminum	0.86	290–000

*After G. Pistoria, O. Bagnarelii, M. Maiocco: J. Appl. Elektrochem. 9, 343 (1979).

The mechanism of such a radiation polymerization may differ depending on the type of monomer used and the reaction temperature. It may be also of free radical, cationic or anionic character.

Studies on the radiation-induced polymerization were carried out mainly on the example of vinyl monomers, which polymerize by free radical mechanism, leading both reactions in the mass of monomer as well as in solution.

The rate of free radicals formation in the polymerization process induced by radiation is directly proportional to the intensity of radiation (I) and depends on the effectiveness of the radiation polymerization.

The quantity indicating the effectiveness of the radiation polymerization is the coefficient G, called also the radiation efficiency. It is equal to the amount of monomer molecules participating in the polymerization reaction after absorption of 1.602×10^{-17} J (100 eV) of radiation energy:

$$G = \frac{w \cdot 6.023 \cdot 10^{23} \cdot 100}{m \cdot D \cdot M \cdot 6.24 \cdot 10^{13} \cdot 10^{6}} = \frac{w}{m \cdot D \cdot M} \cdot 0.97 \cdot 10^{26}$$

where: w – yield of polymer, kg; m – mass of monomer subjected to radiation, kg; D – amount of absorbed dose, M rad (1 M rad = 10^4 J/kg); M – molecular weight of monomer; 6.24×10^{13} – energy equivalent of 1 rad, eV/g (1 eV/g = 1.602×10–^{22} J/kg); 6.02×10^{23} – Avogadro number.

The value of G varies in a wide range, depending on polymerization conditions and the type of monomer. Table 10.15 lists the G values for radiation polymerization of styrene or methyl methacrylate by β and γ radiation at several different temperatures, irradiation power and polymerization rates for both initiation and polymerization processes.

The total activation energy for radiation polymerization varies within the limits of 0–41.8 kJ/mol. A small value of this energy causes that the radiation polymerization reaction rate is temperature independent within a few tens of Kelvins. A characteristic feature of the radiation polymerization is the presence of the retrospective effect (post-effect). It consists on the fact that after the cessation of irradiation the polymerization continues over many hours. This phenomenon concerns primarily the polymerization in solid phase or in solution precipitation. The reason for this phenomenon is the reduced mobility of macroradicals in solid phase and difficulties ending the chains by recombination.

The mechanism of radiation polymerization reaction in the liquid phase can be determined by examination its kinetics. Namely, if the polymerization rate is directly proportional to the square root of the absorbed radiation dose by monomer and the resulting polymer molecular weight is inversely proportional to this value it shows the free radical polymerization mechanism:

$$V = k\sqrt{P_D}$$

$$M = \frac{K}{\sqrt{P_D}}$$

where *PD* denotes the radiation dose.

TABLE 10.15 G Values for Polymerization of Styrene and Methyl Methacrylate

No.	Monomer	Radiation	Power [rad/min]	Temperature [K]	Rate of		G
					poly-merization [m/s•105]	initiation [m/s•108]	
1	Styrene	γ	4100	345	5.84	4.4	0.78
		γ	3200	298	1.01	2.08	0.46
		γ	4400	255	0.24	2.74	0.56
		β		303.5	—	—	0.22
2	Methyl methacrylate	γ	4100	343	49.9	28.4	4.88
		γ	4200	298	21.1	36	5.74
		γ	4400	255	7.25	53.8	7.68
		β		303.5	—	—	3.14

After A. Charles by: Chemia radiacyjna polimerów, WNT, Warszawa, 1962, p. 360.

Additional proofs on the presence of free radical mechanism in the polymerization process are:

- inhibition of the reaction with the introduction of conventional radical inhibitors;
- increase in reaction rate with increasing temperature;
- always positive values of the activation energy (25–41.8 kJ/mol).

If the rate of polymerization is directly proportional to the radiation dose then the mechanism of reaction is of anionic or cationic type:

$$V = kP_D$$

In addition, the course of radiation polymerization running according to the ionic mechanism is confirmed by the lack of effect of radical inhibitors (free radical scavengers) on the process of polymerization and low values of the activation energy.

There are also some known examples of polymerization running along the mixed radical-iononic radiation polymerization mechanism.

An interesting example of the application of radiation polymerization is the polymerization of clathrate (channel complexes) monomer with urea or its derivatives. The channel walls are formed by twisted into a spiral urea molecules, linked together by hydrogen bonds. Inside the channels are arranged in an orderly manner the immobilized monomer molecules associated in a complex manner with urea. Under the influence of radiation, they undergo polymerization with the formation of an ordered polymer structure. At the end of the process urea is dissolved in water and the resulting polymer is isolated by filtration or centrifugation.

Another major application of the radiation polymerization is the use of radiation to initiate the emulsion polymerization. In the case of styrene the radiation polymerization in emulsion runs many times faster than the radiation polymerization in bulk and gives a product with significantly higher molecular weight. The increased efficiency of the polymerization process in emulsion can be also caused, partly, by the formation of additional free radicals in water, which can initiate polymerization of styrene, and partly, by the higher G value for the formation of radicals in this medium. In addition, reducing the rate of termination increases the molecular weight because more monomer molecule per initiating free radical undergoes the polymerization. The activation energy of emulsion polymerization is of 15.5 kJ/mol. Comparative results of the radiation polymerization of styrene in bulk and emulsions are given in Table 10.16.

The radiation polymerization, induced by an intense Co^{60} radiation, was used for emulsion polymerization of vinyl acetate in experiments and on industrial scale.

10.2.7 POLYMERIZATION IN PLASMA

10.2.7.1 Characteristics of Plasma

Plasma is a gas, ionized by any method, in which the inertial motion of electrons prevails over the directed one. The degree of ionization of this gas is always large, reaching in extreme cases 100% of all molecules. In plasma the quantities of molecules charged positively and negatively are equal. It gives almost zero-th spatial charge. The high density of charged particles affects significantly the electrical conductivity of plasma.

Plasma cannot be seen as a normal, electrically neutral mixture of gas like the electron gas. Between the charged and the uncharged particles one observes interaction of forces associated with the electric polarization. For this reason all the neighboring molecules, apart from each other, interact mutually. The plasma properties are not only the sum of the properties

TABLE 10.16 Radiation Induced Polymerization of Styrene in Bulk and in Emulsion

No.	Polymerization	Dose [rad/min]	Temperature [K]	Reaction rate [%/h]	Molecular weight of polymer (×10³)
1	in mass	4400	255	0.1	17–26
		3200	303	0.42	38–100
		4100	345	2.53	165–348
2	in emulsion	1000	298	36	800–1500
		1000	308	45	885–1726
		1000	318	54	1200–2060

After A. Charlesby, Radiation Chemistry of Polymers (Chemia radiacyjna polimerów, in Polish), WNT, Warszawa 1962, str. 365.

of individual molecules, but equally are not the sum of the properties of gases, which compose it.

In order to characterize the energy state of molecules present in the plasma an energetic scale of temperature was introduced. Its basic unit is electronvolt eV (1 eV = 1.602 × 10^{-19} J). The model chosen for such a scale is a gas whose molecules have only two degrees of freedom. For such a gas molecule the energy is given by the following equation:

$$E = k\mathrm{T}$$

where E = energy; k = Boltzmann's constant (1.381 × 10^{-23} J/K); T = temperature (Kelvin).

From this equation we get:

$$T = \frac{E}{k} = \frac{1eV}{k} = \frac{1{,}602 \cdot 10^{-19} J}{1{,}381 \cdot 10^{-23} J/K} \approx 1{,}16$$

The above calculations show that the plasma energy of 1 eV on the energy scale corresponds to the temperature of 11–600 K.

It turned out that the properties of plasma are different at different energy ranges and therefore the concept of high temperature plasma (particles with energy of up to hundreds of electronvolts) and low-temperature plasma (the energy of particles not exceeding a few electronvolts) was introduced.

For the synthesis of polymers only the low temperature plasma, in which the energy of molecules does not exceed a few electronvolts is used. The temperature of electrons in plasma is always significantly larger than the temperature of ionized gas. Therefore, it is possible to run the reaction at a temperature close to ambient temperature where the free electrons have enough energy to break the covalent bonds. For this reason this type of plasma is also called the non-equilibrium plasma.

The most important way of plasma generation, as a medium to conduct the polymerization, is the electric glow discharge. For that a direct (DC) or alternate (AC) current with frequencies ranging from 50 Hz to 5 GHz can be used. A typical plasma polymerization is conducted at low gas pressure of the order of 10^{-1}–10 Pa. The course of process in plasma may vary depending on the method of conducting the reaction. For this reason a distinction is made between:

- plasma polymerization, and
- plasma-initiated polymerization.

10.2.7.2 Plasma Polymerization

The plasma polymerization is a way to obtain specific polymeric materials, which cannot be obtained by other methods. The plasma polymer is significantly different from the polymer formed from the same monomer by using other methods. In the case of a plasma polymer it is impossible to determine the mer unit. For this reason, in order to distinguish this type of compounds from the conventional polymers, the prefix pp (plasma polymerized) before the polymer name is introduced, for example, pp-styrene (polystyrene equivalent).

Almost all organic compounds undergo polymerization in plasma (not only the traditional monomers) as well as a significant number of inorganic compounds.

The chemical reactions occurring in plasma are very complicated and the mechanism of plasma polymerization has not yet been fully understood. The competing reactions running side by side depend always on the level of energy in plasma. These include, in the first order, the molecules decay processes and their re-synthesis. The resulting plasma polymers are often built not from the monomer molecules but from their degradation products, combined with other molecules contained in plasma gas such as nitrogen, carbon monoxide or water.

It is assumed that the primary mechanism for the chain growth in plasma polymerization is a fast gradual polymerization in which dominating are reactions between the active molecules. Such reactions can be also called as the polyrecombination.

In order to obtain polymeric materials with reproducible properties it is necessary to control precisely the reaction parameters such as: concentration of monomer, pressure of reaction mixture, discharge power, type and temperature of substrate, polymerization time and reactor geometry.

In many cases, the plasma polymer deposition process can be described using a complex power parameter:

$$W/FM:$$

where: W – value of the delivered power; F – flow rate (mol/min); M – molecular weight of monomer.

The plasma polymerization runs at large values of W/FM. The yield of polymerization grows always up to a critical value of parameter $(W/FM)_C$.

There exists also a dependence between the critical value of the power parameter $(W/FM)_C$ and the sum of the binding energies of all bonds present in the monomer molecule:

$$(W / FM)c = \alpha\varphi$$

where α is a constant characteristic for the used reactor,

$$\phi = -\frac{\sum (binding\ \ energies)}{\bar{M}}$$

During the plasma polymerization the value:

$$\frac{W / FM}{\alpha\varphi} > 1$$

Depending on the reaction parameters the resulting polymer is emitted in the form of powder, a continuous film, or an oily substance.

From the practical point of view the most important is the way to produce ultra-thin plasma polymer films, which the most frequently are highly branched and cross-linked. Therefore, they are insoluble and forming a good coating material.

10.2.7.3 Polymerization Initiated by Plasma

In this method the plasma state is only used in the initial stage, to initiate the polymerization. The formed active molecules are transferred from the plasma phase to the monomer phase (liquid or solid), in which the chain growth reaction runs already without the plasma participation. As a result of this process one gets a conventional polymer, built of mers, which often exhibits a very high molecular weight. This method is particularly useful for the polymerization of water-soluble monomers such as acrylamide and its derivatives.

10.2.8 TECHNIQUES OF POLYMERIZATION CONTROL

10.2.8.1 Polymerization in Bulk

Polymerization in the bulk, often referred to as the block polymerization, runs in the monomer medium in which the resulting polymer is dissolved. With the increasing degree of conversion increases also the viscosity of the reaction mixture, and then, at a large degree of conversion the gelation occurs and the formed polymer takes shape of the reaction vessel. For this reason the block polymerization is carried out mostly in two steps. In the first step of the process, known as the prepolymerization, a solution of polymer in the monomer of a certain viscosity is obtained, which is then entered into the forms. Examples of such forms used to obtain a transparent organic glass are two plates, assembled from a glass mirror, sealed with a hose made from poly(vinyl chloride) and covered with a thick paper. The space between the plates is poured with the prepolymer. The final polymerization is carried out in the forms at elevated temperatures, the height of which depends on the type of monomer used and on the planned molecular weight of formed polymer.

By increasing the reaction temperature one increases the rate of polymerization, but decreases the molecular weight of the product. Polymerization in bulk has several serious drawbacks, resulting mainly from the high viscosity of the reaction mixture and the resulting poor thermal conductivity of the formed polymer. As a consequence of this fact, within the formed bloc emerge areas with a local overheating. This is due to the large exothermicity of the polymerization reaction. An inhomogeneous distribution of temperature in the block results also in locally different speeds of the polymerization process. Therefore, different degrees of polymerization in different parts of the block (lower inside and higher at its surface) are obtained.

This leads to the formation of internal stresses, causing after a certain time, micro fractures of polymer (so-called silvering). A local overheating may also cause evaporation of monomer. The resulting steam cannot get out due to the high viscosity of the system. It causes formation of bubbles, which lowers the quality of the final product. It shows that the bulk polymerization, apparently very simple, is difficult to implement and requires good skills and experience of technical personnel involved in production. Despite its flaws the bulk polymerization in some cases is irreplaceable (e.g., flooding of organic preparations or minerals with a transparent polymer).

On the industrial scale known are also the semicontinuous or continuous bulk polymerization methods.

The continuous method allows to get a product, which is practically free of monomer. Its molecular weight is higher than in the case of periodic polymerization in molds. It makes also possible the automated management of the process. It ensures the reproducibility of the polymer properties, improves the working conditions, increases productivity of the equipment and lowers the price of the final product.

The most common method of the continuous polymerization is the tower method. The principle of its operation lies in the fact that monomer polymerizes as it flows through the column and is continuously removed by the fresh monomer or solution of prepolymer. Eliminating the possibility of the product mixing at different stages of the reaction allows to obtain a homogeneous polymer with specific, fixed characteristics. This can be achieved, ensuring at the same time a hundred percent yield, by correlating the reaction times with the height of the column (tower) and an appropriate reaction rate.

10.2.8.2 Block-Precipitation Polymerization

The block-precipitation polymerization is a variation of the bulk polymerization during which the formed polymer is insoluble in monomer and precipitates in the form of a sediment.

The technical use of this method has encountered difficulties of the engineering nature. These difficulties are related mainly to the evacuation of the polymerization heat, mixing, and removal of polymer deposited on the inner walls of the reactor.

For polymerization one uses only monomers with the initiator dissolved in. As initiator usually an organic peroxide in an amount of about 0.1% is used.

The liquid monomer is in equilibrium with its vapors and the generated heat of reaction is evacuated, mainly by evaporation and by condensation on the walls of the reactor as well as in the condenser located directly above the reactor.

In the first stage of the reaction a large amount of heat is released. The resulting polymer begins to precipitate in the form of very fine particles of sediment, which already at the monomer conversion of 1% cause clouding of the reaction medium.

The concentration of these particles increases gradually giving rise to the formation of elementary primary grains with diameter of about 0.1 mm. At the higher conversion rate (7–8% of monomer) the primary grains merge into larger aggregates of polymer. Their size depends on the intensity of mixing. The increase of the mixing speed causes a reduction of the average diameter of grains. At this stage of the reaction its medium is still liquid and the grains suspended in polymer have already a shaped structure. During the further polymerization the number of grains is not increasing.

In the second stage of polymerization, with the increasing degree of monomer conversion (from 8% to 90%,), the molecule le diameter increases to about 0.5–1.5 mm. The grains aggregate, creating porous clusters. At that time the initially liquid reaction environmental thickens gradually. At the end of reaction one obtains a polymer in powder form. After reaching the desired degree of conversion (70–90%) the unreacted monomers are driven off, condensed and returned to repolymerization.

10.2.8.3 Polymerization in Gas Phase

The gas phase polymerization is carried out in the case when using gaseous monomers, characterized by a low critical temperature. The best known such monomers are ethylenes and tetrafluoroethylenes. The polymerization is technically difficult to perform, as it requires application of high pressure, of the order of 10 MPa.

The polymerization reaction in a given case is carried out in adiabatic reactors of tube or autoclave type to which one introduces a gaseous monomer, after having pressured it via a two-stage compression to the desired pressure. For this purpose the dry monomer coming from a dry gasometer is mixed with the products coming from degassing liquid polymer. Then it is displaced with a compressor under 25 MPa pressure into a mixer, where it meets the circulating gas. The raw material should be of sufficiently low dew point to prevent clogging of pipes by precipitating crystalline hydrates of ethylene, formed during the cooling and compression of wet ethylene. Then the gas is subjected to the second step of compression to the working pressure, which depends on the kind of produced polymer. The reaction proceeds as follows: cold gas, after mixing it with initiator such as, for example, sprayed peroxides or oxygen, is gradually introduced into reactor on the suction side of the first compressor in such an amount that its content in the reaction zone, depending on the species of the produced polymer, amounts to 0.0045–0.0185 w %. The adiabatic reactor starts by heating its walls with hot air. After initiation of the polymerization process the heating is stopped and the cooling is started.

Immediately after leaving the reactor, the reaction mixture expands in a computer controlled reducing valve to 25 MPa. Then it is headed into the separator, operating at approximately 423 K (150°C). From the separator the gas escapes through the stepwise condenser jacket – tube and cyclone coolers to the fuel gas network or to other production, for example, manufacture of ethyl benzene, ethylene oxide, ethyl alcohol, etc. The liquid polymer flows through a pressure reducing valve to the degassing tank. Then it is extruded in a wire form and cut into pellets with the appearance of rice grains, which are transferred by water jet onto a vibrating sieve. After an initial drying the compressed air the pellets are sent to the departments of averaging, homogenization, dyeing, and expedition. The flowsheet of the polyethylene production process in the gas phase is shown in Figure 10.2.

Another way of gas-phase polymerization is the anionic polymerization of 1,3- butadiene (divinyl). In this case the monomer in gas phase passes over a solid catalyst, which may be a sodium, potassium or lithium. The resultant polymer is deposited in the form of rubber type mass. Compared with the bulk polymerization the benefits in gas phase

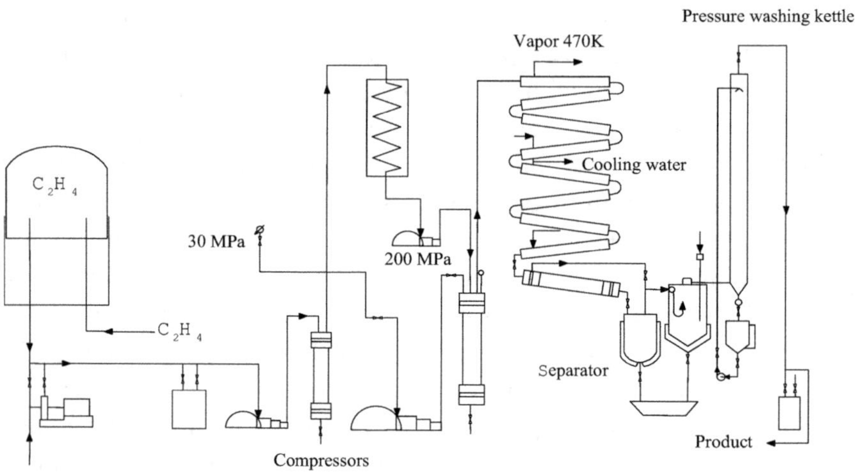

FIGURE 10.2 Scheme of ethylene polymerization in gas phase.

polymerization are: a better heat dissipation and the possibility of obtaining the polymer in large lumps in organic colloidal solution.

10.2.8.4 Suspension Polymerization

The suspension polymerization, also called bubble polymerization or polymerization in suspension, runs under the influence of an initiator dissolved in dispersed in aqueous solution of organic colloids monomer or in aqueous suspension of inorganic salts, being stabilizers, for example, the fixing agents of suspension.

The suspension polymerization process follows a typical bulk polymerization mechanism. Each drop of dispersed monomer behaves like a "microblock." In comparison with the block polymerization it allows to maintain easier a constant temperature and evacuate well the reaction heat.

A much higher rate of suspension polymerization as compared to the block polymerization indicates that there are also important reactions occurring in the surface layers at the interface monomer-emulsifier. For a strongly dispersed monomer the suspension polymerization is similar to the emulsion polymerization.

On the course of the suspension polymerization and the properties of the formed polymer a major impact have: protective colloids, for example, suspension stabilizers, initiators of polymerization and mixing of the system.

The dissolved in the aqueous phase protective colloids, being stabilizers of suspension, counteract the agglomeration of beads during the polymerization. They affect also their size. At the interface water-monomer a protective sheath is created, which prevents the merger (coalescence) of monomer droplets. An excessive reduction in viscosity of the aqueous phase can result in the formation of emulsion, while too high viscosity of this phase impedes the monomer dispersion during the mixing. It can lead to the formation of a coarse-grained polymer.

In the process of suspension polymerization one uses different additives modifying the action of protective colloid and supporting some shortcomings characterizing the given colloid. These additives, known also as auxiliary stabilizers of suspension, improve the hydrophilic-hydrophobic balance of colloid and adjust the surface tension at the interface monomer-water.

The type and the amount of used suspension stabilizer is chosen experimentally as a scientific theory allowing the resolution of this issue in a theoretical way is still lacking. Attempts to generalize in this area rely on introduction of (HLB) number as proposed by Griffin, determining the hydrophilic – hydrophobic or hydrophilic – lipophilic balance of given suspension stabilizer. It takes a value between 0 and 100. The main suspension stabilizer is primarily characterized by solubility or wetting in water, while the supporting stabilizer dissolves better in monomer. The HLB number takes an optimum value when at the maximum concentration of suspension stabilizer the interfacial surface tension on the border of the monomer-water reaches a minimum. As the suspension stabilizers one uses water-soluble organic macromolecular compounds such as poly(vinyl alcohol), copolymer of styrene with maleic anhydride, poly(sodium methacrylate), methylcellulose, methylhydroxypropylocellulose, carboxymethylcellulose, copolymer of allyl alcohol and vinyl acetate, gelatin and starch. One can use also inorganic suspension stabilizers such as calcium phosphate, aluminum and magnesium hydroxides, talc and silicates.

Poly(vinyl alcohol), formed by hydrolysis of poly(vinyl acetate) and used as a protective colloid, should have 8–20% of unsaponified acetate groups. The product containing less than 3% and more than 20% of acetate groups does not have sufficiently good protective properties. The molecular weight of poly(vinyl alcohol) used does not affect in a decisive way on the course of polymerization. A large impact on the size of beads in this case exerts pH of the reaction medium. This stabilizer works more effectively in the acidic medium, where one gets small beads. In alkaline environment, with unchanged other parameters, the formed beads are of larger dimensions. Note that during the polymerization the pH of the aqueous phase is changing. It is a result of a progressive hydrolysis of acetate groups of poly(vinyl alcohol) used and the accumulation of acidic products from the decomposition of benzoyl peroxide and other initiators.

Other macromolecular protective colloids are salts of poly(acrylic acid) or poly(methacrylic acid).

A very good suspension stabilizer is the methylcellulose and its derivatives. The advantage of these compounds is their total non-toxicity, resistance to mold and nonionic nature. The first characteristic determines the possibility of using the polymer for medical purposes, while the latter provides good dielectric properties.

The stabilizing ability of methylcellulose as a protective colloid in the suspension polymerization process depends on its molecular weight. It was found that only the methylcellulose, forming an aqueous solution of low viscosity is able to counter effectively the phenomenon of agglomeration of monomer particles. With increasing concentration of methylcellulose increases the fragmentation of the monomer droplets. At the same time the product benefits of fine-grained structure without a capacity to absorb the softener. Therefore, in order to obtain a porous product one has to keep a low concentration of methylcellulose in the aqueous phase.

There exist a number of patented solutions on the use of kits with different main and subsidiary suspension stabilizers. Noteworthy is the use as a secondary stabilizer the copolymer of vinyl acetate with allyl alcohol, which is more efficient than the well-known secondary stabilizer – glycerol stearate.

The allyl alcohol copolymer with vinyl acetate, used as a self-contained stabilizer during the polymerization of vinyl acetate, forms a porous product with a relatively large dispersion of grain sizes.

Using this copolymer as an auxiliary stabilizer next to the product of oxyethylation and alkylphenol gives a polymer with a porous structure and a homogeneous grain.

The poly(ethylene glycol) used as an independent suspension stabilizer leads to the formation of porous grains with a large dispersion of grain sizes. This phenomenon can be explained by the fact that only the two compounds together create an appropriate hydrophilic-hydrophobic balance. This happens as a result of the simultaneous occurrence in these compounds of hydroxide and ether groups. Therefore, the simultaneous use of both stabilizers in an appropriate proportion allows to get a product with a homogeneous grain.

When applying organic origin stabilizers to the suspension polymerization to some extent the emulsion polymerization takes also place. The post-reaction solution has a white color, which is caused by the presence of the emulsion polymer. To prevent this undesirable phenomenon, causing product loss and contamination, some patents recommend to use polymerization inhibitors, soluble only in aquatic medium, such as, for example, copper salts. An alternative solution is addition of electrolytes to water.

In last years more and more frequently as protective colloids the previously mentioned inorganic substances are used. The behavior of a solid at the border of two mutually not solving liquids depends on its wettability by the liquids. In turn the wetting conditions depend on the size of the surface tension at the interfaces solid – water-monomer ($\delta_{1,2}$), solid-monomer ($\delta_{1,3}$) and water – monomer ($\delta_{2,3}$). When the contact angle θ <90° then the powdered solid is adsorbed by water. In this case the surface tension at the interface solid-monomer is larger than that at the phase boundary solid – water, for example, $\delta 1,3 > \delta 1,2$

Stabilizers of this kind allow to get suspension of a non-polar liquid in water (emulsion of "oil in water" type). If the wetting angle is greater than 90° ($\delta 1.3 < \delta 1.2$) then the powder is adsorbed by the non-polar liquid and one obtains emulsion of "water-in oil" type. For the contact value equal to 90 degrees, depending on such factors as number of separate phases, the way how the phase is dispersed, etc., formation of both types of emulsions is possible.

The size of beads formed during the dispersion of emulsion depends on the type and the quantity of protective colloid used. More of colloid is introduced into the system greater surface will be able to protect. Thus smaller balls will be obtained. The persistence of the protective layers formed from a fixed emulsifier depends on the size of its molecules. Emulsifier of this type must be thoroughly crushed in order to cover completely the surface forming the beads. However, the excessive fragmentation of the emulsifier adversely affects the emulsion stability.

As inorganic stabilizers in t suspension polymerization processes one uses mostly the calcium phosphate in ratio $CaO:P_2O_5$ = 1.35. This is a specially prepared mixture of tribasic calcium phosphate and calcium hydroxide.

The other this type protective colloids include aluminum and magnesium hydroxides. They can be precipitated directly in the aquatic environment through an action on the water-soluble salts of these metals with sodium hydroxide. In this way the cumbersome process of filtration, fragmentation and sieving, is omitted. The advantage of inorganic stabilizers is the ease of their separation from the polymer by dissolving in hydrochloric acid after the polymerization process.

In order to obtain beads of equal size one adds to water 0.001 wt % of potassium persulfate. In practice a mixture of inorganic and organic stabilizers is often used.

Another important factor influencing the course of polymerization and the quality of the formed polymer is the polymerization initiator. As initiator always substances soluble in monomer are used. Usually these are organic peroxides such as benzoyl and lauroyl peroxides. They are chosen depending on temperature and duration of their half-decay. For the polymerization of monomers with a low boiling point such as, for example, vinyl chloride a very effective initiator is the isopropyl dipercarbonate, whose half-life period of decay is 2 hours at 323 K (50°C). The lauroyl peroxide at this temperature has the half-life period of decay 25 times longer. Increasing the initiator concentration in the polymerization system shortens the time of polymerization, but reduces also the molecular weight and the porosity of the product. An excessive shortening of the polymer chains is also caused by the increase of the reaction temperature. The polymer chain length can be adjusted by the application of appropriate molecular weight regulators such as chlorinated hydrocarbons, isobutylene, dienes and mercaptans.

Mixing of components plays an important role in the course of suspensive polymerization, because its intensity influences the size of beads, their shape and the polydispersity. A less rigorous stirring can promote too quickly the agglomeration and the fragmentation of the product. The problem of mixing is particularly important in the initial stage of polymerization, when the amount of reacted monomer is of 10–60%. Stopping the mixer, even for a very short period of time, causes an irreversible agglomeration. The conglutinated beads form a rubbery, extending mass, which cannot be re-dispersed, even by a very intensive mixing.

The advantages of suspension polymerization are:

- possibility of synthesis of a grinded polymer, easy to extract by filtration or centrifugation,
- obtaining of a product with reproducible properties and low degree of polydispersity
- ease of heat dissipation secreted during the polymerization reaction.

The method has found a wide application for obtaining both the linear polymers, which can be then treated by injection molding and extrusion, as well as for the synthesis of cross-linked polymers. The cross-linked products in the form of beads may be subjected to a further chemical modification, leading to the generation of polymeric ion exchangers or polymeric catalysts. The porous polymers can be obtained by adding the liquid hydrocarbons (e.g., decane) to the system.

The suspension method is not applicable for the synthesis of elastomers.

10.2.8.5 Emulsion Polymerization

The principle of emulsion polymerization consists on dispersion of monomer in water using an emulsifier, followed by polymerization at the presence of water-soluble initiators. A polymer dispersion system, with a high degree of fragmentation, is then created in water, often called latex. The advantage of carrying out the emulsion polymerization process is the facilitated reception of the heat generated by the exothermic reaction of the aquatic environment, as well as the opportunity to run the reaction in a continuous way, what allows a substantial automation of the production process and lowering its cost. Due to the low viscosity of the emulsion, even at large concentrations, one can produce by this method viscous and rubbery polymers.

The monomers used in the polymerization process must be of a high degree of purity and be completely free of inhibitor. Water used in the polymerization process has to be demineralized, free of organic compounds and its electrical conductivity should be below 1 μS. It is recommended to remove oxygen from the water. Oxygen inhibits the polymerization reaction and causes formation of unstable, low-molecular weight products. The weight ratio of water to monomer can vary within the limits: 1:1 to 1:2.

As processing aids in the process of emulsion polymerization one uses emulsifiers, emulsion stabilizers, initiators, and regulating substances.

Emulsifiers are the surface-active compounds capable of obtaining a permanent emulsion of monomer in water. They can be divided into four groups:

a) anionic – active,
b) cationic – active,
c) non – ionic, and
d) permanent.

The anionic – active emulsifiers most frequently are the soluble salts of fatty acids, alkyl sulfonates, alkylaryl sulfonates and alkyl sulfates. These compounds are the most frequently used emulsifiers in the emulsion polymerization.

The nonionic emulsifiers are rarely used in the emulsion polymerization. This group includes: esters of glycerol and fatty acids, products of ethylene oxide addition to alcohols, phenols or fatty acids, poly(vinyl alcohol) and others.

To the fourth group of emulsifiers belong: calcium phosphate, talc and oxides of aluminum and magnesium.

Characteristic properties of emulsifiers are given in Table 10.17.

Most of the emulsifier molecules are composed of hydrophobic and hydrophilic parts. The hydrophobic parts of an emulsifier are aliphatic radicals, which give the molecule the ability to dissolve the emulsifier

TABLE 10.17 Properties of Emulsifiers

No.	Emulsifier	Critical micellar concentration		Medium	Molecular weight of micells	Specific surface [Å2]
		[mol/ dm^3]	[%]			
1	potassium laurate	0.0125	0.3	water	11,900	32
2	potassium stearate	0.0005	0.16	water	—	—
3	potassium oleate	0.0012	0.04	water	—	28
4	potassium abietate	0.012	0.39	water	—	25–40
5	n-dodecylosulfonate of sodium	0.0016	0.055	water	8200	32–40
6	n-dodecylosulfate of sodium	0.0087	0.25	water	17,100	—
7	cetyltrimetlyl-amonium bromide	0.001	0.036	0.13 M KBr	61,700	—
8	cetyl-pyridine chloride	0.006	0.21	0.0175 M NaCl	32,300	46
9	dodecylamine hydrochloride	0.014	0.31	water	12,300	26
10	oxyethylated oktyphenol n=9	0.0002	0.012	water	66,700	53
11	oxyethylated octyphenol n=30	0.00025	0.026	water	—	101
12	sorbite monolaurate	0.002	0.067	water	—	36
13	block copolymer (80:20) of ethylene oxide with propylene oxide M = 8200	6.9×10^{-6}	0.006	water	8400	920
14	sodium di(2-ethylhexyl) sulfosuccinate	0.0007	0.03	water	21,300	25

molecule in the organic phase (monomer), while the hydrophilic part is able to dissolve in the aqueous phase. Thanks to that the emulsifier molecules can adsorb at the border surface water – monomer, orienting in such a way that the hydrophilic part heads towards the water and the hydrophobic part to the monomer, respectively.

As the reaction initiators one uses: hydrogen peroxide, cumene hydroperoxide, persulfates and redox systems. The effect of the initiator type on the course of emulsion polymerization is depicted in Figure 10.3.

During the test of the emulsion polymerization mechanism it was found that when the monomer droplets in the emulsion have a diameter of 0.5–10 nm then the grains of the manufactured polymer are ten times smaller. It follows from this fact that the number of original "spores" of grains, contained in 1 cm^3 of emulsion, must be much larger than the number of monomer droplets. The number of polymer grains (in 1 cm^3) increases significantly with the increasing concentration of emulsifier. Therefore, it is assumed that the original spores of grains are formed in micelles of emulsifier. The emulsifier molecules (e.g. sodium palmitate) particles are not separate molecules in aqueous solution, but they are combined into aggregates called micelles.

In this way one can explain the observation that the surface tension of the solution of a surfactant – active molecule is small in comparison with water. But it rises rapidly from the start of polymerization reaction.

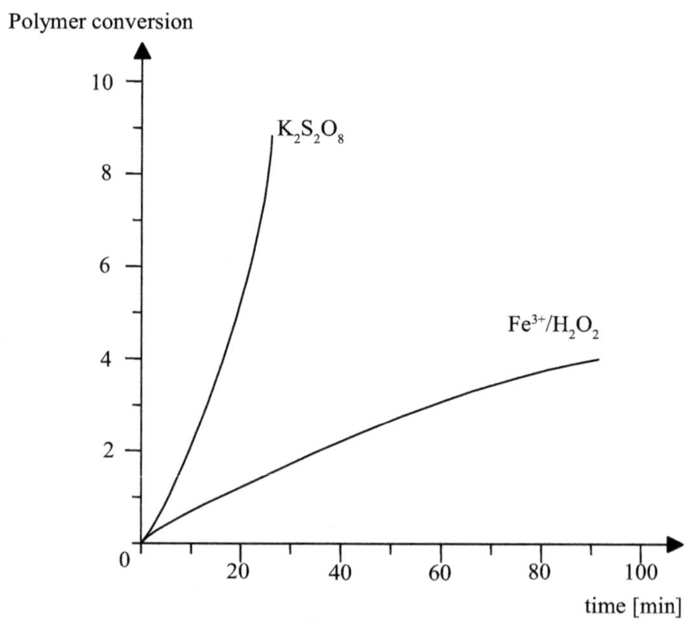

FIGURE 10.3 Effect of the initiator type on the course of emulsion polymerization (after A. S. Dunn, Surf. Inf 1974, 50 (1991)).

It is assumed that the individual emulsifier molecules are arranged in parallel, or directed to inside in such a way that the hydrophilic groups (-COO- or-SO_3-) form the outer layer in contact with water. The course of emulsion polymerization is shown in Figure 10.4.

In inner part of the micelle, where are the hydrophobic groups, may be located also the monomer molecules. In this way one can explain why the small solubility of styrene in water (0.02% at 233 K) increases several times after an addition of 2% of emulsifier.

X-ray diffraction studies of micelles confirm their layered structure. Addition of a micelle to the hydrocarbon system makes diffraction pattern diffuse. This is interpreted as due to the penetration of hydrocarbon inside the core. The size of micells depends on the emulsifier concentration and increases with increasing penetration of water-insoluble monomers inside.

Probably the kinetic chain initiation takes place in micelle formed by absorption of a radical created in the aqueous phase. In micelle, in which a high concentration of monomer is observed, the chain grows proceeds until absorption of a next radical. Due to the very small size of the spore, in which the polymerization started, it seems likely that the second radical will react with the growing chain and will break its growth. Bu it will not start a new chain.

The growing chain takes the needed for its growth monomer molecules first from inside the micelles, and then pulls the monomers dissolved in the aqueous phase. The monomer concentration in the aqueous phase remains constant as long as there exist the droplets. The formed polymer molecules suck monomer and fills like a sponge. Small grains are of gelatinous form, but the pulled through the gel monomer does not dilute it, because it polymerizes immediately. Micelle is disrupted due to the running reaction of polymer growth. Its molecules surround the polymer molecules

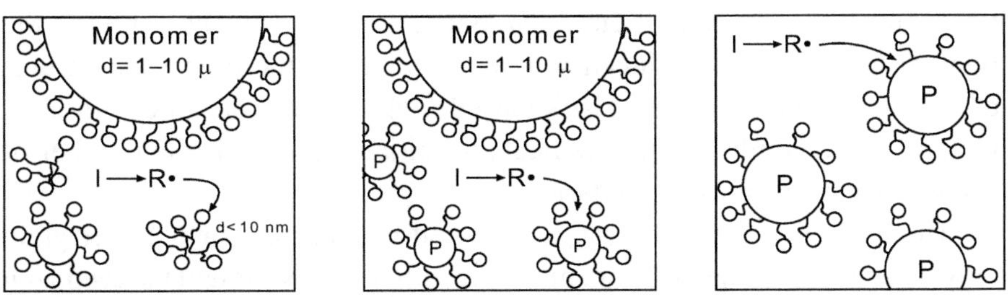

FIGURE 10.4 Chart showing he course of emulsion polymerization.

preventing them from coagulating and conglomerating. Further polymerization takes place in polymer grains. Monomer diffuses into the grains from suspension droplets through the aqueous phase. A characteristic property of the emulsion polymerization and its kinetics is that due to very small size of the polymer grains maximum only one active radical is in. If a second radical arrives during the chain growth it causes its disruption in the recombination reaction. Only after the cessation of the chain growth a new radical entering the grain initiates the growth of a new chain.

The emulsion is composed mainly of polymer grains formed in micelles at the initial stage of polymerization.

For a sufficiently large number of grains and in a given time one half of these molecules possess a radical, whereas the other half not. If we denote by N the number of polymer micelles in 1 cm^3 of emulsion the number of molecules in 1 cm^3 possessing a radical is N/2.

The rate of attachment to the growing polymer radical is given by:

$$-\frac{d[M]}{dt} = k_w[M^\bullet][M]$$

To N/2 radicals is attached at one second $(N/2)k_w$ [M] mole of monomers. As it was shown above, N/2 is the number of radicals in 1 cm^3 of emulsion. This value remains constant during the polymerization process. The rate of polymerization can be described by the formula:

$$V_p = \frac{N}{2}k_w[M]\sqrt{N_A}\left[molecules\right]xcm^{-3}xs^{-1}$$

Because the monomer concentration [M] in the aqueous phase is constant (saturated solution) and practically there is no influence on its concentration in micelles, the rate of emulsion polymerization depends only on the number of grains of produced polymer. However, the number of grains of polymer depends on emulsifier used and on initiator concentrations. Greater is the number of free radicals of initiator greater is the number of primary spores of polymerization.

The initiation of a chain can lead to self-manufacture of polymer grain only when there is a quantity of emulsifier sufficient to surround the seed with a monomolecular protective layer and prohibit connections with other grains of polymer, and thereby protect against coagulation. If there is more polymer grains than it corresponds to the critical amount of emulsifier than their shell is diminishing by combining two or more of them into larger grains. Sometimes, especially at the end of polymerization, coarse grains of coagulum are formed.

In technical conditions at 1% concentration of emulsifier and initiator concentration of about 0.1% in 1 cm^3 the number of molecules (grains) in polymer is of the order of 10^{14}.

A special feature of the emulsion polymerization is the large quantity (N) of polymer grains in 1 cm^3. Therefore, the degree of polymerization (P) can be controlled by the emulsifier concentration, regardless of the concentration of initiator. As result the harmful influence of initiator concentration on the chain length can be compensated. The growing chain radicals are isolated from each other by the action of emulsifier. Even at a large concentration of macroradicals (which takes place at the emulsion polymerization) there is no termination by recombination of growing polymer chains.

The emulsion polymers are characterized by a greater degree of polymerization and higher molecular weight than the block or suspensive polymers formed at the same temperature and at the same concentration of initiator. As an example one can quote the styrene polymerization. In the suspension polymerization of styrene at 80°C one gets a product with an average molecular weight of about 200–000, while the emulsion polymerization at the same temperature and at the same concentration of initiator (0.1%) gives a product with a five orders of magnitude greater molecular weight.

In contrary to the block polymerization where increasing the rate of polymerization by raising the temperature or increasing the initiator concentration leads to a lower degree of polymerization, in the emulsion polymerization it is possible to increase the reaction rate by increasing the concentration of emulsifier, without causing a reduction in the degree of polymerization.

For these reasons the emulsion polymerization was very widely used in the technology of polymers.

10.2.8.6 Polymerization in Microemulsion

In recent years more and more important theoretical and practical importance gains polymerizations carried out in mini- and in microemulsions. The latter are defined as the most complex quaternary systems composed of a monomer, a phase diffuser, a surface-active compound (surfactant) and a cosurfactant. Depending on the degree of hydrophobicity or hydrophilicity of monomer used water or an appropriate hydrocarbon can be used as a phase diffuser.

Microemulsions are characterized by their homogeneity. They are optically isotropic, exhibit a low viscosity and show a much greater stability of emulsion. They are also thermodynamically stable and form spontaneously a continuous phase of the two not mixing ingredients.

There are three types of microemulsions formed in systems in which the quantities of monomer and water are comparable. The three types are:

- oil in water,
- water in oil, and
- bicontinuous strip structure of microemulsion.

The type of formed microemulsion depends primarily on the type of surfactant and cosurfactant used.

The most frequently as cosurfactant one uses a small molecular mass alcohol, containing from three to ten carbon atoms in the chain, which reduces the packing density of surfactant and lowers the interfacial tension.

Bicontinuous microemulsions of water in oil type require the use of a larger amount of cosurfactant than macroemulsion of oil in water type. These microemulsions are distinct by the structure approaching that of liquid crystals, but with a smaller order.

When studying the course of polymerization of styrene it was shown that the formed microemulsions are characterized by transparency and are thermodynamically stable. The diameter of the molecules in microemulsions is between 40 and 100 nm.

There are also so-called mini-emulsions, in contrast to the microemulsions, which form opalescent milky-colored systems and are similar to the traditional emulsions. The diameter of the molecules in mini-emulsions is in 100–400 nm range.

A modification of the micro-emulsion polymerization is polymerization in reverse micelles, called inversed or micellar polymerization. This issue concerns the polymerization of hydrophilic monomers, from which one obtains the water-soluble polymers. Therefore, as the phase diffuser mostly hydrocarbons such as isooctane, heptane, benzene or toluene are used.

In a reverse micelle the aliphatic chains of surfactant are directed to the dispersal phase, while the hydrophilic groups point inside the micelle. The contained in the system water and the monomer molecules penetrate inside the micelle.

The amount of water contained inside the reverse micelles ("water pool") decides on the size and the properties of the formed aggregate. Water contained in the aqueous space of micelle differs from the volume water by physico-chemical properties. The degree of ordering, mobility, viscosity, polarizability, dielectric constant and the chemical activity of water depend on the size of the area occupied by water and change with the distance from the center of the aggregate.

Surfactants, containing a large polar group in molecule (ammonium salts), form small micelles with a diameter of 3–10 nm, characterized by a low water content (up to 3 molecules). Anionic surfactants such as sodium di (2-ethylhexyl) sulfosuccinate, containing a small polar group with two hydrocarbon chains in molecule, form without addition of a cosurfactant almost homodysperse, large aggregates of the size almost independent of the surfactant concentration. At high water content in the system the microemulsions of water in oil are formed.

During the polymerization in inverse micelles initiators soluble in the organic phase, such as dinitryl asoisobutyric acid or organic peroxides, are used.

Polymerization begins when the formed in organic phase free radical penetrates into the "swollen" micelle, in which are monomer molecules. Since in microemulsion only a small number of monomer molecules is present, the chain growth takes place by attachment of monomer molecules located in other micromicelles as result of a monomer collision or diffusion, as shown in Figure 10.5. (Option II).

During the polymerization of acrylamide in microemulsion, in toluene environment and in presence of sodium di(2-ethylhexyl) sulfosuccinate, initiated by UV radiation at a temperature of 10°C, the reaction initialization rate is directly proportional to the radiation intensity. In this process the initial large number of micelles, which in the initial period

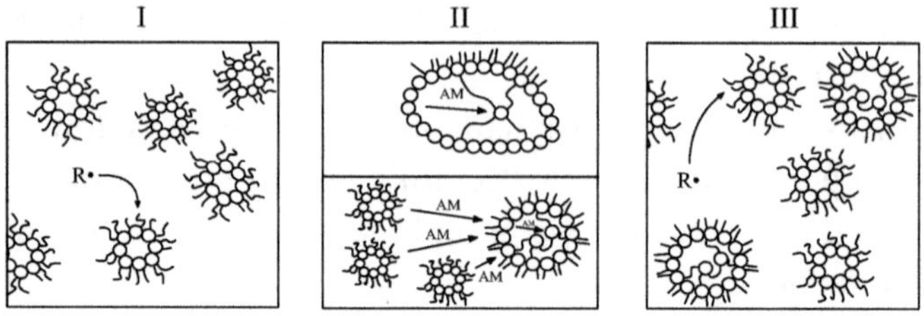

FIGURE 10.5 Flow diagram of micellar polymerization: I – initiation of polymerization, II – increase of the chain length, (a) by collision (fusion), (b) by monomer diffusion, III – complete polymerization (after K.A. Wilk, B. Burczyk, T. Wolf: 7th International Conference Surface and Colloid Science, Compiegne 1991).

of polymerization is of about $10^{21}/dm^3$ decreases to 10 latex molecules in 1 dm^3 of disperse system with dimension of 40 nm, as a result of collisions.

The completion of the polymer chain growth takes place as a result of deactivation by collision with another free radical.

10.2.8.7 Polymerization in Solution

The course of polymerization in solution depends to a large extent on both the type and the amount of solvent used. For some solvents, which dissolve both the monomer and the resulting polymer, the polymerization process proceeds in a homogenous environment. The course of this reaction is strongly influenced by the concentration parameter. With its decline decreases the probability of effective collisions of monomer molecules with the growing macroradical, and simultaneously increases the possibility of chain transfer. This causes the formation of polymer molecules with molecular weight less than obtained during the polymerization in solutions with higher concentration.

The major advantages of the polymerization process in a homogeneous solution is the possibility to obtain polymers with a relatively small dispersion of molecular mass and the possibility of direct use of polymer solutions for manufacturing synthetic fibers by the solution spinning methods. They can also be used for the preparation of solvent adhesives and lacquers.

The use of polar solvents to the polymerization process carried out by ion mechanism improves significantly the efficiency of these reactions.

However, certain difficulties are encountered with the methods of polymer extraction from solutions. The reaction in these systems virtually never runs to the end and a certain amount of unreacted monomer rests always in the solution. Also a part of the solvent is absorbed in the mass of the formed polymer and causes a change in its final properties.

To the methods used to extract the polymer from an aqueous solution belong the steam distillation and the precipitation techniques with the use of diluents miscible with solvent and not dissolving the formed polymer.

During the polymerization reaction in solution possible are also side processes, which sometimes become to be the main processes. An example of such a process is the telomerization reaction. This reaction takes place when the solvent molecules tend to create permanent, stabilized by the resonance, free radicals or ions. Particularly useful for this purpose are halogenated organics, such as carbon tetrachloromethane (tetrachloride), chloroform and ethylene chloride. The solvent molecules can then take part in the chain transfer.

As result of such a reaction telomer molecules are formed, which differ from the oligomer molecules by their ends. These are large substituents coming from the decomposition of the solvent molecule (e.g., Cl and such groups as CCl_3, $CHCl_2$, CH_2CH_2Cl, etc.), which exert a decisive influence on the properties of the final product. For this reason, telomers cannot be equated with oligomers, which do not have reactive terminal functional groups.

The telomerization reaction is often carried out deliberately to get low-molecular weight reactive substances, which serve as raw materials for the synthesis of other polymers or cyclic compounds.

Another way to conduct the polymerization in solution is the solution-precipitation method. It involves the use of a liquid dissolving only the monomer and not dissolving the polymer. Due to that the formed product precipitate from the reaction environment in the form of a very fine powder or suspension, which is separated by centrifugation or filtration. In this process the increasing macroradical, at a certain chain length, ceases to be soluble and precipitates from solution. As result, the resulting polymer exhibits a particularly low polydispersity of molecular weight. This method was successfully applied to the synthesis of solid polymers with a relatively low molecular weight. As examples of such a reaction are polymerizations of styrene in alcohol, methyl methacrylate in saturated hydrocarbons as well as acrylonitrile in water, respectively. The type of solvent used exerts also a great influence on the course of copolymerization in solution.

10.2.9 COPOLYMERIZATION

Copolymerization is a process of joint polymerization of at least two different monomers. Obtained in this way polymers are called copolymers, in contrast to homopolymers, which are produced by polymerization of a single monomer.

Depending on the chemical structure of the starting monomers, their mutual quantitative relationship and the method of conducting the reaction one can obtain copolymers of different composition and different properties. Physicochemical properties of copolymers depend on their composition, but in most cases the relationship is not additive. Often the introduction to macromolecule during the copolymerization of a relatively small amount of another monomer provides the polymer with completely new properties. As an example one can cite a copolymer of isobutylene with a small amount (0.6–3.0%) of isoprene. In this way the formed polymer gets properties of rubber and can be vulcanized with sulfur, while the polyisobutylene does not have such a capacity.

Generally the copolymers are classified in the following way:

- statistical copolymers;
- alternating copolymers;
- block copolymers;
- grafted copolymers.

The statistical copolymers are products characterized by a disordered distribution in formed macromolecules of used in reaction mers A and B.

Alternating copolymers are products of a regular structure and their macromolecules are made up from mers of neighboring A and B mers:

$$\sim\sim\sim ABABABABAB \sim\sim\sim$$

Block copolymers have a chain structure and the macromolecule is composed of separate segments (blocks) made up exclusively of monomer A or monomer B.

$$A_mB_n \text{ or } A_mB_nA_k$$

The block copolymers are the most frequently obtained during the formation of macroradicals in a mixture of polymers. For this purpose one can use the mechano-chemical method involving the rolling of the mixture of polymer or its grounding in a ball mill. However, the post-reaction mixture, in addition to block copolymers, contains in this case always a certain amount of homopolymers. Pure block copolymers can be obtained by using the previously discussed reaction of "living polymers", containing end-groups able to initiate polymerization of another monomer B.

The reaction proceeds according to the scheme:

$$Na^+A^-{-}(A)_n{-}A^-{-}Na^+ + mB \longrightarrow Na^+A^-{-}(A)_{n+1}{-}(B)_{m-1}{-}B^-{-}Na^+ \quad \text{itd.}$$

The grafted copolymers are branched structures. Their main chain is composed of groups of monomers A and the side chains of monomers B:

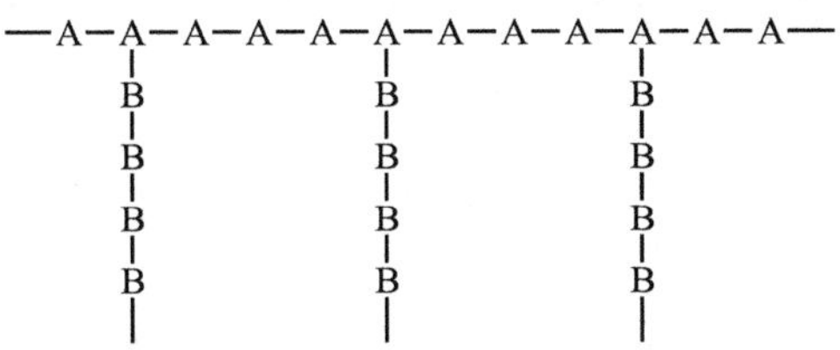

The grafted copolymers are obtained from macromolecules of polymer of A, in which are formed active centers that can then initiate polymerization of another monomer B. The active centers in polymer A may be formed by the reaction with free radicals arising from the decomposition of reaction initiators, and also during the irradiation, oxidation and under influence of ultrasounds, etc.

The copolymerization differs in some particular aspects from the homopolymerization as the initiation of free radical copolymerization of two monomers A and B may run as follows:

$$R^\bullet + A \longrightarrow R{-}H + A^\bullet$$

$$R^\bullet + B \longrightarrow R{-}H + B^\bullet$$

The resulting radicals A • B • may react in the following way:

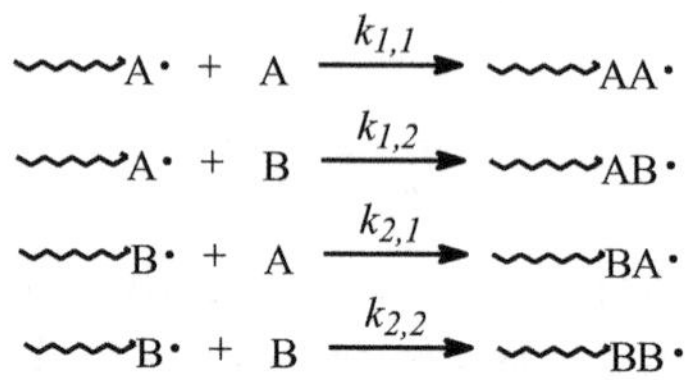

The coefficients $k_{1,1}$, and $k_{2,2}$ denote the reaction rate constants of attachment of monomers A and B to their radicals. The rate constants $k_{1,2}$ and $k_{2,1}$, relate to attachment of these monomers to foreign radicals.

The steady-state concentration of radicals A• and B• does not change. Therefore, under these conditions the rate of the process of the radical transformation A• into B• is equal to the rate of the inverse radical transformation: B• into A•:

$$k_{1,2}[A^\bullet][B] = k_{2,1}[B^\bullet][A]$$

The rate equations of the copolymerization process have the following form:

$$-\frac{d[A]}{dt} = k_{1,1}[A^\bullet][A] + k_{2,1}[B^\bullet][A]$$

$$-\frac{d[B]}{dt} = k_{1,2}[A^\bullet][B] + k_{2,2}[B^\bullet][B]$$

Dividing these equations side by side one obtains:

$$\frac{d[A]}{d[B]} = \frac{[A]}{[B]}\left(\frac{k_{1,1}[A^\bullet] + k_{2,1}[B^\bullet]}{k_{1,2}[A^\bullet] + k_{2,2}[B^\bullet]}\right)$$

Since in the steady state:

$$[A] = \frac{k_{1,2}[A^\bullet][B]}{k_{1,2}[B^\bullet]}$$

then

$$\frac{d[A]}{d[B]} = \frac{[A]}{[B]}\left(\frac{\frac{k_{1,1}}{k_{1,2}}[A]+[B]}{[A]+\frac{k_{2,2}}{k_{2,1}}[B]}\right)$$

This is the copolymerization equation, whose validity was confirmed experimentally.

The occurring in the equation ratios of the reaction rate constants are defined as the monomer reactivity ratios r_1 and r_2:

$$r_1 = \frac{k_{1,1}}{k_{1,2}} \quad r_2 = \frac{k_{2,2}}{k_{2,1}}$$

These ratios are derived from the copolymerization kinetic equations and they determine, respectively:

- r_1 – the ratio of the rate constant of reaction of monomer A with radical A to the rateconstant of reaction of radical A with monomer B;
- r_2 – the ratio of the rate constant of reaction of radical B with monomer B to the rate constant of the reaction of radical B with monomer A.

Depending on the values of parameters r_1 and r_2 the following copolymerization options may take place:

a) when $r_1<1$ and $r_2<1$ – monomer A reacts easier with radical B• than with the own radical A•. Similarly, monomer B reacts more easily with radical A• than with radical B•. Both monomers compile to form a copolymer with no long segments corresponding to only one of the comonomers;
b) when $r_1>1$ and $r_2<1$, facilitated is the reaction of both radicals A• and B• with monomer A. As result the formed polymer is enriched in monomer A in comparison to the initial monomer mixture;
c) when $r_1>1$ and $r_2>1$, a mixture of homopolymers is formed;
d) when r_1 x r_2 = 1 then the composition of the resulting copolymer is directly proportional to the composition of the starting monomer mixture. Such a system is called a perfect one;

e) when $r_1 = r_2 \approx 0$, alternating copolymers are formed;
f) when $r_1 = r_2 = 1$, the copolymer composition is identical with that of the initial comonomer mixture.

The selected reactivity ratios r_1 and r_2 for different monomers are given in Table 10.18.

TABLE 10.18 Reactivity Coefficients for Selected Monomer Systems Used in the Synthesis of Copolymers

No.	Monomer 1	Monomer 2	r_1	r_2
1	acrylamide	akrylic acid	0.48	1.73
2	acrylamide	acrylonitrile	1.04	0.94
3	acrylamide	vinyl chloride	19.6	0
4	acrylamide	methyl methacrylate	0.44	2.6
5	acrylamide	Styrene	0.3	1.44
6	methyl acrylate	vinylidene chloride	0.8	0.5
7	methyl acrylate	2-vinylpirydine	0.19	0.23
8	methyl acrylate	Styrene	0.17	0.77
9	1,3-butadiene	butyl acrylate	0.99	0.08
10	1,3-butadiene	acrylonitrile	0.18	0.03
11	1,3-butadiene	vinylidene chloride	1.9	0.05
12	1,3-butadiene	2-vinylpirydine	0.75	0.85
13	1,3-butadiene	styrene	1.4	0.5
14	ethylene	butyl acrylate	0.2	11
15	ethylene	vinyl chloride	0.6	1.85
16	ethylene	vinyl acetale	0.7	3.7
17	N-phenylmaleimide	vinyl chloride	4.37	0.03
18	N-phenylmaleimide	methyl methacrylate	0.183	1.022
19	N-phenylmaleimide	vinyl acetale	1.269	0
20	N-phenylmaleimide	styrene	0.047	0.012
21	metmethyl acrylate	styrene	0.35	0.35
22	metmethyl acrylate	2-hydroxyethyl methacrylate	0.296	1.054
23	vinyl acetate	vinyl chloride	0.23	1.68
24	vinyl acetate	vinylidene chloride	0	3.6
25	vinyl acetate	styrene	0.01	55
26	styrene	maleic anhydride	0.04	0.015

TABLE 10.18 Continued

No.	Monomer 1	Monomer 2	r_1	r_2
27	styrene	N-(4-chlorophenyl) maleimide	0.22	0.03
28	styrene	2-vinylpirydine	0.56	0.9
29	styrene	vinylidene chloride	2.1	0.45
30	styrene	N-vinylcarbazol	5.6	0.062

*After R.Z. Greenley: J. Macromol. Sci. Chem. A 14, 445 (1980).

There are various methods of determining the coefficients r_1 and r_2. As described above, the copolymerization equation assumes that:

$$a = \frac{[A]}{[B]} \quad i \quad b = \frac{d[A]}{d[B]} = \frac{-[A]-}{-[B]-}$$

where [A] and [B] indicate the contents of mers A and B in copolymer, respectively.

After introducing the coefficients r_1 and r_2 the copolymerization equation takes the form:

$$b = a\left(\frac{r_1 a + 1}{r_2 + a}\right) = \frac{r_1 + 1}{\frac{r_2}{a} + 1}$$

A graphic image of this equation is a straight line.

The most frequently used methods of determining the reactivity ratios are:

- Mayo-Lewis method; and
- Finneman-Ross method.

The Mayo-Lewis method consists on determining the value of b in a series of successive copolimerization tests carried out at different contents of copolymerizable monomers used: A and B. The obtained values are entered into the copolymerization equation, which is solved in order to get r_1 and r_2.

The solution of the equation for r_2 is:

$$r_2 = a\frac{r_1 a + 1}{b} - a = r_1\frac{a^2}{b} + \frac{a}{b} - a$$

After substituting into the equation the *a* and *b* values obtained from measurements, one gets a series of straight lines corresponding to different dependencies $r_2 = f(r_1)$. Substituting the rising value of r_1 (e.g., from 0 to 1) and calculating r_2 one obtains a family of straight lines intersecting at one point (Figure 10.6). At this intersection point all obtained values of *a* and *b* correspond to the values of coefficients r_1 and r_2.

The Finnemana-Ross method consists on the solving the copolymerization equation with respect to a and b parameters:

$$a - \frac{a}{b} = r_1\frac{a^2}{b} - r_2$$

Then a graph, showing the dependence:

$$a - \frac{a}{b} = f\left(\frac{a^2}{b}\right)$$

is drawn.

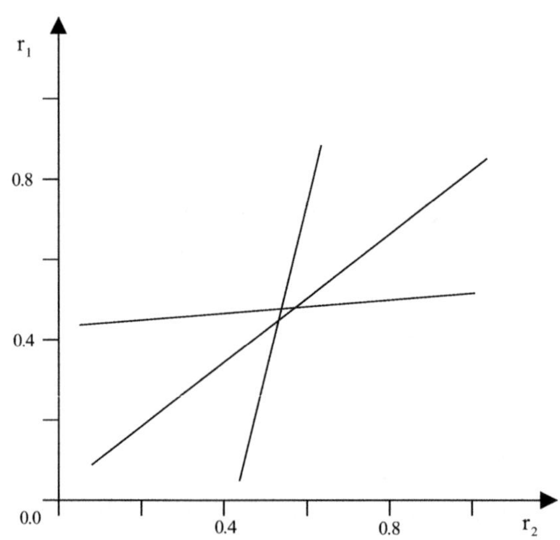

FIGURE 10.6 Graphical determination of reactivity ratios r_1 and r_2 by the Mayo-Lewis method.

The value of r_2 is the abscissa on $(a–a/b)$ axis and r_1 is the slope of the curve. The Finneman-Ross method is simpler than the Mayo-Lewis method, because it requires obtaining of only one curve, describing the above dependence, only. But it is not always accurate (cf. Figure 10.7).

In order to simplify the way used to determine the reactivity ratios Alfrey and Price proposed a semiempirical relationship in which these factors is expressed as a fixed characteristic of the monomer, but independent of the comonomer. In this way it is not necessary to determine experimentally the relationship for each pair of individual monomers in the copolymerization reaction.

This relationship is called Alfrey-Price's Q-e system:

$$r_1 = \left(\frac{Q_A}{Q_B}\right) e^{-e_A(e_A - e_B)}$$

$$r_2 = \left(\frac{Q_B}{Q_A}\right) e^{-e_B(e_B - e_A)}$$

where Q_A and Q_B are the measures of the reactivity of monomers A, B, respectively and are connected by the resonance stabilization of monomer. Constants e_A and e_B are the measures of the polarity of monomers.

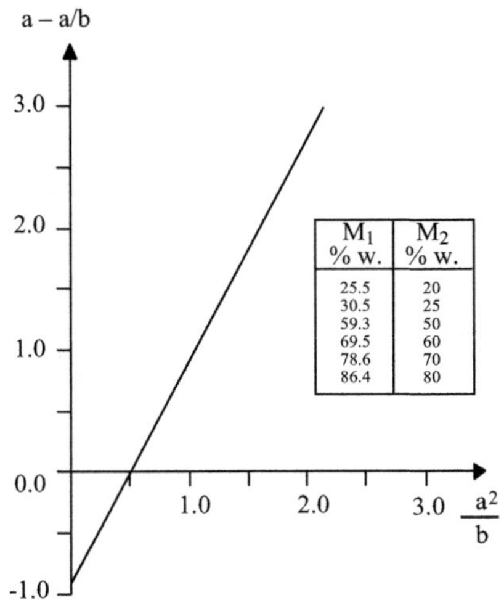

FIGURE 10.7 Graphical determination of the reactivity ratios r_1 and r_2 by the Finneman-Ross method for copolymerization of acrylic acid with acrylonitrile.

Styrene monomer was chosen as standard with assigned values of Q = 1.00 and $e_A = -0.80$.

The Q values increase with the increasing resonance stabilization, while the values of e becomes to be less negative when the groups attached to atoms forming a double bond attract stronger electrons.

Q and e values for selected monomers are given in Table 10.19.

It should be noted that the values given in Table 10.19 are only approximate and in the best case they provide only semi-quantitative informations. The latter are, however, very useful, as they help to predict the value

TABLE 10.19 Q and e values for Selected Monomers

No.	Monomer	Q	e
1	acrylaldehyde (propenal)	0.8	1.31
2	acrylamide	0.23	0.54
3	acrylonitrile (vinyl cyanide)	0.48 ± 0.07	1.23 ± 0.08
4	ethyl acrylate	0.41	0.55
5	octyl acrylate	0.63	2.01
6	acrylic acid	0.83 ± 0.47	0.88 ± 0.23
7	1,3-butadiene	1.7	–0.5
8	vinyl chloride	0.056	0.16
9	vinylidene chloride	0.31	0.34
10	ethylene	0.016	0.05
11	1-hexene	0.035	0.92
12	indene	0.13	–0.71
13	diethyl maleate	0.053	1.08
14	maleic anhydride	0.86	3.69
15	maleimide	0.94	2.86
16	N-phenylmaleimide	2.81	3.24
17	butyl methacrylate	0.88	3.7
18	ethyl methacrylate	0.76	0.17
19	methyl methacrylate	0.78 ± 0.06	0.40 ± 0.08
20	propylene	0.009	–1.69
21	styrene	1	–0.8
22	N-vinylcarbazole	0.26	–1.29
23	2-vinylpirydine	1.41	–0.42
24	4- vinylpirydine	2.47	0.84

*After R.Z. Greenley: J. Macromol. Sci. Chem. A 14, 427 (1980).

of the reactivity coefficients in the situation when no other experimental data are available.

The reactivity coefficients are affected by the sum of steric, resonance and polar effects.

Resonance in the monomer molecule affects the reactivity of the intermediate radical. A large resonance stabilization of the monomer reduces its reactivity in the chain growth reaction. For example, it is difficult to copolymerize styrene with vinyl acetate. This happens because the resonance stabilized styrile radical shows no tendency of addition to vinyl acetate, since the resulting radical would not be stabilized.

On the contrary, styrene copolymerizes readily with methyl methacrylate, which forms like styrene, a resonance stabilized radical and exhibits the acceptor-donor effect.

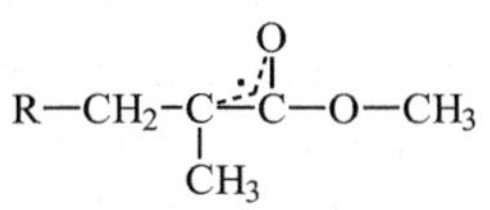

In order to clarify the cases in which there exists a strong tendency to form alternating copolymers (r_1 and $r_2 \approx 0$) two theories were proposed. The first one takes into account the polar effects. The second explore the mutual interactions connected with the charge transfer.

Effect of polar effects can be seen on the example of the styrene copolymerization with maleic anhydride. Attachment of the maleic radical to styrene leads to a transition state, which can be stabilized by the resonance structure formed as a result of the electron transfer:

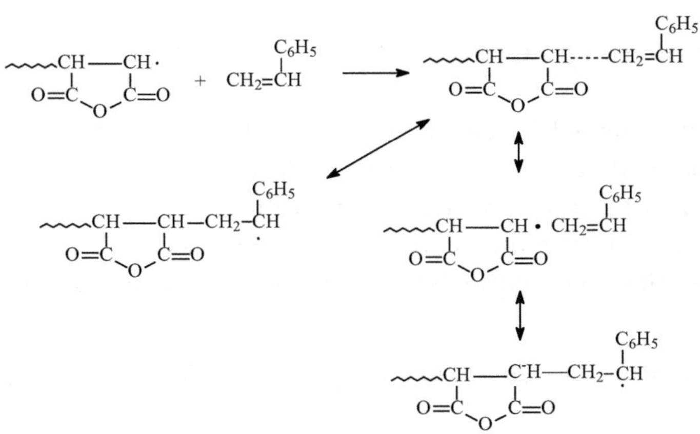

This type of structure may arise when a monomer includes numerous electronegative substituents, as it is the case of maleic anhydride.

The attachment of the styrile radical to the maleic anhydride gives rise to a transition state, which is stabilized in a similar way. In sum, these interactions take such an effect that the styrile radicals prefer to join the maleic anhydride and the maleic radicals prefer to combine with styrene.

A newer theory of the formation of alternating copolymers assumes the formation of complexes between comonomer molecules, associated with a charge transfer.

According to this theory the alternating copolymerization reaction is in fact a homolymerization or an alternate copolymerization of formed complex with the charge transfer, which reacts favorably with other charge transfer complexes. This is confirmed by the observed increasing tendency to form alternate copolymers when the reactants increasing the acceptor properties of one of the monomers are introduced to the system. This phenomenon is observed, for example, when the ethyl aluminum sesquichloride is added to the styrene maleic anhydride or the zinc chloride to the butadiene-acrylonitrile system. This is also confirmed by the fact, according to which after the introduction of other highly reactive monomers the latter do not react. It was also found that the molecular weight regulators have only an insignificant influence on the synthesis of alternating copolymers.

Other examples of compounds which copolymerize with vinyl monomers are: carbon monoxide, forming polyketones and sulfur dioxide which reacts forming a polysulfone. It should be noted that sulfur dioxide forms complex compounds with alkenes in the 1:1 ratio.

10.3 VINYL POLYMERS

The vinyl group CH2=CH– is present in ethylene and its derivatives, in which the hydrogen atom has been replaced by an atom or a group of atoms. As a result of polymerization of this type of compounds, called vinyl monomers, the vinyl polymers are formed.

Depending on the type of substituent the vinyl polymers can be divided into:

- Polyolefins, resulting from the polymerization of alkenes (olefins);

- Styrene polymers containing aromatic substituents;
- Poly(vinyl halides), containing halogen atoms;
- Poly(vinyl amines);
- Poly(vinyl ethers);
- Poly(vinyl ketones);
- Poly(vinyl esters);
- Acrylic polymers, containing groups CN, $CONH_2$, COOH and COOR.

10.3.1 POLYETHYLENE (PE)

Polymerization of ethylene is very difficult and therefore the first synthesis of polyethylene was carried out accidentally. During the test of the reactivity of chemical compounds under pressure a mixture of ethylene and benzoic aldehyde was introduced into the reactor. The reaction product did not contain aromatic groups and proved to be polyethylene. Further tests of the ethylene polymerization under elevated pressure were initially unsuccessful and only the discovery of the reaction initiation by atmospheric oxygen enabled the development of this synthesis on an industrial scale. The small molecular weight of polyethylene and its excellent insulation properties preferred it to be used as a raw material in the manufacture of submarine cable under the La Manche channel.

The first charge of technically produced polyethylene was obtained on 01.09.1939, coinciding with the start of World War II. The lightness and the dielectric properties of polyethylene have enabled the construction of light aircraft radars, which contributed to the sinking of several German submarines, and thereby to break the blockade of the sea around England, influencing in this way the outcome of the war.

The ethylene polymerization process is an exothermic one and the heat of reaction is of 93 [kJ/mol].

The reaction is carried out in adiabatic autoclave or tubular type reactors. Instead of the previously applied oxygen the hydroperoxides are used as initiators. The conversion degree of ethylene during one cycle is about 23%. The unreacted ethylene is recycled to the circulation, and the molten polyethylene is extruded in strips, cooled and granulated.

The discovery of organometallic coordination catalysts by Ziegler and Natta allowed to obtain high-density polyethylene, which is characterized by a greater degree of crystallinity, worse transparency and better mechanical properties in comparison with the high-pressure polyethylene, called a low-density polyethylene.

The high-density polyethylene is obtained using also the technique introduced in U.S. in 1957. The method is called the Philips medium pressure technique. As catalyst the chromium trioxide, fixed on a support consisting of a mixture of silicon dioxide and aluminum oxide, is used. In this method a 5–7 percent solution of ethylene in the aromatic or the aliphatic hydrocarbons from pentane to decane is heated at a temperature of 150°C and at a pressure of 3.5 MPa. Under these conditions ethylene polymerizes virtually in a single cycle.

The structure of polyethylene depends on the method used for its production. During the high-pressure free radical polymerization the grafting side reactions take place which cause side branching in polyethylene molecules. It makes difficult its crystallization and lowers the density. The products of ethylene polymerization obtained by Ziegler-Natta or Phillips methods are practically free of branching.

Polyethylene dissolves in hot aromatic and halogen derivative aliphatic hydrocarbons and exhibit good chemical resistance, except powerful oxidizing agents and concentrated mineral acids.

10.3.2 POLYPROPYLENE (PP)

The free radical polymerization of propylene proceeds unsatisfactory. The cationic polymerization at a temperature of –50°C and in the presence Friedel-Crafts catalysts leads to the formation of an amorphous polypropylene with molecular weight of 1000–10000 and with appearance of wax, which can be used to improve the viscosity of oils.

The use of Ziegler-Natta catalysts for propylene polymerization allowed to obtain isotactic polypropylene with superior utility performances, enabling a widespread application of polypropylene products for the manufacturing of construction plastics and synthetic fibers.

The properties of polypropylene depend largely on its molecular weight, degree of polydispersity, its tacticity and crystallinity.

The average values are as follows:

density [g/cm^3]	0.90–0.91
melting temperature of crystallites [°C]	70–75
glass transition temperature [°C]	–35
tensile strength [Pa]	(1–3.6) x 10^7
bursting strength [Pa]	3–4 x 10^7
drawing elongation [%]	5–15
ultimate elongation [%]	500–700
compressive strength [Pa]	(0.7–1) x 10^8
specific heat at temperature of 23°C [J/g K]	1.92
dielectric strength [kV/m]	350–800
surface resistance [cm/cm]	5 x 10^{13}
oblique resistance [cm × cm]	10^{15}–10^{17}
dielectric loss factor [10^4 Hz]	0.0019
water absorption after 24 hours [%]	0.03

Polypropylene is characterized by a high chemical resistance. It shows no sensitivity to water, solutions of strong acids, bases and inorganic salts. However, it is susceptible to the action of strong oxidants and non-polar liquids. Polypropylene is a thermoplastic material, which is suitable for injection and extrusion moldings as well as for the production of foils and fibers.

The products made of polypropylene retain the shape up to the temperature of 423 K (150°C). However, the value of the elasticity modulus at 413 K (140°C) is only 10% of that at room temperature. Therefore, in practice, polypropylene is used at temperatures below 408 K (135°C).

10.3.3 POLYISOBUTYLENE (PIB)

Polyisobutylene is obtained by the cationic polymerization of isobutylene in the presence of Friedel-Craffts catalysts. The most widely used catalyst is the bromine trifluoride and t-butyl alcohol as co catalyst. The polymerization reaction proceeds according to the following scheme:

$$BF_3 + R\text{-}OH \longrightarrow H^{\oplus}[BF_3OR]^{\ominus}$$

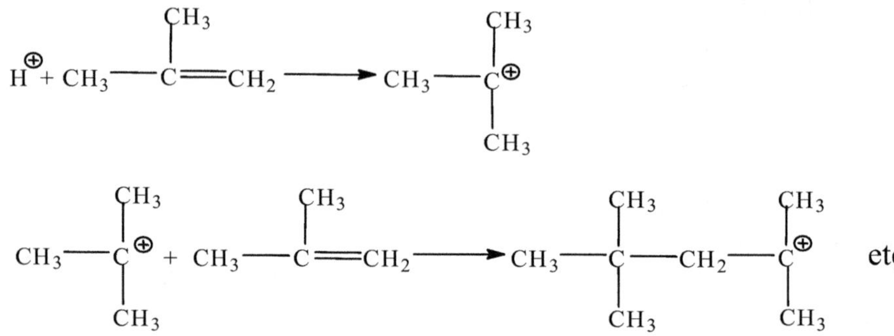

Completion of the chain growth takes place by the transfer of proton to a new monomer molecule or a negative ion.

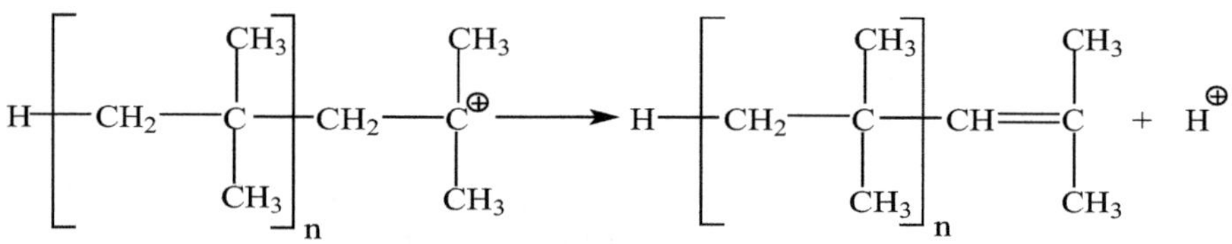

Polymerization inhibitors are sulfur, mercaptans, hydrogen sulfide, hydrogen chloride, etc. The bromine trifluoride is a very energetic catalyst in this process. The reaction is one of the fastest in the chemistry of macromolecular compounds and is highly exothermic (heat emission is of 54.4 kJ/mol). It impedes greatly the process, because it requires rapid removal of the generated heat. For this reason the isobutylene polymerization reaction is carried out in solution.

The course of polymerization is highly influenced by the way how monomer is mixed with catalyst and by temperature of the conducted process. Products with high molecular weight are obtained by polymerization of isobutylene at low temperatures (below –70°C), while the increase in temperature leads to the formation of polymers with a low molecular weight.

The properties of polyisobutylene vary depending on its molecular weight. Low molecular weight products (M < 5,000) are oils, while polymers with an average molecular weight of 100,000–500,000 are typical elastomers, characterized by a very low glass transition temperature (199 K) and a low tendency to crystallize.

Some properties polyisobutylene are as follows:

density [g/cm^3]	0.91–0.93
glass transition temperature [°C]	–74
brittle point [°C]	–50
bursting strength [Pa]	(2–13.5) x 10^6
elongation [%]	1000–2000
water absorbability [%]	0–0.05
dielectric breakdown [kV/mm]	23–25
cross-resistance [Ω/cm]	10^{15}–10^{16}
dielectric loss factor [tan δ]	0.0004–0.0008
dielectric constant [at 10^6 Hz]	2.3–2.2

The polyisobutylene is relatively easily soluble in aromatic hydrocarbons, carbon disulfide and halogenated derivatives. It is insoluble in alcohols, ketones, esters and other polar solvents.

Polyisobutylene is characterized by a high chemical resistance and a resistance to the water action. At ambient temperatures polyisobutylene is insensitive to the action of almost all acids, even the nitric acid, and in some cases resists to the aqua regia too. It is also resistant to bases and to halogens. High chemical resistance of polyisobutylene originates from its saturated character. It belongs to the family of weakly polar polymers. Due to that polyisobutylene is a good dielectric.

In the presence of oxygen and under the UV light polyisobutylene oxidizes with formation of oily decay products. The introduction of active fillers (soot, talc, etc.) enhances its resistance to oxygen and to radiation.

Polyisobutylene has found application in various fields of technology and in particular for wire and cable insulation, as an anti-corrosion lining as well as for obtaining protective coatings and adhesives.

The polyisobutylene oils have found application as brake fluids and additives for lubricating oils to improve their viscosity characteristics.

10.3.4 POLY(METHYLPENTENE)

The poly(methylpentene) is formed by the polymerization of 4-methyl-1-pentene on Ziegler-Natta coordination catalysts:

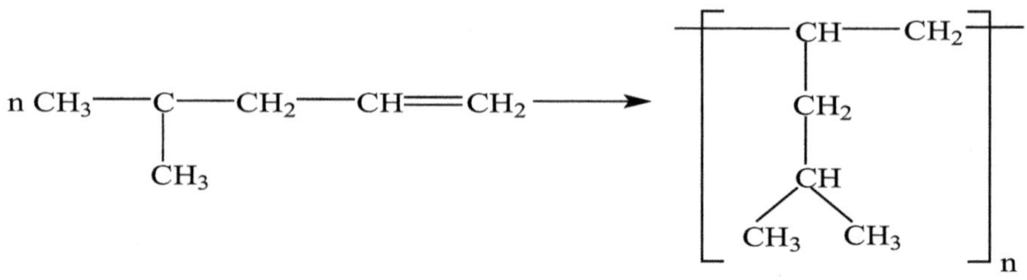

Poly(methylpentene) is the lightest plastic material with the following characteristics:

contents of the crystallites [%]	40–65
density [g/cm^3]	0.83
tensile strength [MPa]	28
elongation at rupture [%]	15
softening temperature (Vicat) [° C]	179
glass transition temperature [° C]	–40
melting temperature of crystallites [° C]	242
dielectric constant	2.12

Poly(methylopentene) is produced since 1968 by the English company ICI in a variety of transparent and opaque materials. It can be sterilized and utilized up to the temperature of 170°C.

10.3.5 POLYSTYRENE (PS)

Polystyrene is one of the oldest known thermoplastics.

It is obtained by radical polymerization of styrene (vinylbenzene) in bulk or in suspension. The reaction proceeds according to the scheme:

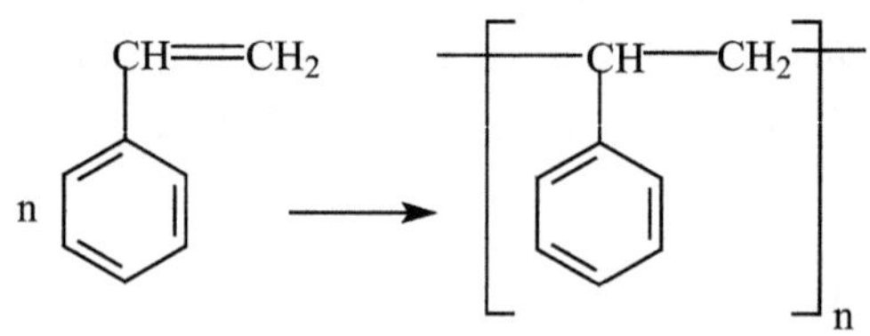

Polystyrene is characterized by a linear structure with "head to tail" arrangement of molecules and by a high molecular weight. The material is hard, brittle and transparent, soluble in aromatic hydrocarbons, chlorinated organic derivatives, esters, ketones and carbon disulfide. The mechanical properties of polystyrene depend on its molecular weight as well as on temperature and diminish when approaching its softening point.

The average values characterizing the properties of polystyrene are as follows:

density [g/cm^3]	1.05
tensile strength [MPa]	55
modulus of elasticity [MPa]	3.4×10^3
bending strength [MPa]	105
impact strength by Charpy [J/m^2]	
• without notch	25×10^3
• notched	2.5×10^3
cross-resistance [Ω cm]	10^{18}
dielectric breakdown [kV/mm]	60
dielectric permeability	2.5
dielectric loss factor (tan δ)	0.0003

The specific heat of polystyrene is lower than that of majority of other polymers and is 1.34 [J/gK]. The coefficient of linear thermal expansion of polystyrene is of 8×10^{-6} [K^{-1}]. It diminishes with addition of fillers and increases as a result of the softeners addition.

The visible light transmittance of polystyrene is about 90%, so that it can be used in optics. Indeed, polystyrene is widely used in the manufacturing of optical fibers. Its refractive index n_D^{20} = 1.5916–1.5927 and the critical incidence angle of light is of 38° 55″. Thanks to that the phenomenon of the total internal reflection of light occurs even at a small radius of curvature. Major disadvantage of polystyrene for application in optical elements is its low durability and easy scratching of surface.

The biggest disadvantages of polystyrene are its fragility and combustibility. In order to counter these shortcomings, limiting application of polystyrene, new polystyrene plastics, with improved properties, were developed.

High impact polystyrenes (HIPS), containing physically (type K) or chemically (type G) bound rubber by the grafting reaction are characterized by good mechanical properties. The addition of decabromodiphenyl to polystyrene plastic makes it so much self-extinguishing that the material can be used to manufacture the rear walls of TV receivers.

The polystyrene foam with a cellular structure, known as Styrofoam, is used as insulating and packaging material.

10.3.6 STYRENE COPOLYMERS

Styrene is a highly reactive comonomer and forms easily series of copolymers, whose properties are often much better than those of polystyrene. Depending on the reactivity of the second comonomer Q and the double bond polarization e different reactivity ratios r_1 and r_2 in the copolymerization process, which determine the structure of the resulting copolymer, are observed.

The monomers used in the copolymerization reactions with styrene can be divided into three main groups.

Group I includes monomers for which the products of reactivity coefficients with styrene are equal to zero. These monomers copolymerize easily with styrene and show a distinct tendency for alternating in the growing chain. They are characterized by the presence of electronegative groups in the molecule. This type of monomers includes: maleic anhydride, methyl fumarate, fumaronitrile, acrylonitrile, methacrylonitrile, methylvinyl ketone and p-cyanostyrene.

Group II of monomers, among the others, includes methyl methacrylate, substituted in the ring styrene derivatives and butadiene These monomers copolymerize easily with styrene but show a weak tendency to alternate.

Group III includes monomers, which hardly copolymerize with styrene. A privileged reaction is that of styrene homopolymerization. Example of this type of monomers are: vinylidene chloride, methyl maleate, trichlorethylene, vinyl chloride and vinyl acetate.

10.3.7 POLY(VINYL CHLORIDE) (PVC)

Poly(vinyl chloride) is produced by the free radical polymerization of the vinyl chloride:

$$n\ CH_2{=}CHCl \longrightarrow -\!\!\left[CH_2-CHCl\right]_n\!\!-$$

The rate of polymerization of vinyl chloride in the presence of peroxide initiator increases gradually to about 40% of the reacted monomer and then remains stable up to about 80% of conversion. At the end of the reaction the rate decreases, what is due to the lack of solubility of the formed polymer in the monomer. Further polymerization takes place in the headed nodules of polymer. The completion of the growth of poly(vinyl chloride) macromolecules takes place mainly by the transfer of chain reaction onto monomer or polymer molecules. The latter reaction is followed by the branching of the chain.

The average molecular weight of poly(vinyl chloride) depends on the method of polymerization used and decreases with increasing initiator concentration and reaction temperature.

Poly(vinyl chloride) is a thermoplastic polymer, resistant to the action of water, acids, bases, mineral oils, hydrocarbons, oxygen and ozone. It dissolves in esters, ketones, chlorinated organic derivatives, tetrahydrofuran, pyridine, and carbon disulfide.

In order to characterize the properties of poly(vinyl chloride) a constant called number K, known also as the Fikentscher's number, was introduced. This constant depends on the size and the shape of macromolecules and is proportional to the molecular weight.

The relationship between the relative viscosity and the K constant is given by the equation:

$$\log\frac{\eta}{\eta_o} = \frac{75K^2}{1+1.5c} + K$$

where η_0 – viscosity of cyclohexanone at a temperature of 25°C; η – viscosity of 1 g PVC solution in 100 cm^3 of cyclohexanone at 25°C; K – Fikentscher's constant; c – concentration.

The K number is derived from viscosity measurements. It is read from special nomograms or from tables for the measured viscosity. Therefore, it is unnecessary to solve the above cited equation to get the value of K. Most often the K values are in the range from 50–80. It corresponds to the average molecular weight between 40,000 and 129,000.

Average values characterizing the properties of PVC:

density [g/cm^3]	1.38
bursting strength [MPa]	45–60
tensile elongation at break [%]	8–20
glass transition temperature [° C]	80
softening temperature [°C]	145–170

The products made from poly(vinyl chloride) can be found in commerce in form of vinidur (hard poly(vinyl chloride)) or viniplast, containing 20–60% of plasticizer. Vinidur is processed at the temperature of 160°C. A small amount of hydrogen chloride is then split off. It catalyzes the decomposition, while causing browning of the material. To avoid degradation, thermal stabilizers are added to PVC in order to bind the evolved hydrogen chloride. Among them the metal carboxylates, for example, lead, cadmium, or calcium, stearates, organotin compounds and organic phosphates are most frequently used.

10.3.8 POLYTETRAFLUOROETHYLENE (PTFE)

The discovery of polytetrafluoroethylene occurred by chance. Dr. Roy J. Plunkett, who worked for Du Pont company, intended to use the tetrafluoroethylene as a non-toxic cooling agent in the manufacturing of compressors for refrigerators. But it turned out that in one container, instead of the gaseous tetrafluoroethylene he found a white powder with a wax appearance. This powder was characterized by a very good chemical resistance and lack of solubility in all known solvents. This discovery coincided with the World War II and the studies conducted on the fission of uranium. General Leslie R. Groves, responsible on behalf of the US army for construction of the atomic bomb, was looking for material to seal the apparatus for the distillation of highly corrosive uranium hexafluoride. It was needed to obtain the uranium isotope with the mass number of 235, necessary to produce the uranium bomb. Because of the high importance of the case he funded the research on polytetrafluoroethylene and implementation of its production.

Currently, known as Teflon, the polytetrafluoroethylene is obtained by free radical, emulsion polymerization of tetrafluoroethylene in the presence of peroxide initiators. This polymerization reaction takes place more easily than the polymerization of ethylene and the emitted heat of reaction is of 196 [kJ/mol].

The reaction proceeds according to the scheme:

$$n\,CF_2{=}CF_2 \longrightarrow \mathrm{-\!\!\!}[CF_2-CF_2]_n\mathrm{\!\!\!-}$$

Polytetrafluoroethylene is a linear polymer with a high degree of crystallinity (80–85%). The polymer is thermoresistant and can be used up to the temperature of 250°C, and in some circumstances even up to 327°C. At this temperature the crystalline structure disappears and the material becomes to be amorphous and transparent.

Physical properties of polytetrafluoroethylene:

density [g/cm^3]	2.1–2.3
glass transition temperature [°C]	–120
bursting strength [MPa]	14–20
bending strength [MPa]	11–14
impact strength [J/m^2]	10^5
cross-resistance, [Ωcm]	10^{18}
dielectric breakdown, [kV/mm]	25–27
dielectric loss tan δ	0.0002
coefficient of friction	0.06–1

Polytetrafluoroethylene is processed by sintering at temperatures of 360–380°C. It is used in aviation, chemical, radiotechnic, radio engineering refrigeration, food and pharmaceutical industries, as well as a coating and fiber-forming material.

One of the unconventional uses of polytetrafluoroethylene are its reactions with metals such as aluminum, magnesium and titanium, which are accompanied by the secretion of large amounts of heat energy in a very short time. This allows to produce very high temperatures, allowing for cutting metal or other materials.

Another application of Teflon powders are the manufacturing methods of submicron silicon carbide or tungsten carbide during powders during the self propagating reaction between silicon or tungsten and the powdered graphite with the addition of Teflon powder.

10.3.9 POLY(VINYLIDENE FLUORIDE) (PVDF)

The poly(vinylidene fluoride) is obtained by the free radical polymerization of vinylidene fluoride.

$$n\, CF_2{=}CF_2 \longrightarrow -\!\!\left[CF_2-CF_2\right]_n\!\!-$$

The polymerization process is carried out by emulsion method in water as well as in aqueous solutions of acetone, alcohols (e.g., t-butyl) or ethers. The polymerization reaction proceeds with good yield at temperatures of 50–130°C and at a pressure of 2.1–8.4 MPa. As initiators potassium persulfate, hydrogen peroxide as well as peroxides of dicarboxylic acids are used. Tetrachlorophthalic and pentachlorobenzoic acids as well as the ammonium salts of chloro- or fluorocarboxylic acids are used as emulsifiers.

A method of suspensive polymerization of vinylidene fluoride in the presence of 0.01–3 wt% of dicarboxylic perdicarbonates at a temperature of 0–50°C at and 4 MPa pressure was also elaborated. In this method the methylcellulose in an amount of 0.1% is used as the suspension stabilizer. Acetone, 2-butanone or saturated linear hydrocarbons containing 3–12 carbon atoms serve as molecular weight regulators. The suspensive polymerization reaction proceeds with the yield of 98%.

The poly(vinylidene fluoride) polymer is thermally stable up to temperatures of 300–350°C (573–623K). Its melting point is 170–180°C. It enables its processing using the typical methods for thermoplastics within the temperature range of 200–250°C. Heating of poly(vinylidene fluoride) to a temperature above 430°C causes its pyrolysis combined with its degradation into short chain fragments of polymer and decomposition with emission of hydrogen fluoride. The poly(vinylidene fluoride) decomposition takes also place during its irradiation with gamma rays in the presence of oxygen. The action of these rays in the absence of oxygen (in vacuum) can cause a cross-linking of polymer.

The poly(vinylidene fluoride) resists to the action of acids, bases, strong oxidizers, halogens and a very large number of organic compounds.

It is not resistant to polar solvents, in which the poly(vinylidene fluoride) dissolves as well as to acetone, ileum and other sulfonating agents at elevated temperature.

The poly(vinylidene fluoride) dissolves in dimethylformamide, dimethylacetamide and dimethyl sulfoxide. The polymer dissolution is carried out at 40–50°C.

PVDF is the hardest plastic material with the highest mechanical strength of all currently known fluoropolymers and can be compared with polyamides and with Penton. It breaks under the tension of 50–60 MPa at room temperature. Its relative tensile elongation at break is of 200–300%. The deflection temperature of poly(vinylidene fluoride) is of 150°C and 98°C under the load of 0.46 MPa and 1.86 MPa, respectively.

The poly(vinylidene fluoride) is currently used as a construction and lining material. It is utilized in the form of tubes and foils in the chemical industry, machine building, construction, precision industry, and medicine.

The polymer is recommended for use in the building of chemical equipment, more particularly for making bearing shells, valves, pipes, cocks, gaskets, carpets for autoclaves, tanks and pumps. It may be also used for fabrication of elements the apparatus resistant to bromine action. It is also used for the manufacture of sealing joints in the rocket devices. The poly(vinylidene fluoride) is also used to make thermoshrinkable hoses (trade name Termofit) designed for the isolation of electronic circuits and electro heating elements.

Metal tapes coated with poly(vinylidene fluoride) are used in architecture as cladding element of buildings for various destinations. Strong foils of PVDF found application for packing chemical reagents, pharmaceuticals and medical instruments.

10.3.10 POLY(VINYL ACETATE) (PVAC)

Poly(vinyl acetate) is obtained by the free radical polymerization of vinyl acetate

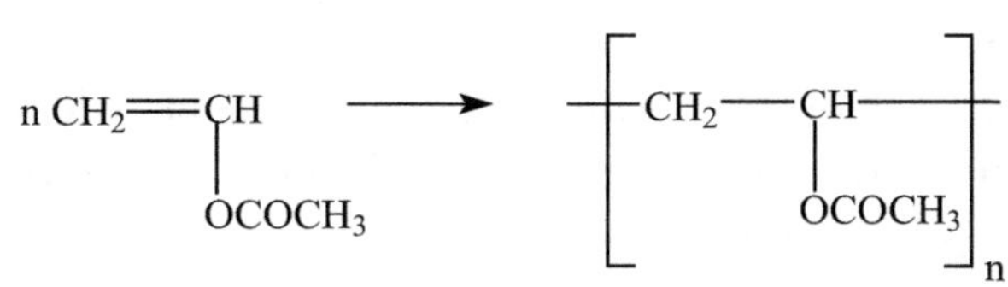

The emitted heat of reaction is of 88 kJ/mol. The period of induction and the kinetics of the process depends on the impurities contained in the vinyl acetate which inhibits the polymerization.

The presence of a tertiary carbon atom in the molecule of vinyl acetate may cause the formation of a branched product at high temperature and high degree of conversion. Branching can occur at both carbon atom of vinyl group and the rest of acetate.

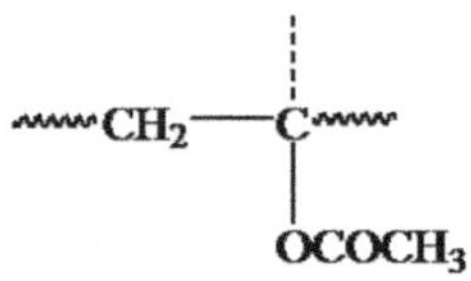

or

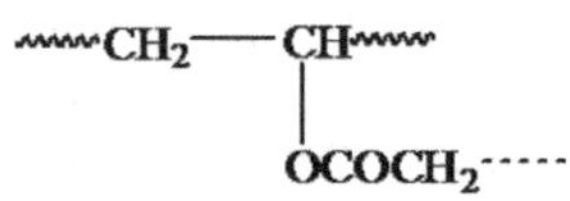

The properties of poly(vinyl acetate) depend on its molecular weight. Polymers of vinyl acetate with a low molecular weight are soft and resinous. With its growth they become to be hard and brittle.

Typical characteristics of PVAC:

refractive index n_D^{20}	1.467
density [g/cm3] (at 20°C)	1.191
specific heat [J/gK]	1.63
breaking strength [Pa]	0.1–0.34
softening temperature [°C]	44–86

Poly(vinyl acetate) is used as an ingredient in adhesives, plastics, coatings, binding masses, as a raw material for the manufacture of poly (vinyl alcohol).

10.3.11 POLY(VINYL ALCOHOL) (PVOH)

PVOH is a vinyl polymer, which is not obtained by polymerization, but by the hydrolysis of poly(vinyl acetate). Under the term poly(vinyl alcohol) one understands also its copolymers with vinyl acetate of low content of acetate groups.

As a result of the ionic polymerization of acetaldehyde, carried out at low temperatures, one obtains poly(vinyl alcohol). However, this method has not found yet a practical use.

The hydrolysis of poly(vinyl acetate) is carried out in a methanol solution in alkaline or acidic medium at 50–70°C. The polymer precipitates spontaneously from the solution at the time when approximately 60% of acetate groups are hydrolyzed.

The reactions proceed according to the scheme:

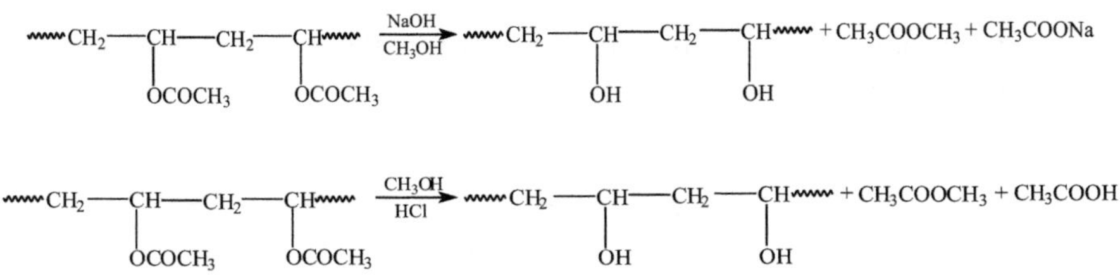

The hydrolysis process under the acidic conditions is slow. The product contains still over a dozen percent of acetate groups. For this reason the hydrolysis reaction is carried out most frequently in an alkaline environment, in which it runs much faster and the degree of hydrolysis depends on the catalyst concentration.

Due to the presence of a large number of hydroxyl groups poly(vinyl alcohol) is soluble in water and insoluble in most of organic solvents.

The solubility of poly(vinyl alcohol) and the properties of its aqueous solutions depend on the percentage of unhydrolysed acetate groups. In the case of atactic polymers the best solubility exhibit the poly(vinyl alcohol) species containing 11–13% of acetate groups. A further degree of hydrolysis makes easier the formation of hydrogen bonds, which in turn cause the formation of crystalline forms and hinder the solubility.

The glass transition temperature of poly(vinyl alcohol) is of 80°C. A further heating leads to changes in the polymer structure, which are due to intra- and intermolecular dehydration reactions leading to the formation of lactones and double as well as ether bonds.

The poly(vinyl alcohol) through the presence of hydroxy functional groups shows a high chemical reactivity. It is relatively easy for performing esterification reactions, etherification, acetalisation, oxyethylation, cyanoethylation, dehydration, chlorination, bromination and reaction with isocyanates.

Poly(vinyl alcohol) is used in the manufacturing of adhesives, finishes, protective colloids, photographic emulsions, as a thickener and gelling ingredient in cosmetics and pharmaceuticals, pipes, tubes, plates, aprons and gloves resistant to gasoline and other solvents, surgical sutures, coatings and membranes, which are only in a small extent permeable for gases. Large amounts of polyvinyl alcohol is processed into polyacetals. PVOH was also used to obtain synthetic fibers.

10.3.12 POLYACRYLAMIDE (PAM)

The polyacrylamide is obtained during the radical polymerization of acrylamide in aqueous solution.

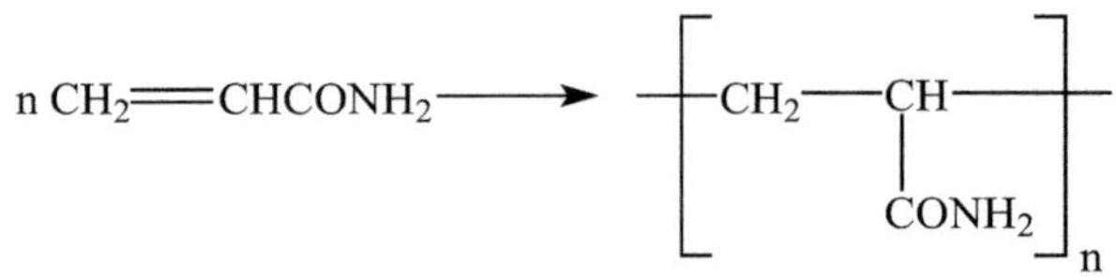

The emitted heat of reaction is of 81.7 kJ/mol. The molecular weight of the formed polymer can be adjusted by choosing the ratio of oxidant to reductant concentrations, used to initiate the reaction in the oxidation-reducing system as well as by the addition of regulators such as the isopropyl alcohol or the sulfur compounds.

A very important parameter in the acrylamide polymerization reaction is temperature. The process should be conducted at a temperature below 60°C, since at higher temperatures the hydrolysis of amide groups to carboxylic ones takes place and as a reaction product one can obtain copolymers of acrylamide with ammonium acrylate.

The industrial process of polyacrylamide synthesis is carried out in a reactor with a mixer fitted with a heating – cooling system in an aqueous medium or in water – isopropanolic solution. As initiators a redox system consisting of potassium persulfate and sodium hydrogen sulfate is used, It allows to carry out the reaction at an appropriate temperature during 5–7 hours at pH = 7–85. The course of reaction is determined by following the amount of the emitted heat. The cessation of the temperature growth in reactor signals the end of the process.

Is also possible to carry out the polymerization of acrylamide in the presence of potassium persulfate at the temperature of 75°C within two hours. The product separates from the aqueous solution by precipitation with acetone or methanol.

Polyacrylamide is a white powder. Because of its polar groups it is easily soluble in water and is insoluble in acetone, alcohols and hydrocarbons. It mixes well with surface – active substances as well as with many other water – soluble polymers. Similarly as poly(vinyl alcohol) the polyacrylamide can be plastified with glycerine. Heating polyacrylamide to a temperature above 373K (100°C) leads to changes in the structure combined with emission of nitrogen.

Polyacrylamide is a reactive compound and can be chemically modified. After a treatment with formalin or glyoxal the polymer becomes to be insoluble in water. As a result of Hofmann degradation under the influence of sodium hypobromite 94% of amide groups convert into amino groups with the formation of poly(vinyl amine).

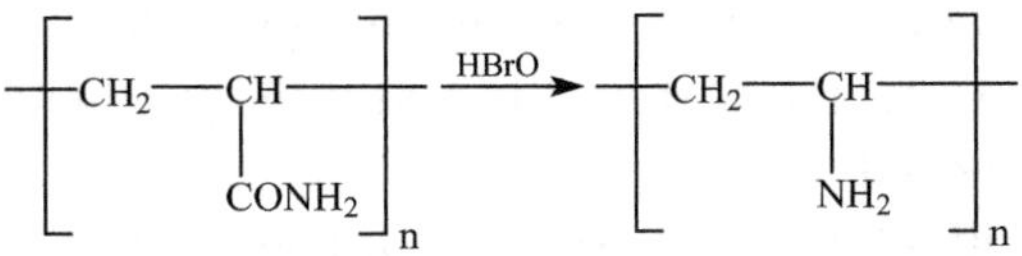

During the reaction with 80% of hydrazine the polyacrylic hydrazid is formed, whose degree of conversion is of 85%, as determined by the iodometry method

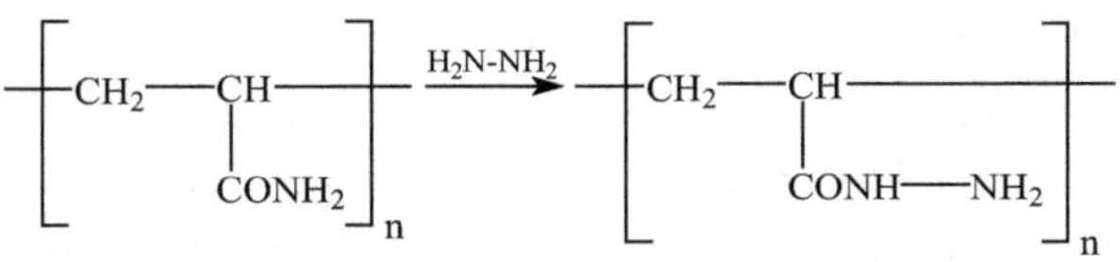

By the condensation of polyacrylamide with formaldehyde one obtains a hard product with a high melting temperature. It is suitable to use as a binder for brake discs.

Acrylamide polymers and copolymers are used as flocculation means serving for suspensions deposition in the process of clarification of waste water and improving the quality of drinking water. They are also applied in soil conditioning, to improve its structure, in the manufacture of adhesives, dispersants, as auxiliaries for textile, thickening agents as well as to stabilize the natural rubber, latex and poly(vinyl acetate) emulsion. Polyacrylamide is a better emulsion stabilizer than poly(vinyl alcohol).

10.3.13 POLY(METHYL METHACRYLATE) (PMMA)

The poly(methyl methacrylate) is obtained by a free radical polymerization of methyl methacrylate. The polymerization reaction is exothermic and proceeds according to the scheme:

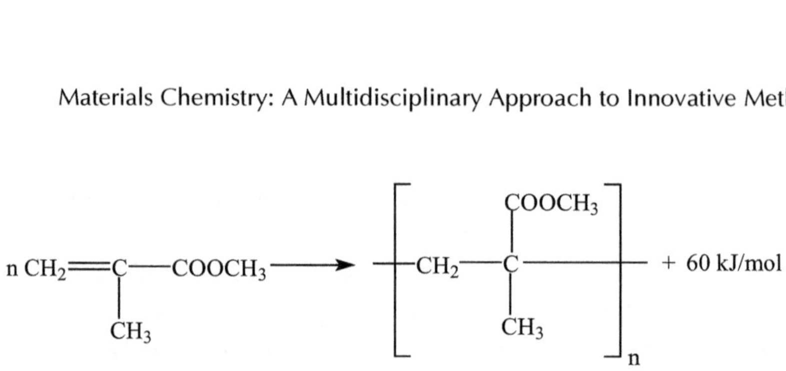

A characteristic property of poly(methyl methacrylate) is its good transparency. It allows its use as an organic glass, known under the commercial name of Plexiglass. For this reason its production technology is tailored to those needs.

The bulk method of polymerization of methyl methacrylate is carried out in a periodic way. In the first stage one obtains a prepolymer, formed by the initial polymerization of methyl methacrylate at the presence of benzoyl peroxide or by dissolution of pieces of poly(methyl methacrylate) waste in monomer. The content of polymer in prepolymer is of 5–10%.

The obtained prepolymer is used to fill the forms made from plates of mirror glass, separated by dividers, carefully cleaned and covered by a thick paper. Forms are filled to 98–99% of their volume. Then the inlet is sealed and the prepolymer is subjected to polymerization in these cells by raising gradually the temperature from 298 to 363 K (25–90°C). After completion of the polymerization the forms are exposed to hot water which detach the paper and facilitates separation of glass panels from the formed sheets of poly(methyl methacrylate). Obtained in this way polymer plates are subject to a technical control in order to eliminate the products with blisters. Then both sides of the PMMA plates are wrapped with thin paper, protecting them against scratches.

Currently, on the industrial scale, this method has been displaced by the method of suspensive polymerization of methyl methacrylate and the organic glass plates are obtained by extruding the polymer in extruding machines with a nozzle slot.

Properties of poly(methyl methacrylate) are as follows:

density [g/cm^3]	1.18
refractive index n_D^{20*}	1.58–1.69
absorptiveness of water [%]	0.25
tensile strength [MPa]	60–70

softening temperature (Vicat) [°C]	105–110
flow rate, [g/10 min]	0.8
cross-resistance [Ω m]	10^{16}
dielectric loss factor	
tan δ at 50 Hz	0.07

*index of refraction at the wavelength of sodium doublet line (589 nm) and at 20°C.

10.3.14 POLY(N-VINYL-PYRROLIDONE) (PVP)

Poly(*N*-vinyl-pyrrolidone) is formed as a result of the free radical polymerization of N-vinylpyrrolidone in aqueous or alcohol solution. Its average molecular weight is in the range 10,000–360,000.

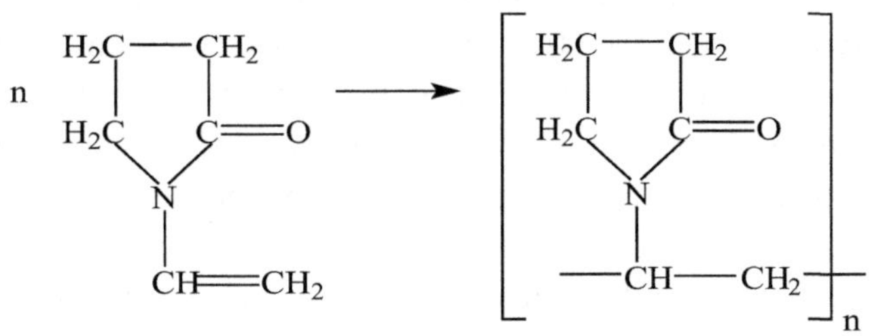

In industry the polymerization process of N-vinylpyrrolidone is most frequently carried out in the 30-percent aqueous solution at a temperature of 50–70°C and in the presence of hydrogen peroxide as the initiator and ammonia as the activator (regulator). One uses 0.1 wt% of hydrogen peroxide and 0.1 wt% of ammonia with respect to the amount of monomer. The reaction proceeds at this temperature during 2–3 hours.

Poly(N-vinyl pyrrolidone) is isolated from the post reaction solution by precipitation in acetone, salted out with sodium sulfate or spray-drying in an oven.

Poly(N-vinyl pyrrolidone) is a colorless, amorphous, hygroscopic polymer with the density of 1–190 kg/m^3. It is well soluble in water, alcohols, higher ketones, organic acids and in aromatic and chlorinated hydrocarbons. It does not dissolve in ethers as well as in aliphatic and alicyclic hydrocarbons.

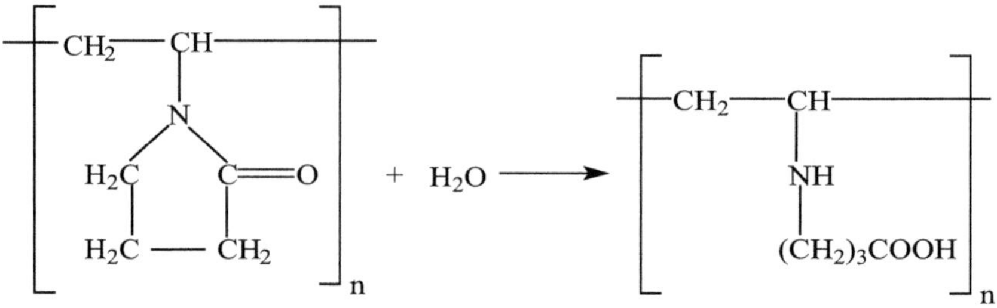

The polymer softens at temperature 140–169°C, but it becomes insoluble and decomposes at the temperature of 230°C. The aqueous solutions of this polymer are weakly acidic (pH ~ 5) and are characterized by persistence in a neutral and weakly acidic environment, even when heated to the temperature of 100°C.

In an alkaline medium poly(N-vinyl pyrrolidone) undergoes a hydrolysis associated with the formation of the poly-N-vinyl-γ-aminobutyric acid.

Poly(N-vinyl pyrrolidone) forms complexes with many inorganic and organic compounds, in particular with dyes, vitamins, medicines and poisonous substances. It is characterized by the miscibility with many resins, polymers and plasticizers.

One of the main applications of poly(N-vinyl pyrrolidone), produced under the trade names Subtosan and Periston is the production of synthetic plasma (blood plasma substitute) It is possible due to its non-toxicity and suitable speed of excretion from the body. This preparation was extensively tested and used on a large scale in hospitals in Germany during the World War II. The poly(N-vinyl pyrrolidone) solutions are also used to treat the poisoning occurring in children with diphtheria, diabetes insipidus, a fatty degeneration of the kidneys and other ailments. The Periston type solutions have particular advantages in the treatment of shock caused by heavy burning, when the blood may be concentrated due to the loss of water and of plasma by the walls of damaged capillary cells. These solutions serve also to strengthen the injections of insulin, penicillin, novocaine, hormones, salicylates and other drugs.

Poly(N-vinyl pyrrolidone) is also used as a binding agent in the pharmaceutical industry in the tableting of drugs, the production of hair lacquers, to bleach badly dyed fabrics and as an activator to improve the capacity of synthetic fibers for staining. It is also used as a component of adhesives.

There exists also a cross-linked form of PVP known as cross-linked polyvinyl pyrrolidone (PVPP), crospovidone or crospolividone. PVPP is insoluble in water but still absorbing it. It swells very rapidly, generating a swelling force. Therefore, PVPP is used as a disintegrant in pharmaceutical tablets. It is also used to bind impurities to remove them from solutions. In beer processing PVPP is used to remove polyphenols to clear them and to stabilize foam. Because of its binding properties PVPP is also applied as a drug against diarrhea.

10.3.15 POLY(N-VINYLCARBAZOL) (PVK)

The poly(N-vinylcarbazol) is obtained by the polymerization of n-vinylcarbazole in the presence of ionic polymerization catalysts and radical initiators.

The polymerization proceeds according to the following scheme:

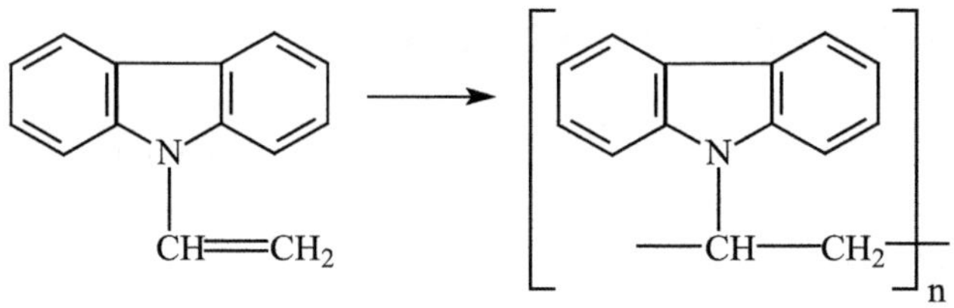

The rate of polymerization reaction depends on the purity of the monomer and on the amount of the added initiator or catalyst.

Poly(N-vinylcarbazol) is a chain polymer, in which predominates the "head to tail" type structure. It is characterized by a high degree of crystallinity, suggesting a low branching and a more symmetrical structure. Due to the presence of large chain rings that prevent a free rotation of chain segments its glass transition temperature is high, of about 426 K (173°C).

Poly(N-vinylcarbazol) is characterized by a high heat resistance. It is weakly polar, hydrophobic, chemically resistant and exhibits good insulating properties. PVK is soluble in ketones, esters, aromatic and chlorinated hydrocarbons.

PVK is used in electrical engineering as a substitute for mica and asbestos. In chemical industry it is used for the manufacture of chemical equipment resistant to bases, acids and fluorinated compounds at a temperature of 393 K (120°C).

Vinylcarbazole copolymers, with acrylonitrile, replace the printing metal in printing industry.

The recent applications of PVK base on the use of the electric, photoelectric and photoconducting properties of charge transfer complexes of poly(N-vinylcarbazol) with a photosensitizer like trinitrofluorene (TNF). The PVK–TNF complexes are used in electrophotography, the manufacture of thermistors, phototermistors, as elements of thin film electronic components, photorefractive materials and as elements of optical memories. These are the prospective applications of poly(N-vinylcarbazol) in which the production costs and the product price are not crucial, for example, the development of manufacturing technologies is not hindered by the economic considerations. The production of this polymer shows an upward trend.

10.3.16 POLY(VINYL ETHER) (PVE)

In contrary to alkyl-vinyl and aryl-vinyl esters the alkyl-vinyl ethers do not polymerize in the presence of free radicals but polymerize by the cationic mechanism.

The reaction proceeds according to the scheme:

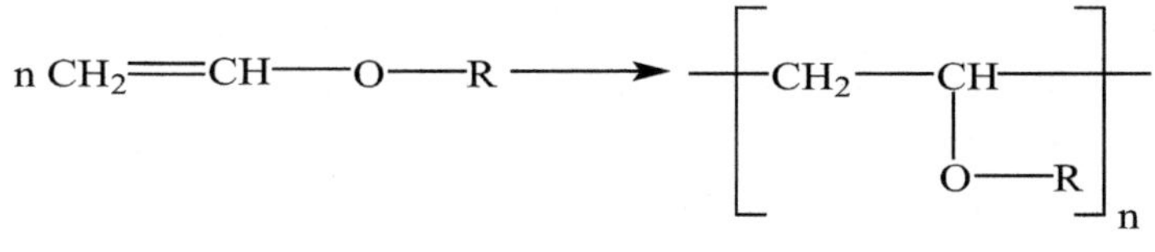

As catalysts metal fluorides and chlorides, such as BF_3, $SnCl_2$, $SnCl_4$, $FeCl_3$ and $AlCl_3$, are used.

During the room temperature polymerization of alkyl-vinyl ethers liquid or viscous oligomers are formed. At lower temperature, below 263 K (–10°C) one obtains solid, similar to rubber, polymers. The polymerization reaction is the most frequently carried out in solution, in liquid propane.

The molecular weight of vinyl polymers depends on the type of catalyst, its concentration and on the reaction temperature.

In industry the polymerization of vinyl ethers is carried out by the batch method. The reaction proceeds in an apparatus fitted with stirrer, heating-cooling jacket and a reflux condenser. In the polymerization of n-butyl ether-vinyl the ferric chloride (III) solution in butyl alcohol, in an amount of 0.03–0.15 wt% is used as catalyst. After reaction the formed polymer is detached from the catalyst and dried.

During the low temperature (about 200 K) bulk polymerization of alkyl-vinyl ethers stereoregular polymers with high mechanical strength are formed.

Polymers of vinyl ethers are produced both as concentrated solutions in hydrocarbon solvents, in aliphatic and aromatic hydrocarbons, as well as stabilized solids.

Known are also vinyl ether copolymers with various vinyl monomers, which are obtained by radical polymerization initiated with peroxides.

The copolymerization is conducted mostly by the suspensive emulsion method.

The poly(vinyl ethers) are well soluble in organic solvents except ethyl alcohol. Poly(methyl vinyl ethers) are soluble even in cold water. PVE's are resistant to the action of aqueous solutions of acids and bases. They adhere to glass, metal, wood, fabrics and other materials. They are used in production of paints and lacquers, in treatment and impregnation of fabrics for the manufacture of artificial leather, adhesives, plastyfying additives and as thickeners. Poly(butyl vinyl ether), known under the trade name of Szostakowski balsam, is used in medicine for the treatment of burns.

Copolymers of vinyl ethers with maleic acid are used as additives in lacquers and impregnating agents of electrical insulation materials. Copolymers of vinyl ethers with methyl methacrylate can be used as adhesive layers in the production of safety glass.

10.3.17 IONOMERS

The term ionomers denote polymers, which in their macromolecule contain a small (up to 8%) number of ionic groups.

The first ionomers, which have found industrial importance, are copolymers of ethylene and sodium acrylate with trade name "Surelen A".

Polymers of this type are obtained by a partial neutralization of α – olefin copolymers with acrylic acids by metal compounds of I, II and III group of periodic table. Due to that only a partial ionization of carboxyl groups, which is a measure of the degree of ionization of the system, takes place.

In ionomer molecules the non-neutralized carboxyl groups form hydrogen bonds and the metal cations ionic bonds with carboxyl anions, respectively. It causes, in particular, the strengthening of intermolecular bonds. The degree of ionization of such a polymer, for example, the fraction of ionized carboxyl groups, contributes substantially to the properties of the product. Introduction to the chain of the ionic poly(α-olefin) groups have a particular impact on the morphology of crystal structures of these polymers. With the increasing content of carboxyl groups the structure changes from the spherulites type to the amorphous one.

Selected properties of poly(α-olefin) ionomers:

density [g/cm^3]	0.93–0.96
hardness (after Shore) [MPa]	60–65
tensile strength, [kG/cm^3]	245–390
relative elongation at break [%]	200–600
modulus [kG/cm^3]	1,800–5,200
shrinkage under pressure [%]	0.3
friction coefficient for the foil	0.6–0.9
softening temperature (after Vicat) [°C]	71–96
[K]	344–369
maximum processing temperature [°C]	330
[K]	603
dielectric constant	2.5
dielectric loss factor tan δ	0.0015
specific resistance [Ωcm^3]	0.5×10^{17}
Dielectric breakdown [kV/cm]	2–250

Ionomers have a higher tensile strength of the starting copolymers and have also good chemical resistance. At room temperature they are insoluble in organic solvents. At elevated temperatures they dissolve partially only. In some cases to obtain a ionomer solution one can use a mixture of such solvents as decaling and dimethylacetamide. The extremely high resistance of these polymers to the action of oils and greases make them a good material for packaging. Ionomers are characterized by absorption of water. The moisture deteriorates their mechanical properties.

Ionomeric materials are suitable for the manufacture of numerous products by extrusion, injection, blow molding and vacuum forming.

The olefin ionomers have found the largest application in production of packaging for food and for medicines. The packaging materials made from ionomers are transparent and highly resistant at low temperatures. They are resistant also to abrasion and they are more resistant to puncture with a sharp object than films made from other polymers.

The large adhesion of ionomers to metals and nonmetals allows to produce the laminates, which are characterized by durability and elasticity.

The ionomers of this type can be used to fabricate the protective glass, fancy goods, shoe soles, gaskets, helmets, syringes, insulating materials, etc.

Ionomers with α-olefins are characterized by easily connecting to other polymers, pigments and fillers. A color effect of ionomer with a simultaneous transparency can be obtained by a proper selection of ions or of mixtures of metal ions used to neutralize the carboxyl groups. The miscibility of ionomer with fillers is so good that the fivefold increase in material stiffness does not decrease its toughness. However, when selecting the fillers, antioxidants, stabilizers and other additives a special attention has to be paid to avoid that they react with the cross linking ionic bonds.

10.3.18 TERPOLYMERS ETHYLENE/PROPYLENE/CARBON OXIDE

Terpolymers are obtained as a result of copolymerization of ethylene and propylene with carbon monoxide. Its structure is that of an aliphatic polyketone in which the carbon monoxide mers in chain are arranged alternately with the olefin mers as presented in the following scheme:

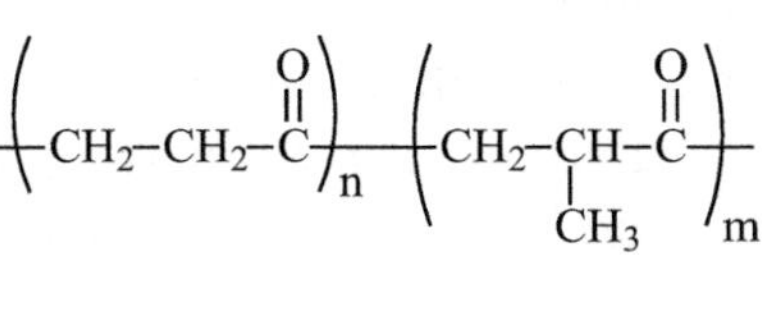

The obtained terpolymer is partially crystalline with a density of 1.235 g/cm^3 (density of the amorphous phase is 1.206 g/cm^3), the melting temperature of 493K (220°C) and the glass transition temperature of 288 K (15°C). The polymer contains 30–40% of crystalline areas. The packing density of atoms in the amorphous phase is larger than in other polymers of this type. Thus the terpolymer exhibits good barrier properties. This material is resistant to most organic solvents and to not highly concentrated acids as well as bases.

The poly(ethylene/propylene/carbon monoxide) is a new plastic that can be used to build the plant for fuel in cars, for the manufacture of parts of mechanisms (including gears), and in fabrication of electrical and electronic equipment. It is produced by the company "Shell Chemical Company" under the trade name "Carilon."

10.4 OTHER POLYMERIZATION METHODS

10.4.1 MIGRATION POLYADDITION

In contrary to the previously discussed polymerization methods of vinyl compounds the migration polyaddition is not based on the polarization of the unsaturated bonds. It is due to the displacement of the mobile hydrogen atom originating from another reactive chemical compound to this bond.

A typical example of such a reaction is the synthesis of polyurethane from diisocyanate and glycol. The reaction of the isocyanate group with the hydroxyl group is as follows:

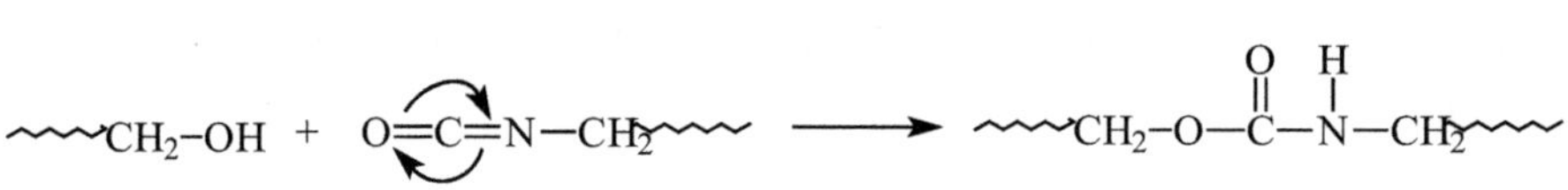

The reaction is a gradual process. It means that the molecular weight of polymers obtained by this method increases steadily during the subsequent reaction steps and the growing macromolecule after each act of addition is completely stable.

However it is not a reversible reaction. The addition of successive monomer molecules to the growing chain is very fast. Therefore, the separation of the simple addition products such as dimers and trimers is very difficult. For this reason the reaction is very similar to the previously discussed reaction of polymerization. Migration polyaddition catalysts are salts, acids, bases (III-row amines) and water. Linear polymers of high molecular weight can be obtained only in the case of equimolecular ratio of reactants, namely diisocyanate and glycol. In the case of excess of one of them the macromolecules with the same end groups are formed. It limits or even prevents the further growth of the chain.

The use in the reaction of monomers containing at least three functional groups results in formation of polymers with cross – linked spatial structure.

Other types of migration polymerization include:

- Synthesis of polyurethanes from diisocyanates and diamines;

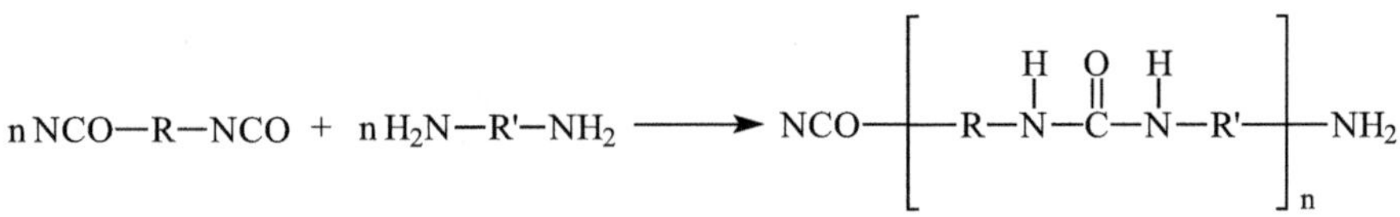

- Synthesis of polyamides by polymerization of N-substituted acrylamide derivatives;

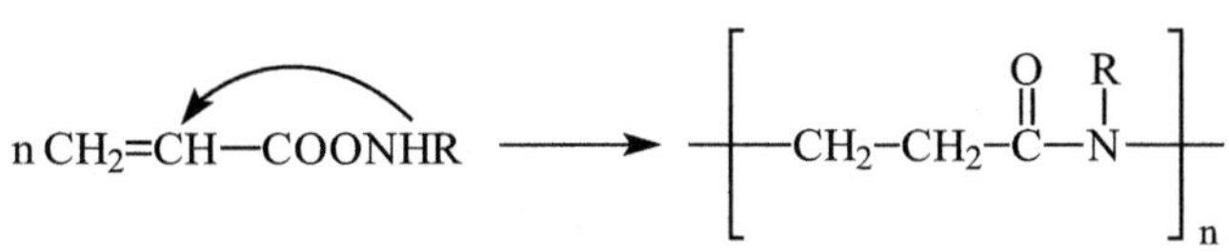

- Synthesis of polyamides by a reaction of carbon suboxide with diamines;

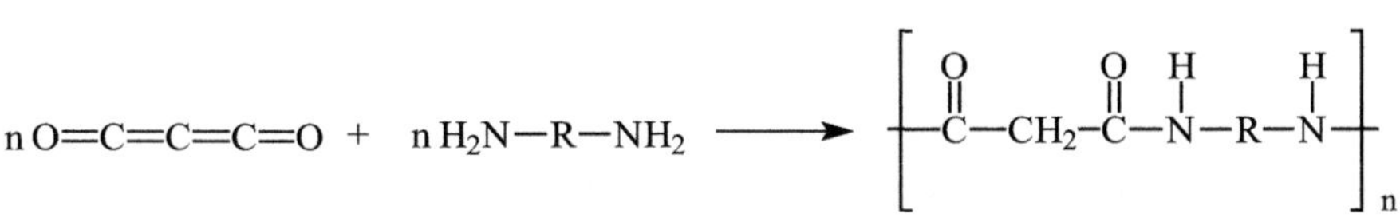

- Polyquaternization of pyridine derivatives

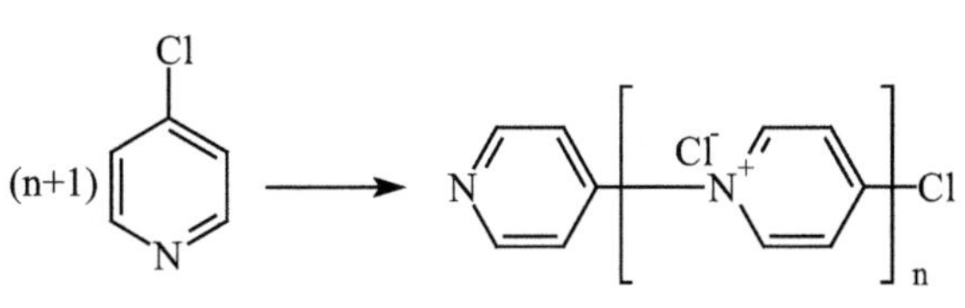

10.4.2 CYCLOPOLYMERIZATION

The cyclopolymerization is a special case of tetrafunctional polymerization of unsaturated compounds containing two double bonds in which there is no cross-linking reaction and a linear polymer containing rings in the main chain is formed.

A classic example is the cyclopolymerization reaction of methyl is the polymerization of N-diallyl diethyloammoniom bromide conducted in the presence of persulfates, and iodine as initiator:

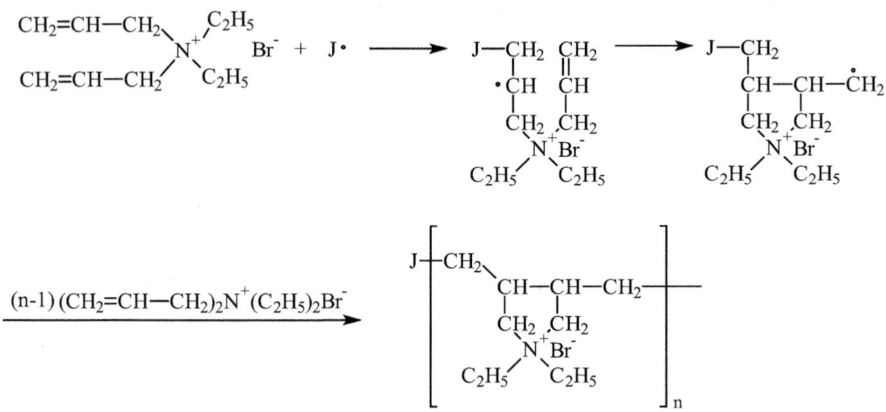

In a similar way runs the cyclopolymerization of anhydrous methacrylic acid:

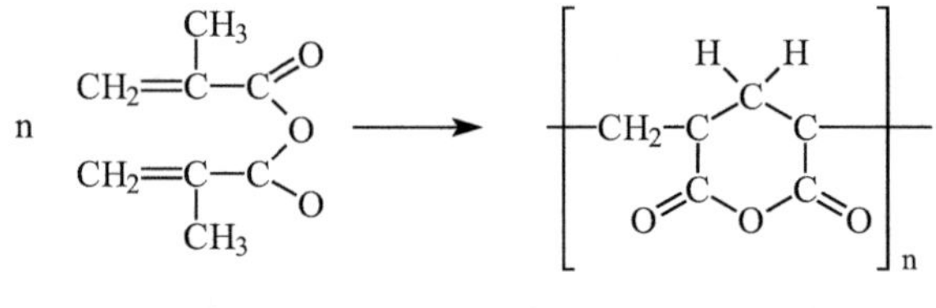

Some of dienes containing isolated double bonds may also polymerize with a formation of cyclic polymers. This reaction takes place when there is a possibility of formation of rings with five or six members.

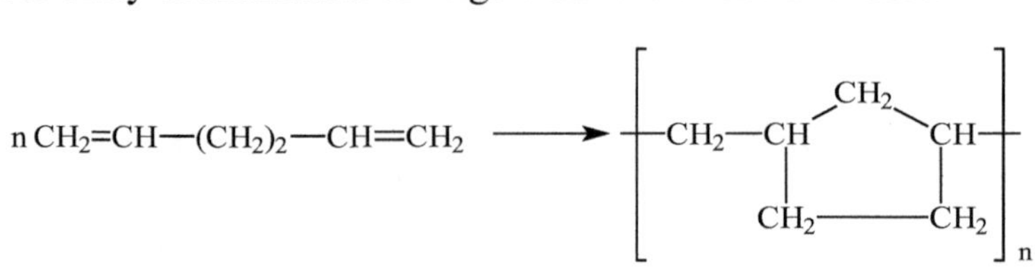

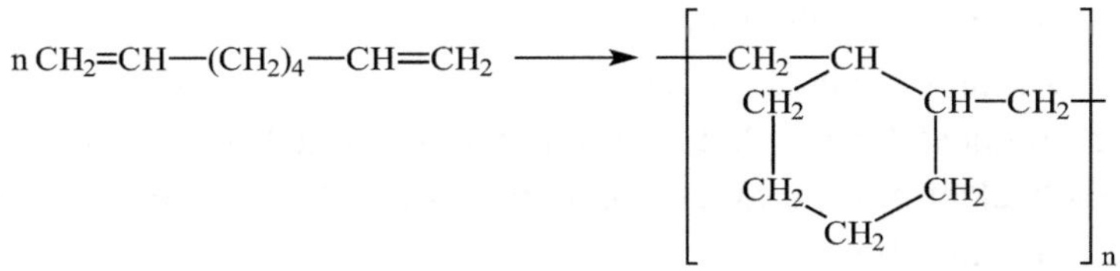

Dienes with isolated double bonds containing less than five methylene groups are sporadically and incompletely cyclizated. The formed polymers contain in the chain a number of multiple member rings and linear segments with the unsaturated bonds in side chains. Such polymers can undergo a cross-linking reaction.

An interesting example of cyclopolymerization is the reaction of divinyl ether with the maleic anhydride:

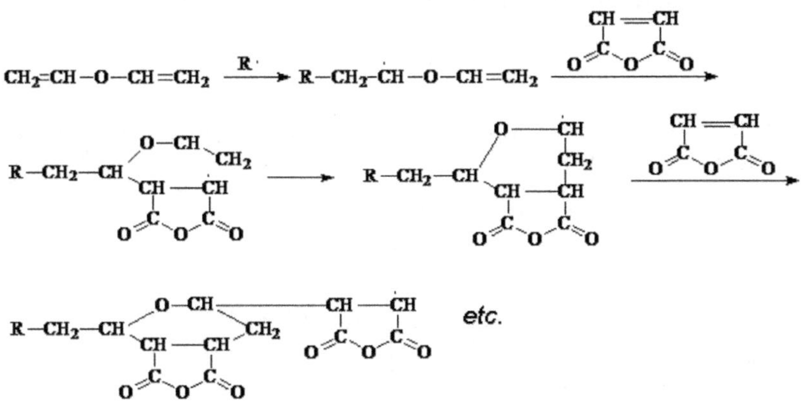

and of divinyl ether with divinyl sulfone that passes spontaneously through the stage of a charge transfer complex:

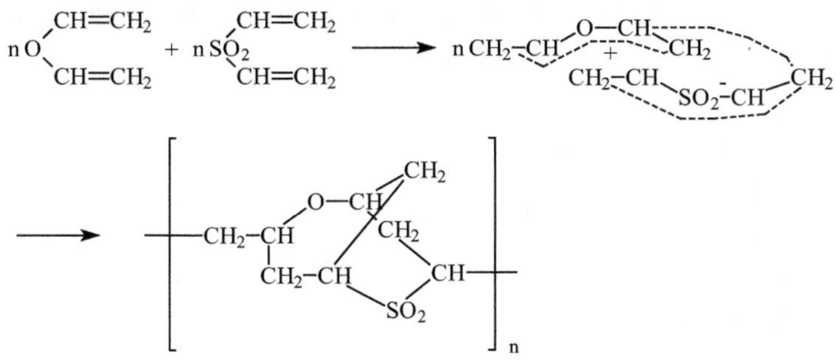

An important method of cyclopolymerization is the cycloaddition polymerization of Diesel – Alder.

The classical Diesel-Alder reaction is the reaction of a nucleophilic diene with electrophilic alkene, called dienophile, with the formation of the cyclohexane ring.

In practice, the most spectacular are dienophiles with double electron attracting substituents to the atoms forming double bonds. The diene reactivity increases if the monomer is in a more reactive cisoidal configuration.

Of course the dienophile used for the synthesis of polymers cannot be a simple alkene, but must be a ternary functional monomer, for example, it has to contain two unsaturated bonds.

The Diesel-Alder polymerization allows for the formation of the ladder polymers. An example of such reaction is the cycloaddition of 2-vinyl-buta-1,3-diene to p-benzoquinone:

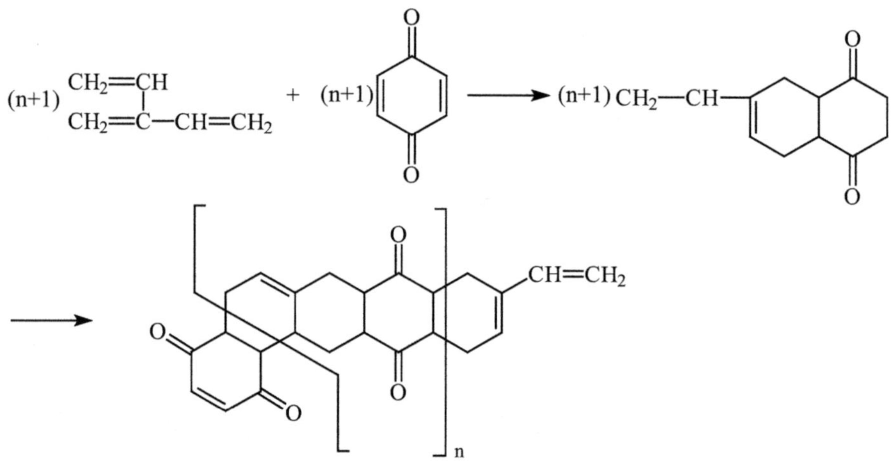

In a similar way reacts N, N'-alkylene-bis (cyklopentadiene) with p-benzoquinone or a suitable bismaleimide.

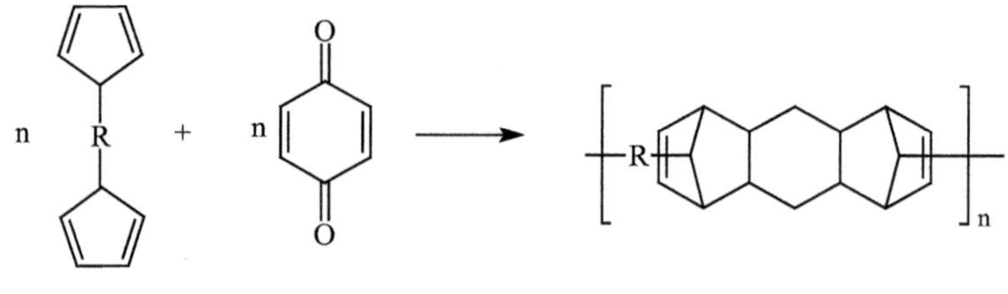

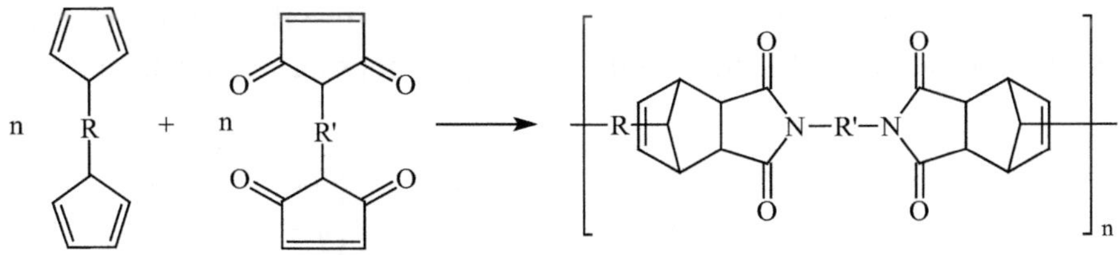

In the cyclopolymerization reaction of Diesel – Alder one can also use monomers being at the same time both a diene and a dienophile. An example of such a reaction is the spontaneous cyklopolimerization of biscyclopentadiene:

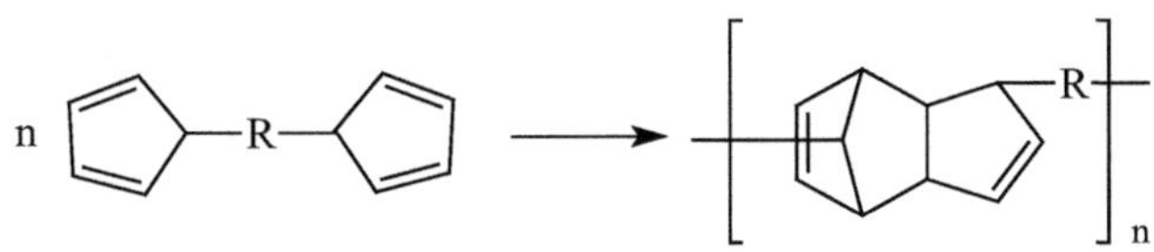

The above-presented structure of the emerging polymer is a probable one. In fact the structure of poly(dicyclopentadiene) is more complex because in the Diesel – Alder reaction different ways of addition or of cross-linking are possible.

10.4.3 OXIDATIVE POLYMERIZATION

The oxidative polymerization is a new method of synthesis of polymers from phenols or from aromatic amines. As it turned out, the oxidation of phenols leads to the formation of free radicals which react together to form a poly(phenylene oxide) (PPO) or its derivatives. This reaction is carried out especially easy when using phenol derivatives having substituents at positions 2 and 6. As oxidizing agents mild oxidants such as potassium ferrycyanide (hexacyanoferrate (III)), lead (II) oxide and silver oxide may be used. However, the most frequently the oxidation is performed with oxygen from the air in the presence of catalysts derived from copper (II) chloride and pyridine or other tertiary amines. Probably the alkaline salt of copper, obtained by oxidation of copper (I) chloride and coordinatively bound with two amino groups can be used as an active catalyst too.

The resulting catalytic complex reacts with the dialkylphenol. The reaction is associated with the formation of an appropriate phenolate:

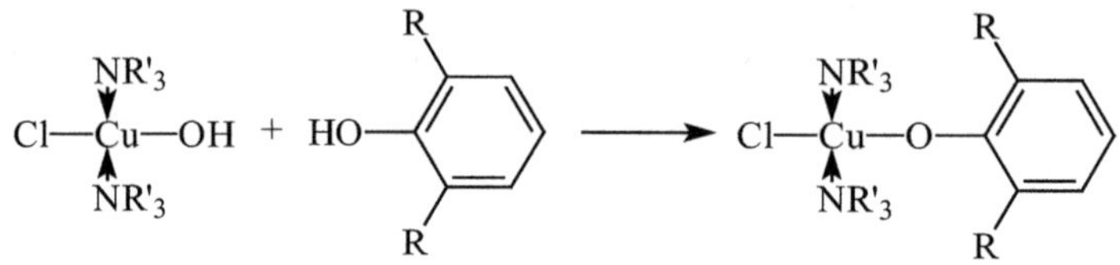

The formed phenoxyl anion is oxidized in a reaction to two mesomeric radicals, which recombine together with the formation of a dimer:

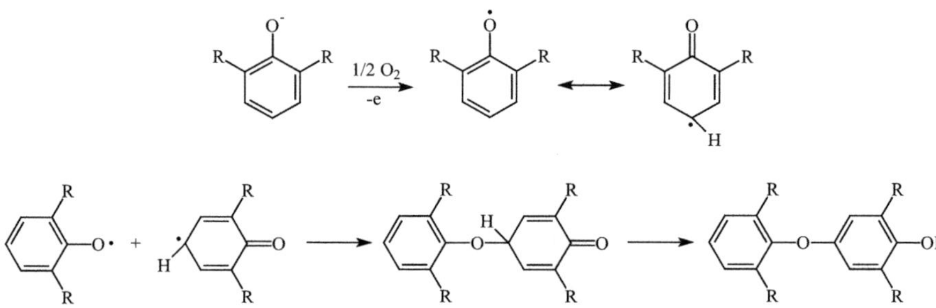

The resulting dimer reacts again with the catalytic complex and the reaction is repeated successively, until the formation of polymer. So the chain is created in a progressive manner and a large distribution of molecular weight is result of the transfer of hydrogen atom reaction.

Known are also the ways to carry out the oxidative copolymerization of different phenols. For example, it was shown that during the oxidative copolymerization of 2,6-diphenylphenol with 2,6-dimethylphenol one can obtain block copolymers as a result of the initial oxidation of 2,6-diphenylphenol and after its exhaustion the next introduced portions of 2,6-dimethylphenol. Changing the order of addition of these monomers or use of their mixtures in the process leads to the formation of a statistical copolymer.

In recent years a number of research works were devoted to the oxidative polymerization process of aromatic amines. In fact the resulting polymers react readily with acids, forming conducting ionic polymers.

Aniline and its derivatives, substituted in the phenyl ring, can polymerize by oxidizing both chemically and electrochemically. Most frequently the oxidation is performed with the solution of ammonium persulfate in an acidic environment.

The reaction yield and some properties of ionic polymers obtained by the oxidation of aniline and its derivatives are given in Table 10.20.

Molecular mass of synthesized polymers depends on the reaction conditions and on the nature of monomer used.

Table 10.21 compares the molecular mass of polyanilines obtained by the chemical and electrochemical oxidation method. Polymers formed in the electrochemical process are characterized by a lower molecular mass. Similarly, the increase in volume of the substituent and its distance from the amino group lowers the efficiency of polymerization products with a simultaneous decrease of the molecular mass of the formed polymer.

TABLE 10.20 Reaction Yield and Some Properties of Ionic Polymers Formed During the Oxidation of Aniline and Its Derivatives by Ammonium Persulfate

No.	Monomer	Yield [%]	Content of ionic systems Cl/N	Electrical conductivity [Ω^{-1}cm^{-1}]
1	Aniline	82	0.44	5
2	2-methylaniline	80	0.65	0.3
3	3- methylaniline	29	0.7	0.3
4	2-ethylaniline	16	0.68	1
5	3-ethylaniline	1	—	0
6	2-n-propylaniline	2	—	0

*After M. Leclerc, J. Guay, L.H. Dao: Macromolecules 22, 649 (1989).

TABLE 10.21 Molecular Mass of Polyanilines Obtained by the Methods of Chemical and Electrochemical Oxidation

No	Polymer	Molecular mass of polymers	
		Oxidized chemically	Oxidized electrochemically
1	Polyaniline	80,000	9,000
2	poly(2-methylaniline)	7–000	4,300
3	poly(3-methylaniline)	—	4,000
4	poly(2-ethylaniline)	5,000	—

*After M. Leclerc, J. Guay, L. H. Dao: Macromolecules 22, 649 (1989).

The study of the reaction mechanism of oxidative polymerization of aniline has shown that, in the first stage, a split of proton takes place, associated with the formation of the nitrene radical cation ($C_6H_5N_2.^+$). As a result of the electrophilic substitution of the formed t nitrene cation to the next molecule of aniline the substrates dimerization to p-aminodiphenylamine takes place. Then, similarly as in the case of oxidation of phenols, the process is repeated several times until the formation of the polymer. The course of the synthesis of polyaniline may be shown schematically as follows:

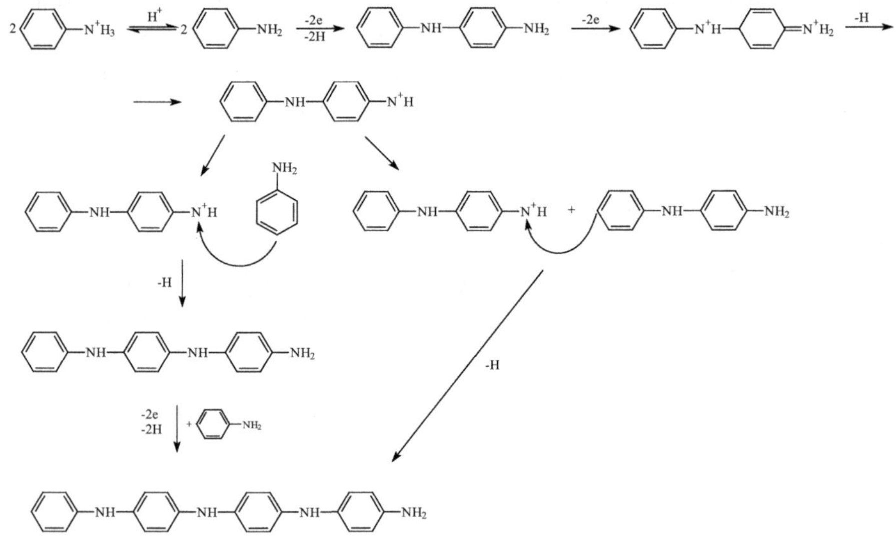

The oxidative polymerization method was also applied to the synthesis of conductive poly(heterocycles) like: polypyrrole, polythiophene, polyfuran, polyselenophene and polypiridazine.

For example, during the oxidative polymerization of piridazine, a good oxidizing agent may be iodine. The reaction is conducted in a polar solvent such as, for example, acetonitrile and at room temperature or at 0°C.

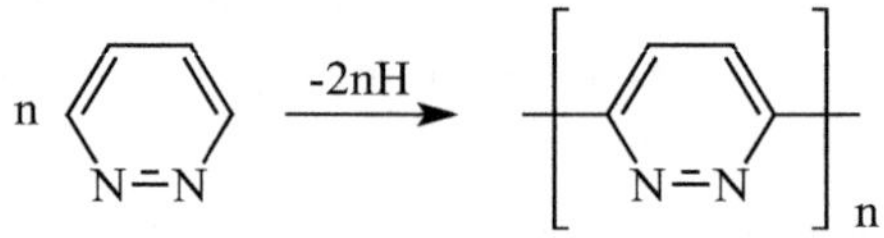

The resulting polymer forms a charge-transfer type complex with iodine. Complexes of poly(polypiridazine-J) are black with electric conductivity of 2–3 S/cm.

10.4.4 POLYMERIZATION OF ALDEHYDES

Aldehydes are highly reactive compounds, undergoing easily the oxidation and reduction reactions. In particular, equally in acidic and in alkaline environment the aldehyde molecules react with each other and, depending on the process conditions, they are subject to aldol condensation, cyclization, oligomerization or polymerization.

An example of the aldol condensation is the synthesis of crotonaldehyde, running through the stage of dimerization of the acetaldehyde as shown below:

$$2\ CH_3{-}CHO \longrightarrow CH_3{-}CHOH{-}CH_2{-}CHO \longrightarrow CH_3{-}CH{-}CH{-}CHO + H_2O$$

In an acidic environment the aldehydes containing up to 10 carbon atoms in the molecule undergo easily the cyclization process and most frequently the cyclotrimerization:

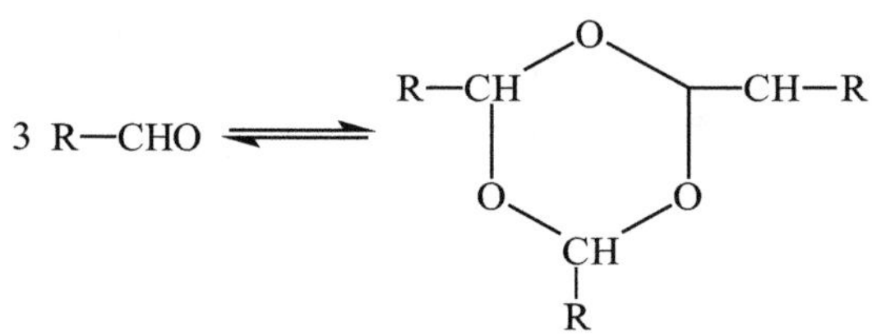

A pure trimmer of this type doesn't undergo either depolymerization during the distillation or during the storage.

However, after adding an acid catalyst it can depolymerize to the starting aldehyde state.

At temperatures below 273 K in acidic environment the aldehydes polymerize with the formation of crystalline oligomers (most frequently the tetramers) called meta aldehydes. Aldehyde oligomers heated to the temperature above 100°C depolymerize to the starting compound.

By using appropriate catalysts and the conduction process conditions one can obtains from many aldehydes polymers with high molecular weight. The resulting polymers can be made resistant to temperature by blocking the terminal hydroxyl group by esterification, most frequently with the acetic anhydride.

Polymerization of aldehydes takes place by opening the double carbon–oxygen bond, whose energy is almost four times smaller than that of the double carbon–carbon bond and is about 20.9 kJ/mol.

Polymerization of aldehydes takes always place below the limit temperature T_c, at which the free energy of polymerization is equal to zero.

The characteristic thermodynamic values of the enthalpy and the entropy changes at the limit temperature measured during the polymerization of some aldehydes are listed in Table 10.22.

The aldehyde polymerization reaction process is always exothermic. The heat of the most frequently performed formaldehyde polymerization is of 71.23 kJ/mol.

The mechanism of this process in an acidic environment consists on addition of a proton to the oxygen atom of carbonyl group with the formation of an active carbocation. This carbocation reacts with other aldehyde molecules leading to the formation of polymer molecule:

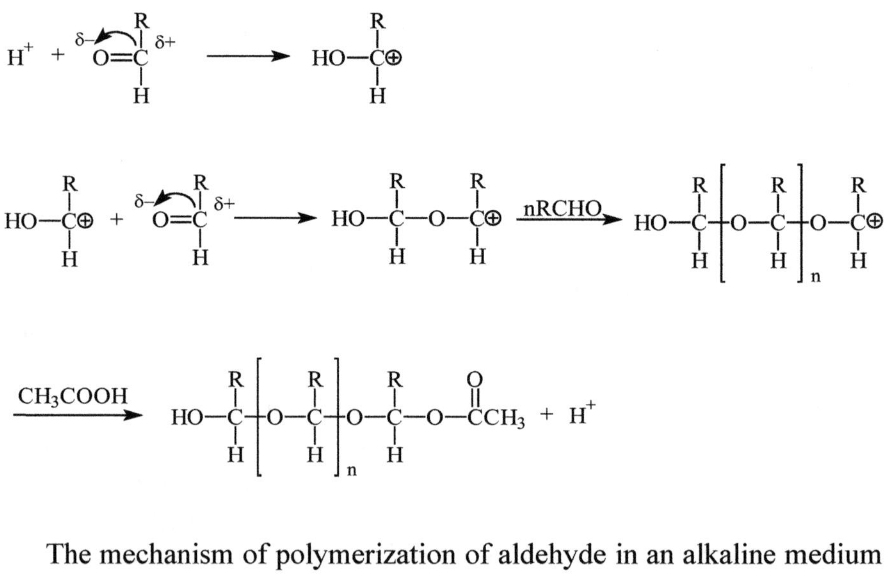

The mechanism of polymerization of aldehyde in an alkaline medium is based on the attachment of the Nu- nukleofile to carbon atom and creating in this way a negative charge on the oxygen atom. The resulting anion reacts with the next aldehyde molecules with formation of a macroanion, which can be stabilized in reaction with the acetic anhydride.

TABLE 10.22 Thermodynamic Data for Polymerization of Aldehydes

No.	Aldehyde	Chemical formula	Solvent	–ΔH [J/mol•10^{-3}]	–ΔS [J/mol•K]	Tc* = $\Delta H_{pol}/\Delta S_{pol}$ [K]
1	acetalaldehyde	CH3CHO	—	27.6	118	234
2	butyraldehyde	C3H7CHO	hexane	35.5	122	—
3	isobutyr aldehyde	C3H7CHO	—	46	—	—
4	isobutyr aldehyde	C3H7CHO	tetrahydrofuran	15.5	74	—
5	valeralaldehyde	C4H9CHO	—	22.6	97	231
6	methoxypropionic aldehyde	CH3OCH2--CH2CHO	—	19.7	82	238
7	trichloroacetic aldehyde	CCl3CHO	tetrahydrofuran	14.6	52	284
8	trichloroacetic aldehyde	CCl3CHO	toluene	37.8	134	282
9	trifluoroacetic aldehyde	CF3CHO	toluene	54.9	155	354
10	tribromoacetaldehyde	CBr3CHO	toluene	19.1	100	196

*Tc – limit temperature at which the free energy of polymerization is equal to zero (from P. Kubis, K. Neeld, J. Starr, O. Vogl: Polymer 21, 1433 (1980)).

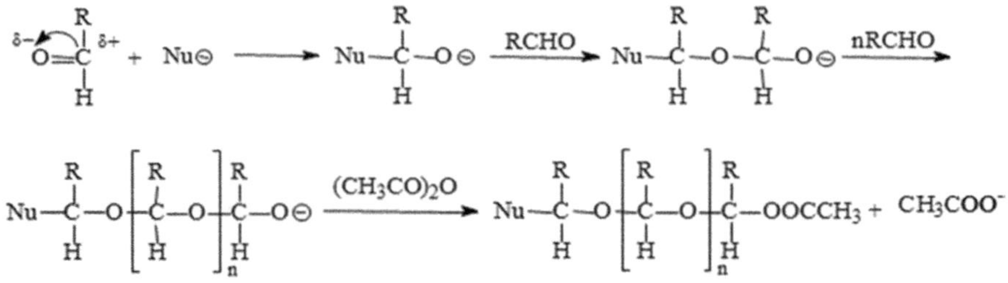

where Nu Θ – nukleofile factor.

Similarly as in the case of anionic polymerization of vinyl compounds, there is no a proper termination reaction here. "Living" polymers, suitable for the block copolymerization, can be formed too. In the reaction with acetic anhydride one obtains the hydrophobic chain termination and in this way a stabilized polymer with increased thermal resistance.

The low limit temperature in aldehydes in the polymerization is utilized in the synthesis of poly(trichloroacetic aldehyde) (polichloral), characterized by non-combustibility and good mechanical properties.

In order to obtain this product chloral is mixed with a catalyst at a temperature above the limit temperature, which amounts to 58°C. Then the mixture is poured into the mold and cooled to the temperature of about 10°C to obtain a polymers with high molecular weights:

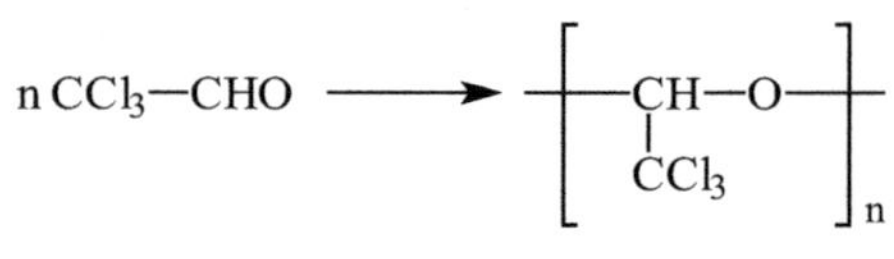

In the discussed group of polymers a great importance has also the stereospecific polymerization of aldehydes on coordination catalysts.

As it was shown in previous studies only the isotactic polymers are obtained. The reaction is assumed to run as follows:

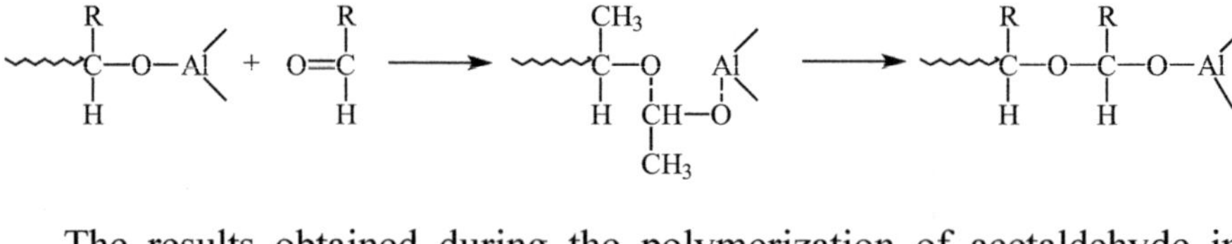

The results obtained during the polymerization of acetaldehyde in the presence of the triethylaluminum complex with different complexing compounds are given in Table 10.23.

TABLE 10.23 Polymerization of Acetaldehyde in the Presence of Triethylaluminum Complex

No.	Complexing factor	Chemical formula	Yield* [%]	Stereospecificity indicator ** [%]
1	cycloheksanon	(CH2)5CO	70	96
2	2-pyrrolidone	OC(CH2)3NH	49	85
3	acetamide	CH3CONH2	87	95
4	methylacetamidee	CH3CONHCH3	69	88
5	phenylacetamide	CH3CONHC6H5	87	95
6	phenylbenzamide	C6H5CONHC6H5	85	92

* Fraction insoluble in hexane.

** Ratio of the fraction insoluble in chloroform to the fraction insoluble in hexane; (after H. Tani, Polymer Advances in Science, 11, 1957 (1973)).

During the polymerization of dialdehydes the formed polymers contain cyclic ether groups. For example, during the polymerization glioxal in the presence of sodium naphthalene, performed in tetrahydrofuran at −78°C, a poly glyoxal with the following structure is formed:

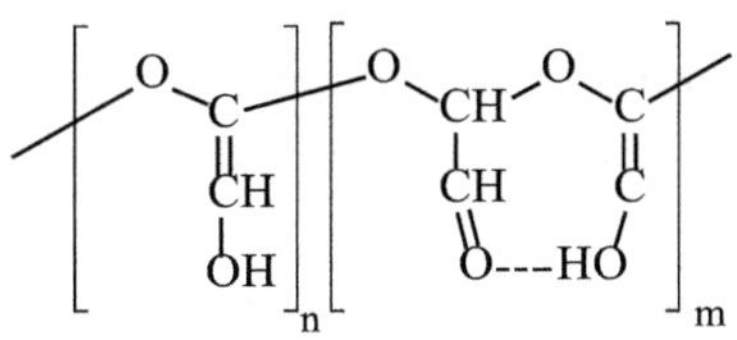

n > m

Aldehydes such as glutaric, succinic and phthalic, polymerized in the presence of Lewis acids, give amorphous polymers with low molecular weights. The chemical structure of the polymer formed from glutaraldehyde is the following:

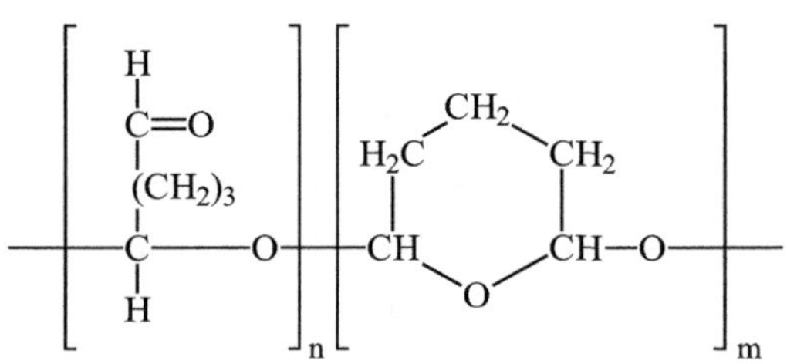

Terephthalic and isophthalic aldehydes engage in the process of polymerization only one aldehyde group, whereas the o-phthalic aldehyde undergo a partial cyclopolymerization at 0°C. It cyklopolimerizes almost exclusively at the temperature of −78°C according to the following scheme:

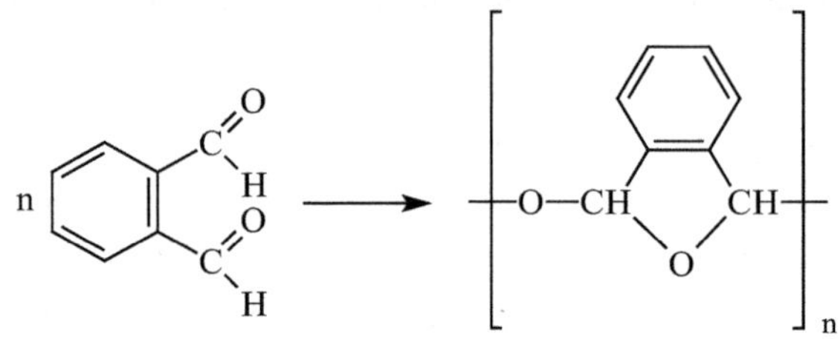

10.4.5 POLYMERIZATION OF KETONES

Ketones are less reactive compounds than aldehydes and harder to polymerize. The polymers formed from them have no practical significance.

Acetone polymerize at low temperatures at the presence of Ziegler-Natta catalysts or under the γ irradiation. The obtained in this way polyacetone is unstable. More stable is a block copolymer of acetone with propylene.

Significantly better monomers are ketenes. Depending on the reaction environment various polymers are formed during the polymerization of dimethylketenes.

In polar solvents formed are polyethers. In nonpolar solvents and in the presence of lithium, magnesium and aluminum ions formed are polyketones. In nonpolar solvents and in the presence of sodium and potassium ions one obtains polyesters:

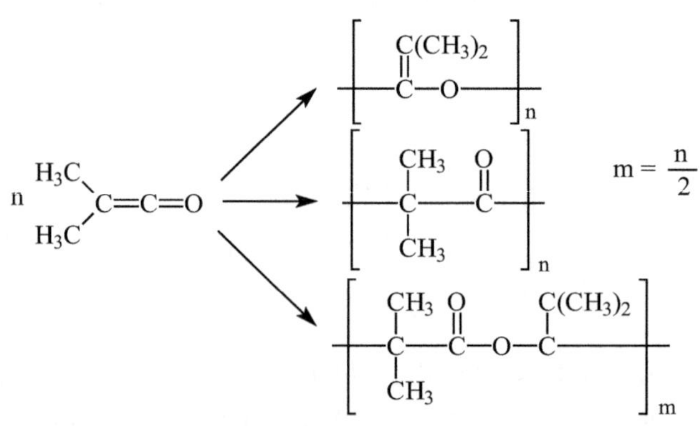

10.4.6 POLYMERIZATION OF HETEROCYCLIC COMPOUNDS WITH RING OPENING

Some heterocyclic compounds undergo relatively easy the ring-opening polymerization resulting in the formation of linear polymers.

As catalysts for these reactions acidic (usually Lewis acids), alkaline or coordination compounds may be used.

The method of the ring-opening polymerization was widely used for the synthesis of polyethers from oxacyclic connections, polyamides from lactams, polyesters from lactones and polyorganosiloxanes.

10.4.7 POLYMERIZATION OF CYCLIC ETHERS

The most commonly used oxacyclic monomers include the following compounds or their derivatives:

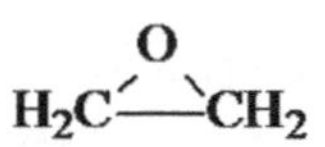

oxirane, epoxyethane (ethylene oxide)

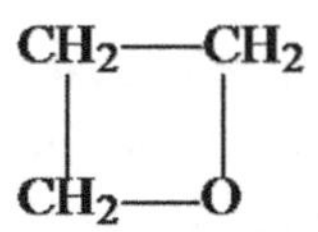

oxetane (oxacyclobutane)

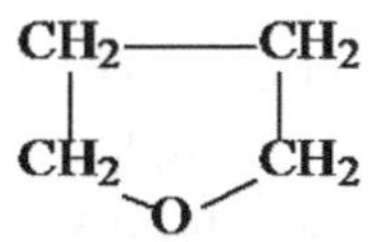

tetrahydrofuran

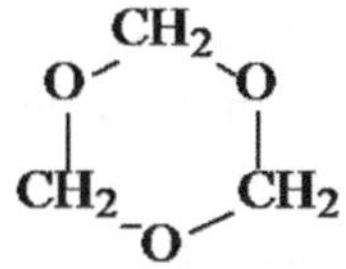

trioxymethylene

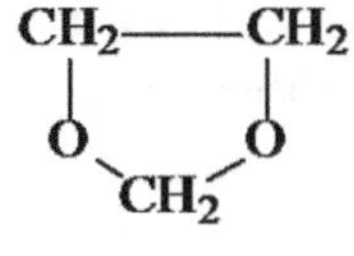

1,3 - dioxalon

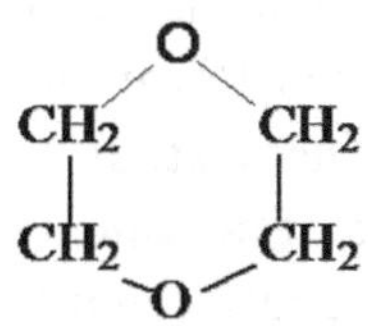

1,4 dioxane

10.4.7.1 Cationic Polymerization of Cyclic Ethers

Polymerization of cyclic ethers by cationic mechanism is triggered by various protic catalysts and Lewis acids. In the event of proton donors the polymerization reaction proceeds through the stage of an oxo ion, but as a result of this reaction mostly oligomers or polymers with small molecular weights are formed. The course of polymerization of ethylene oxide under the influence of acid or boron trifluoride in the presence of water is given by the following scheme:

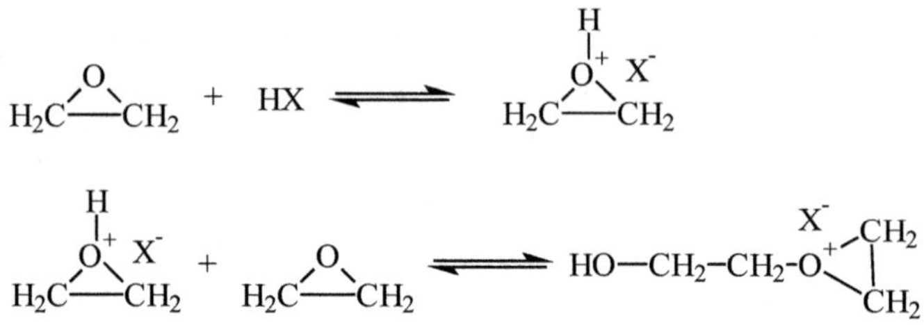

The completion of the chain growth takes place by the reaction of macrocation with a water molecule:

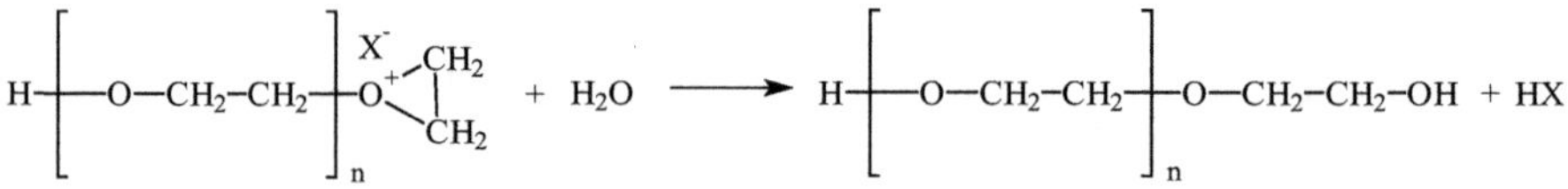

In the case of using other cationic catalysts the polymerization mechanisms are often complex and not completely understood. Some of these catalysts most likely react with monomer, forming metal alkoxides. It means that the polymerization proceeds according to the coordination anionic mechanism.

During the polymerization of ethylene oxide in the presence of tin tetrachloride the reaction mechanism differs from the mechanism initiated by boron(III) fluoride and leads to the high molecular weight polymer. The kinetic studies indicate that each particle of tin chloride initiates two chains of polymer.

The reaction may be represented schematically as follows:

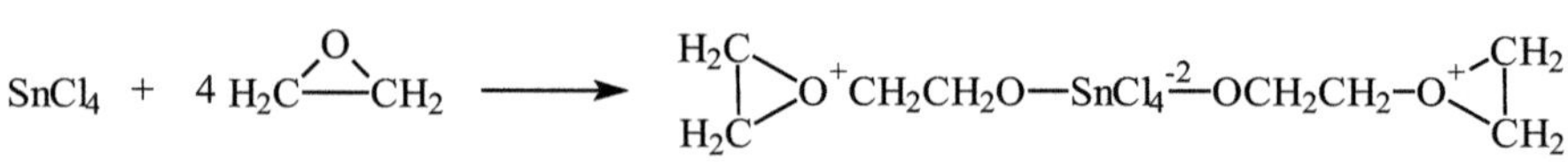

The "living" polytetrahydrofuran is formed by the reaction of alkyl halide with the silver hexafluoro antimony:

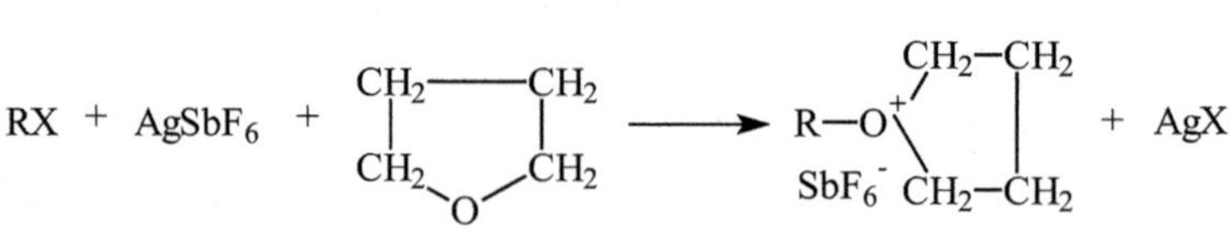

The chain growth completion can take place in reaction with the triphenylphosphine:

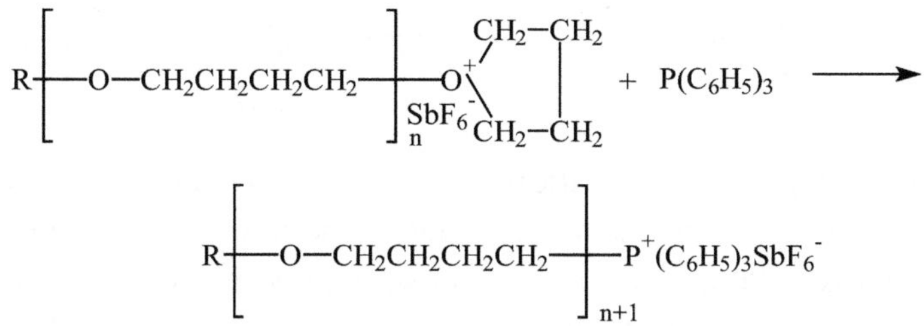

The above-mentioned reaction has found also application in the synthesis of graft copolymers.

The hexachloro antymony of triphenylmethyl has proved to be a good catalyst, enabling the cationic copolymerization of cyclohexene oxide with styrene oxide. The initiation reaction in this case consists on formation of the carbonion cation:

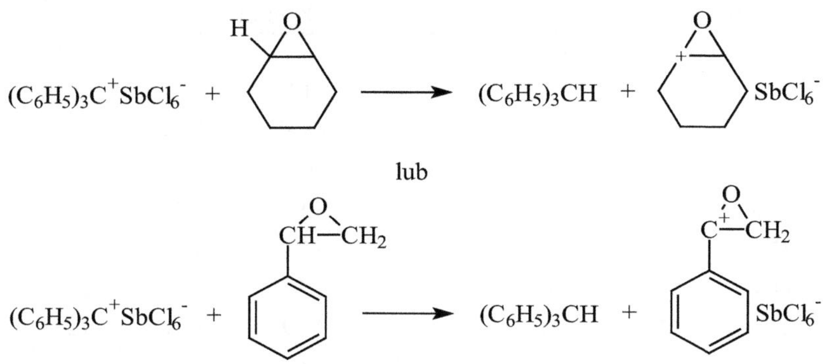

The further increase of the copolymer chain proceeds through the stage of the oxonium ion:

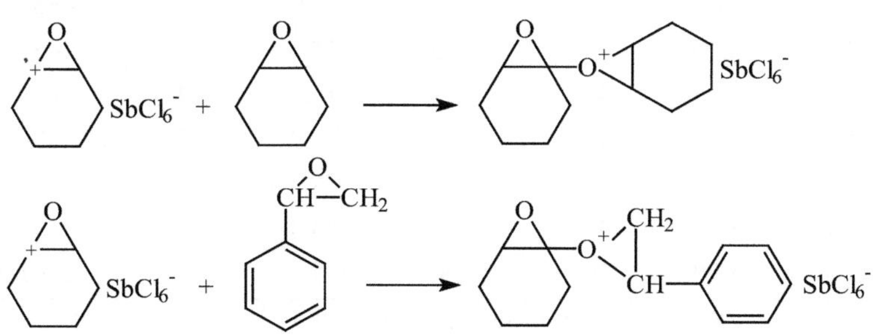

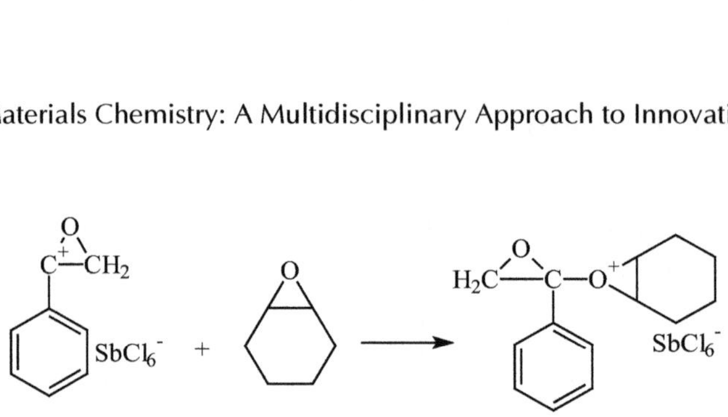

The resulting oxonium cations react with next monomer molecules, with ring opening. As consequence linear copolyethers are formed.

The results obtained during the course of cationic copolymerization of cyclohexene oxide with styrene oxide and the measured reactivity ratios are given in Tables 10.24 and 10.25.

TABLE 10.24 Course of Cationic Copolymerization of Cyclohexene Oxide (M1) with Styrene Oxide (M2) Within 1 Hour and at the Given Initiator Concentration

No.	Composition		Reaction temp. [°C]	Yield [weight %]	Copolymer composition		Molecular mass of copolymer Mn*102
	M1 [% mol]	M2 [% mol]			M1 [% mol]	M2 [% mol]	
1	11.1	88.9	0	1	30.7	69.3	901
2	22.5	77.5	0	1.3	51.7	48.3	1048
3	38.7	61.3	0	1.1	68.6	31.4	1296
4	53.5	46.5	0	1.48	80.4	19.6	1188
5	66.6	33.4	0	2.2	90.5	9.5	1550
6	82.1	17.9	0	3.8	95.1	4.9	1278
7	11.1	88.9	30	16.4	27	73	1858
8	22.5	77.5	30	18	46.9	53.1	1298
9	38.7	61.3	30	16.3	67	33	1137
10	53.5	46.5	30	17.8	78.9	21.1	889
11	66.6	33.4	30	18.1	88.7	11.3	778
12	82.1	17.9	30	17.6	94.9	5.1	668
13	11.1	88.9	46	6.3	25.3	74.7	1655
14	22.5	77.5	46	9.1	45.1	54.9	1672
15	38.7	61.3	46	13.4	66.8	33.2	1490
16	53.5	46.5	46	13.7	78.9	21.1	1256
17	66.6	33.4	46	15.6	86.3	13.7	1159
18	82.1	17.9	46	14.6	95	5	1481

*From W.M. Pasika, D.J. Chen: J. Polym. Sci. A, 29, 1457 (1991).

TABLE 10.25 Reactivity Ratios of Cyclohexane Oxide (r_1) with Styrene Oxide (r_2)

No.	Temperature [°C]	Kelen–Tudos method		Fineman–Ross method	
		r1	r2	r1	r2
1	0	4.24	0.4	4.4	0.35
2	30	4.19	0.42	4.15	0.4
3	46	3.94	0.46	3.9	0.45

*From W.M. Pasika, D.J. Chen: J. Polym. Sci. A, 29, 1457 (1991).

One of the cationic polymerization of cyclic ethers method is the polymerization initiated by a constant electric current. It can be applied to both: oxirane derivatives and tetrahydrofuran. Polymerization oxirane derivatives are performed in the dichloromethane solution using a platinum electrode. As electrolyte the tetrabutylammonium hexafluorophosphate with concentration of 0.1 mol/dm^3 is used.

The results obtained during the electroinitiated cationic polymerization of some epoxy compounds are presented in Table 10.26.

10.4.7.2 Anionic Polymerization of Oxirane and Its Derivatives

Oxiranes polymerize by the anionic mechanism too. Catalysts for these reactions are usually hydroxides or alkoxides of sodium and potassium. The oxirane ring opening reaction proceeds by S_N2 substitution with the formation of an alkoxylane ion. As consequence the chain growth takes place either by the nucleophilic rearrangement of the newly formed alkoxylane ion or in a gradual process, consisting on transfer or termination following the addition of alcohol, as illustrated by the following reaction scheme:

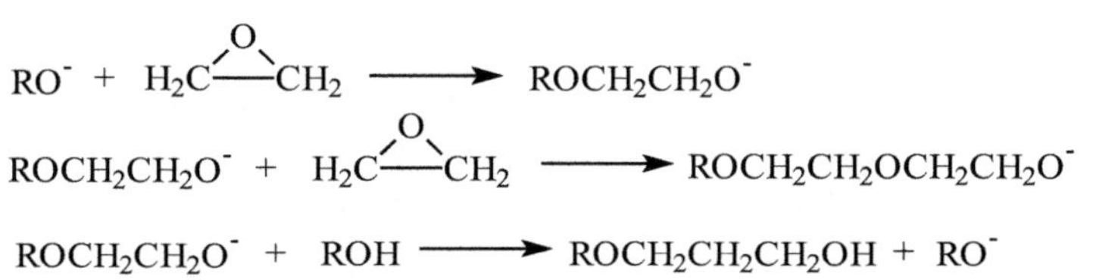

The initiation reaction takes place in this case faster than the growth chain reaction. It causes the formation of polyethers of low molecular

TABLE 10.26 Electroinitiated Cationic Polymerization of Some Epoxides

No.	Monomer	Anode poten. [V]	Poten. polym. [V]	Max. current inten. [mA]	Temp. [°C]	Reaction time [min]	Monomer conversion degree. [%]	Polymer melting temp. [°C]	Polymer solution viscosity in benzene 0.1 dm^3/g at 35°C
1	1,2-epoxy-4-epoxy-ethyl cyclohexane	2.7	2.3	2	-37	5	5	>340	—
2	epoxycyclohexane (cyclohexene oxide)	3.01	2.3	4.5	-37	90	68.8	93	0.071
3	epoxycyclopentane (cyclopentane oxide)	2.92	2.2	5	-35	90	69	-10	0.13
4	Epoxyethylbenzene (styrene oxide)	2.39	2.2	5	-20	75	>95	15	0.092

*After U. Akbult et al., Macromol. Chem. Rapid Commun., 4, 259(1983).

weight. The curse of this reaction is largely influenced by the type of alcohol used during for its initiation. When using a tertiary alcohol a more reactive primary alcohol is formed in the stage of initiating. It causes growth rate advantage over the rate of chain reaction initiation. It allows thus to obtain polymers with higher molecular weight. The course of the anionic polymerization of epoxides is also affected by the type of solvent used and the temperature at which the reaction is conducted.

It was shown that during the reaction carried out in dimethylsulfoxide a methylsulfinyl the methanide anion is formed in the reaction environment,

$$[CH_3-\underset{\underset{O}{\|}}{S}-CH_2]^-$$

which participates in the polymerization process and affects the kinetics of the reaction.

This anion is in equilibrium with the alkoxyl anion:

$$[CH_3-\underset{\underset{O}{\|}}{S}-CH_2]^- + ROH \rightleftharpoons RO^- + CH_3-\underset{\underset{O}{\|}}{S}-CH_3$$

The results of kinetic studies of anionic polymerization of oxirane and its homologues in the dimethyl sulfoxide environment are displayed in Table 10.27.

TABLE 10.27 Kinetic Data of Oxirane and Its Homologues Polymerization in Dimethyl Sulfoxide

No.	Monomer		Temp. °C	Order with respect to		References
				Monomer	Initiator	
1	oxirane (ethylene oxide)	tert-butylan of potassium	25	1	1	Polym. 10, 653, (1969)
2	oxirane	tert-butylan of potassium	20	1	1	J. Am. Chem. Soc. 88, 4039 (1966)
3	methyloxirane (propylene oxide)	1-ethoxy-3-oxapentylan of cesium	50	1	1.9	Makromol. Chem. 176, 3107 (1975)

TABLE 10.27 Continued

No.	Monomer		Temp. °C	Order with respect to		References
				Monomer	Initiator	
4	methyloxirane (propylene oxide)	tert-butylan of potassium	30	1	1	J. Am. Chem. Soc. 88, 4039 (1966)
5	methyloxirane (propylene oxide)	tert-butylan of potassium	40	1	1.7	J. Polym. Sci. A-1 10, 1353 (1972)
6	ethyloxirane (1,2-butylene oxide)	tert-butylan of sodium	30–60	1	1.8	J. Polym. Sci. Polym. Chem. Ed. 10, 3089 (1972)

In a completely different way runs the anionic polymerization of oxirane derivatives in the presence of 4-ethylopyridine. In that case no polyethers are formed, but aromatic type polymers containing pyridine groups in the main chain.

The reaction of 4-ethylopyridine with styrene oxide can be presented as follows:

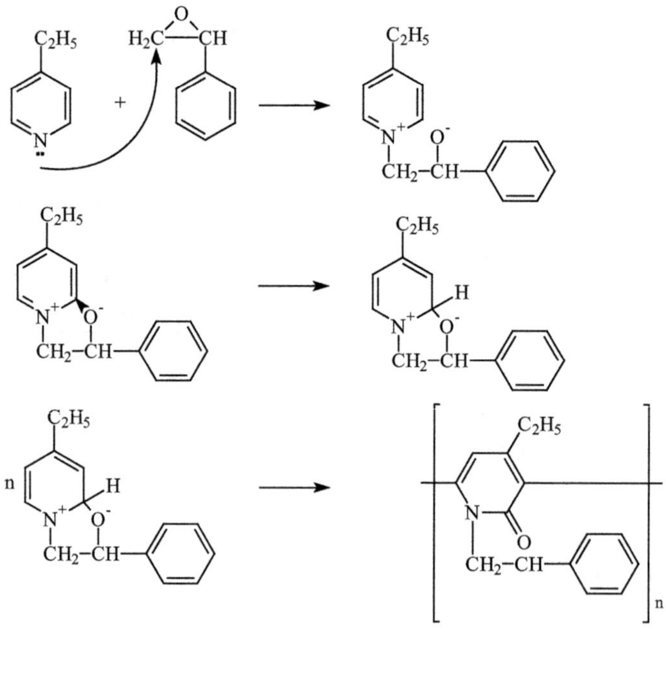

Known are also methods of anionic polymerization of propylene oxide in the presence of crown ethers and co polymerization of epoxides with carbon dioxide in the presence of zinc salts of carboxylic acids.

10.4.7.3 Stereospecific Polymerization of Cyclic Ethers

During the polymerization of cyclic ethers in the presence of coordination catalysts formed are polyethers with high molecular weights and a stereoregular spatial structure. The mechanism of formation of such polymers is similar to the already described mechanism for obtaining stereoregular vinyl polymers with Ziegler-Natta catalysts.

The stereoregularity effect depends on the type of catalyst and the reaction conditions.

The coordination catalysts used for polymerization of cyclic ethers are chelates of β-hydroxyesters or β-diketones, calcium ethoxy amide, Ziegler-Natta catalysts, trialkylaluminum, dialkylzinc and a complex of ferric chloride with 1,2-epoxypropane.

Some coordination catalysts (mainly alkylmetals) increase their activity in the presence of cocatalysts such as water or ketones.

The addition of a cocatalyst changes the structure of the basic catalyst and renders it more active. For example, the addition of water to diethylzinc causes likely its partial hydrolysis with the formation of a compound of the following chemical structure:

$$C_2H_5{-}Zn{-}O{-}Zn{-}C_2H_5$$

The formed catalyst may react with the propylene oxide in the following way:

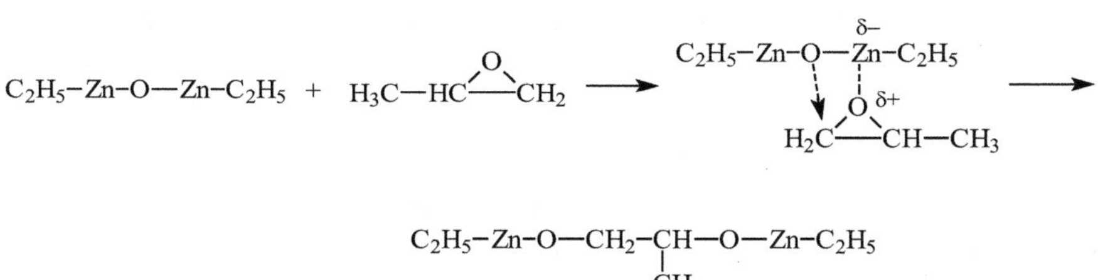

A further chain growth takes place as a result of an intramolecular rearrangement. The mechanism of this reaction has not yet been clearly determined and requires a further research. The polymerization reaction of oxirane and its derivatives in the presence of dialkylzinc is conducted mostly in a benzene solution.

The results obtained during the polymerization of ethylene oxide in benzene at the temperature of 60°C and in the presence of water activated by zinc diphenyl are presented in Table 10.28. The presented there data show that the increase of the amount of ethylene oxide with respect to the catalyst and the extension of the reaction time increases the efficiency of the reaction and favors the formation of a polymer with a higher molecular weight. The optimal molar ratio of water to zinc diphenyl is 1: 1.

The molecular weight of formed polyethers was determined by the viscosimetry technique in water at 298 K using the following formulas:

$$[\eta]=3.97\times10^{-4}\left(\overline{M}_v\right)^{0.686} \quad \text{for ethylene oxide}$$

$$[\eta]=1.12\times10^{-4}\left(\overline{M}_v\right)^{0.77} \quad \text{for propylene oxide}$$

Tetrahydrofuran and 2-methyl-2-oxazoline can polymerize through the formation of complexes such as "charge-transfer" complex (CTC).

TABLE 10.28 Polymerization of Oxirane in the Presence of $(C_6H_5)_2Zn - H_2O$ in Benzene at 60°C

No.	Zinc diphenyl [mmol]	Oxirane (C6H5) 2Zn [mol/mol]	Reaction time [h]	Yield [%]	Intrinsic viscosity [0.1 dm^3/g]	Molecular mass MV×10^{-5}	Polymer flow tempera ture [°C]
1	0.285	199.7	2	1	—	—	—
2	0.285	199.7	4	6	3.04	3	62–64
3	0.285	199.7	6	24.7	1.44	1.5	61–63
4	0.285	199.7	12	30.3	3.7	6.1	63–65
5	0.285	214	24	38.2	3.28	5.1	59–63
6	0.278	200	36	56.8	3.07	4.7	54–60
7	0.285	199.7	48	63.4	5.44	10.7	54–59
8	0.285	199.7	72	83.9	4.93	9.3	52–63
9	0.143	214.6	120	93.3	5.98	12.3	59–63
10	0.278	200	144	97.1	6.96	15.4	57–61

*After F.M. Rabagliati, C. Bradley: Eur. Polym. J. 20, 571 (1984).

An example of such a reaction is the polymerization of tetrahydrofuran in the presence of tetracyanoethylene (TCNE):

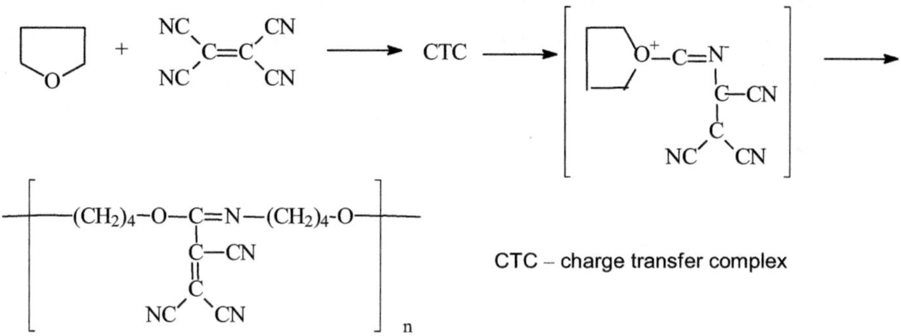

CTC – charge transfer complex

This method can be also used to obtain polymers from 2-methyl-2-oxazoline with molecular mass between 2–200 and 18–800.

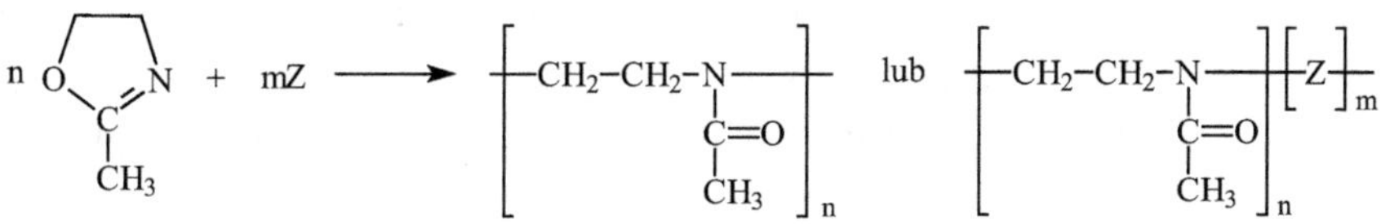

Z - acceptor CTC

where, Z – acceptor CTC.

10.4.8 POLYMERIZATION OF CYCLIC IMINES

Cyclic imines polymerize similarly as oxirane and its derivatives. Propyleneimine copolymerize with carbon dioxide without the use of a catalyst. As a result of this reaction formed are polyurethanes containing 10–35 mol % of urethane groups. The amount depends on the temperature of conducting the process. The structure of the resulting copolymer can be represented by the following formula:

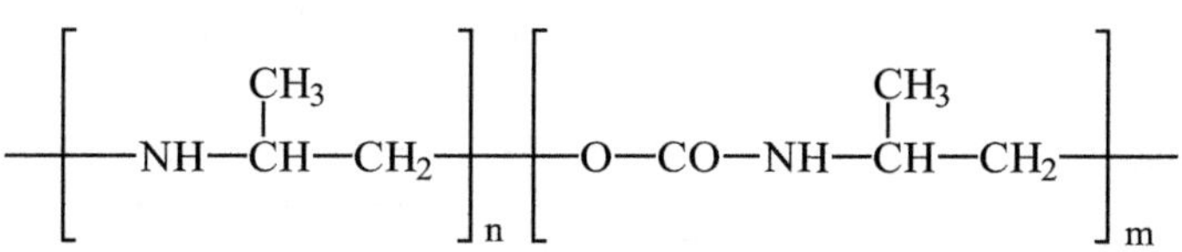

Table 10.29 displays the results showing the influence of temperature and the reaction time on the course of copolymerization of propylenimine with carbon dioxide at the molar ratio of reagents 1:5.

From the obtained results one can conclude that increasing the temperature and prolonging the reaction time results in the increase of the degree of substrates conversion and of the molecular weight of formed polymer, as it follows from the observed increase of the limit viscosity value.

10.4.9 POLYMERIZATION OF LACTAMS

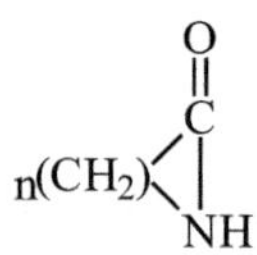

Lactams, similarly as cyclic ethers, can polymerize with ring opening using acidic or alkaline catalysts. The reaction is carried out in an anhydrous medium to prevent the hydrolysis of the ε – aminocarboxylic acid.

TABLE 10.29 Influence of Temperature and Reaction Time on the Course of Copolymerization of Propylenimine with Carbon Dioxide

No.	Temperature [°C]	Reaction time [H]	Conversion rate	Intrinsic viscosity [g/0.1 dm3]
1	–78	3	0	—
2	–20	3	27.4	—
3	0	3	32.9	0.114
4	0	4.5	90.9	0.238
5	0	16	91.7	0.282
6	20	3	58.1	0.285
7	50	0.5	46.7	0.232
8	50	1	52.1	0.208
9	50	3	78.5	0.434
10	50	5	70.1	0.312
11	80	3	75.9	0.528
12	100	3	93.9	0.515

*After Soga K., W. Y Chiang, S. Ikeda: Polym J.. Sci. Polym. Chem. Ed. 12, 121 (1974).

10.4.9.1 Cationic Polymerization of Lactams

During the polymerization reaction of caprolactam in the presence of hydrogen chloride a tautomeric equilibrium between two forms of cationic caprolactam is fixed: the mesomeric stabilized imide form (A) on one side and the form containing the amide cation form (B) on the other one.

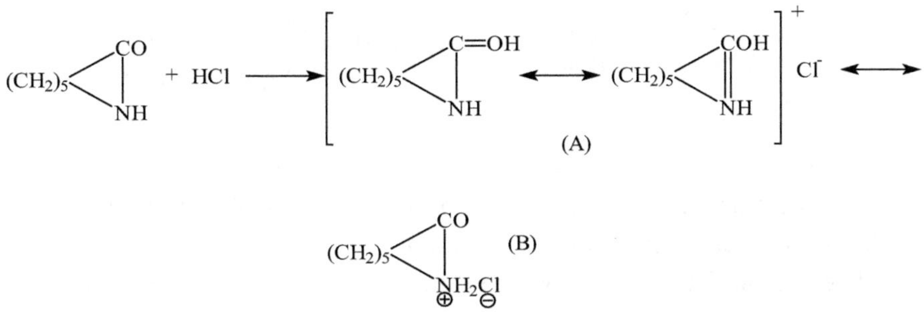

Form (B) is more active because of the absence of mesomeric stabilization. It reacts with the next caprolactam molecule with formation of hydrochloride of amino caprolactam, which initiates a further polymerization:

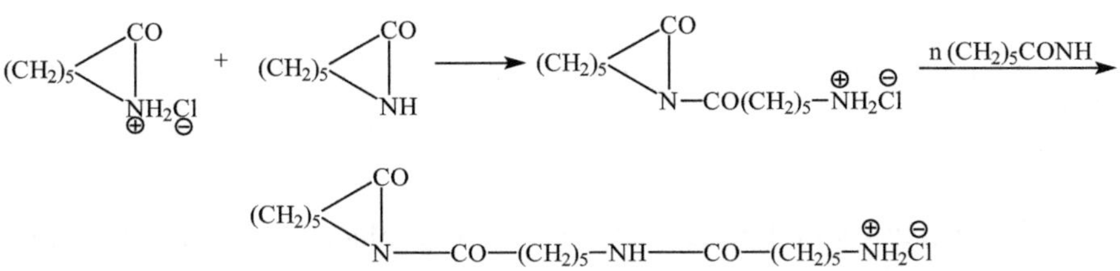

The cationic polymerization of lactams has no practical significance. Industrial application has found the anionic polymerization of lactams.

This method gives from caprolactam the polyamide with a molecular weight many times higher than that obtained during the polycondensation of ε-aminocaproic acid.

10.4.9.2 Anionic Polymerization of Lactams

Typical catalysts for anionic polymerization of lactams are alkali metals and their hydroxides or alkoxides.

The mechanism of anionic polymerization of lactams has been thoroughly tested on the example of caprolactam. Lactams are very weak acids. Acidity of the hydrogen atom of the lactam amide group is a little greater than the acidity of the hydrogen atom of water or alcohol. For this reason in the reaction of caprolactam with the metal hydroxide or alkoxide an equilibrium state is reached:

The reaction of anionic polymerization of lactams is a complex process and can be carried out only in an anhydrous and alcohol-free medium in order to prevent the hydrolysis of lactam.

The mechanism of chain growth during the anionic polymerization of caprolactam is as follows:

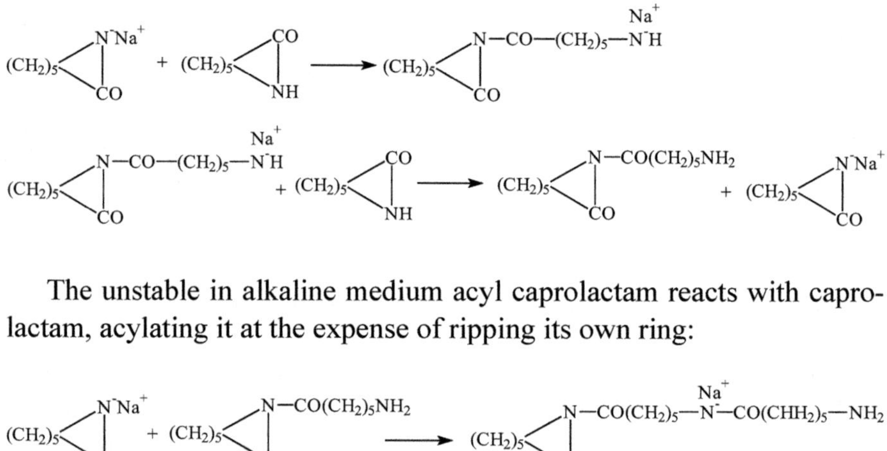

The unstable in alkaline medium acyl caprolactam reacts with caprolactam, acylating it at the expense of ripping its own ring:

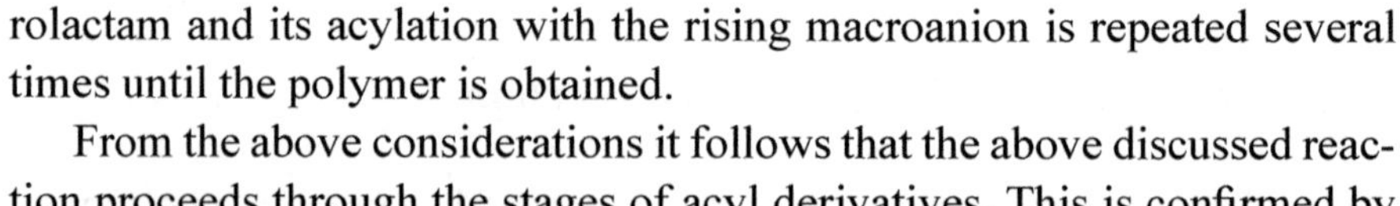

The cycle of sodium cation exchange with the next molecule of caprolactam and its acylation with the rising macroanion is repeated several times until the polymer is obtained.

From the above considerations it follows that the above discussed reaction proceeds through the stages of acyl derivatives. This is confirmed by experimental data which show that the added to the system chlorides and acid anhydrides are activators of the process, because they shorten the induction time and allow the reaction to run at lower temperature. The anionic polymerization of caprolactam without using an activator is carried out at temperatures above 190°C.

Polymerization of five- and six-member lactam rings is difficult. However, under the influence of catalysts one can obtain polymers at temperatures below 60°C. For example, polyamide-4 is obtained by anionic polymerization of 2-pyrrolidone (butyrlactam), catalyzed by silicon tetrachloride.

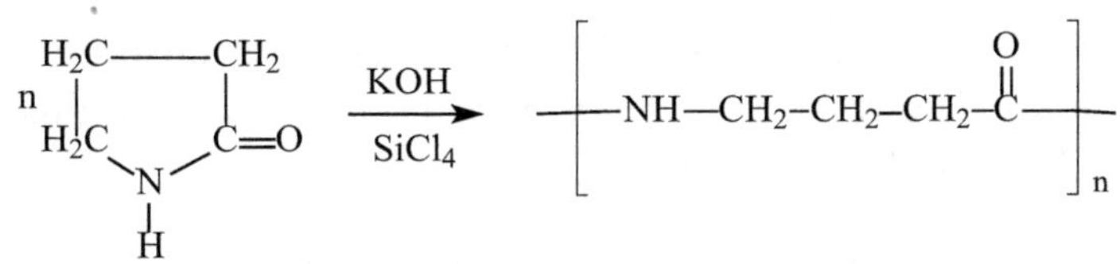

The substituents in the lactam ring, particularly those connected to the nitrogen atom, contribute in reducing the activity of monomer.

10.4.10 POLYMERIZATION OF CYCLIC SILOXANES

During the hydrolysis of organic derivatives of dichlorosilane, beside the linear polyorganosiloxanes, formed are also cyclic siloxanes with six- or more members. The cyclic siloxanes polymerize easily by ionic mechanism, both by cationic and anionic, with formation of polyorganosiloxanes of large molecular weights.

10.4.10.1 Cationic Polymerization of Cyclic Siloxanes

The cationic polymerization of cyclic siloxanes is usually performed in the presence of acids: sulfuric, chlorosulfon, phosphoric or boric, and also in systems containing the Lewis acids such as tin tetrachloride, aluminum chloride, boron trifluoride and zinc chloride.

The relatively well-studied reaction of cationic polymerization of hexamethylcyclotrisiloxane proceeds along the following scheme:

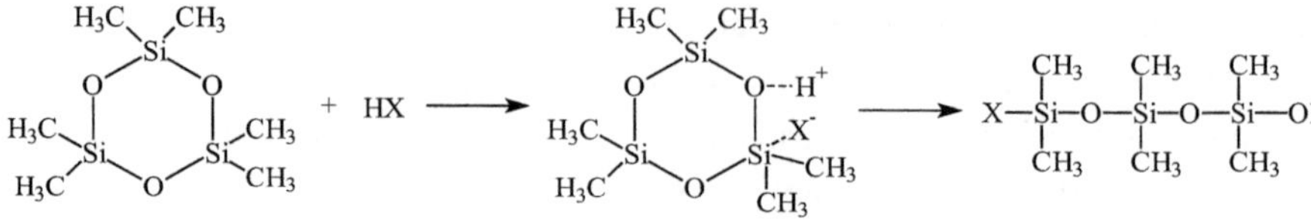

The resulting linear compound reacts with the next molecules of hexamethylcyclotrisiloxane forming a linear polymer.

10.4.10.2 Anionic Polymerization of Cyclic Siloxanes

The cyclic siloxanes polymerize by an anionic mechanism in the presence of bases. The catalytic activity of metal hydroxides in this process decreases in the following order:

$$Ca(OH)2 > KOH > NaOH > LiOH$$

An example of such a reaction can be the anionic polymerization of hexamethylcyclotrisiloxane in the presence of potassium hydroxide. At the first stage of this process as a result of the nucleophilic attack of hydroxyl anion on the silicon atom in cyclosiloxane a transition compound is formed, in which breaks the weakened silicon – oxygen bond of siloxane:

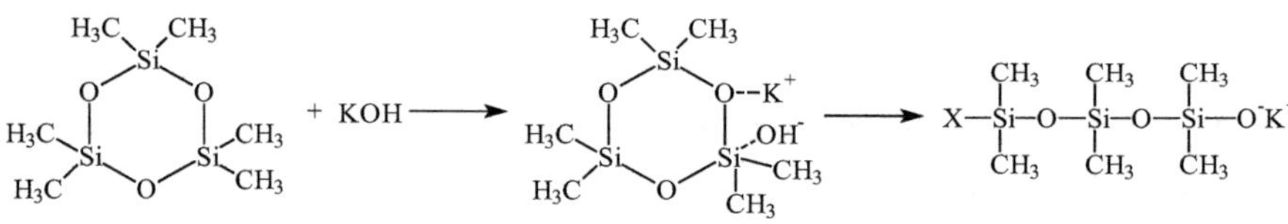

The formed active anion reacts sequentially with the next hexamethylcyclotrisiloxane molecules, causing a rupture of the ring and chain increase with the creation of a new active center at the end of the chain. The reaction proceeds until obtaining a polymer with high molecular weight, which is a "living polymer".

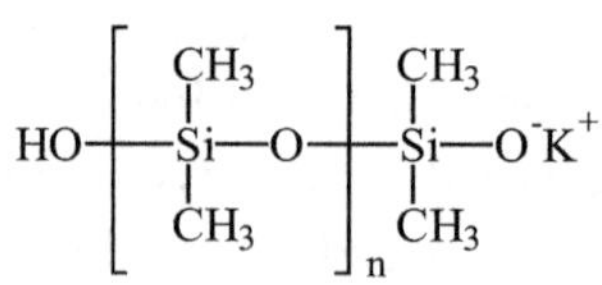

10.4.11 POLYMERIZATION OF ALKYNES (POLYACETYLENE)

Acetylene is the simplest representative of the homologous series of alkynes. The first attempts to carry out the polymerization of acetylene resulted in obtaining low molecular weight compounds. During the transmission of acetylene through the glass tubes, heated to the red heat, acetylene undergo a trimerization process, which results in the formation of benzene:

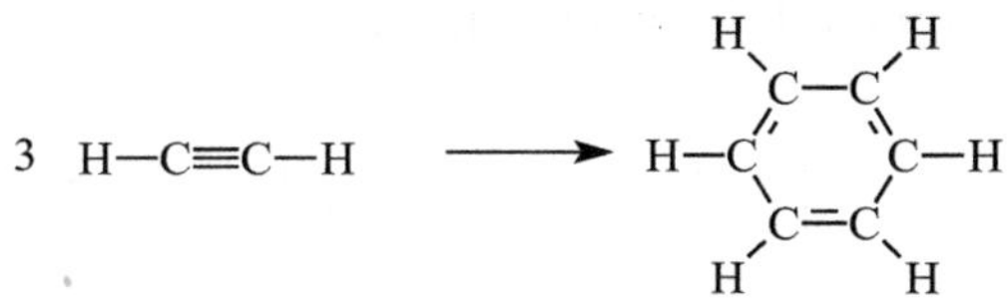

In the presence of Cu_2Cl_2 and NH_4Cl catalysts at 50°C two molecules of acetylene dimerize with the formation of vinylacetylene:

$$2\ H{-}C{\equiv}C{-}H \longrightarrow CH_2{=}CH{-}C{\equiv}CH$$

It was also found that acetylene passed through a copper sponge polymerize to a mass similar to that of cork. This form didn't find practical significance. Research over the course of the polymerization of acetylene and other alkynes have gained importance after finding in 1977 by H. Shirakawa and C.K. Chiang that the electrical conductivity of polyacetylene, after an appropriate doping, increases by 11–12 orders of magnitude and is comparable to that of metals.

The use of Ziegler-Natta catalysts, modified by Shirakawa, allows synthesis of different species of polyacetylene with an average molecular weight ranging from several to tens of thousands. For this purpose one uses most commonly the soluble coordination catalysts prepared with prepared from tetrabutoxytitanium and triethylaluminum with the chemical formula:

Ti(OC4H9)4—Al(C2H5)3, where the ratio Al:Ti = 4.

Thin films of polyacetylene are obtained on catalysts, which are complexes of titanium triacetylacetonate with triethylaluminum or chromium triacetylacetonate with triethylaluminum.

During the study of the catalytic activity of metal complexes of metal acetylacetonate complexes with triethylaluminum the following order was found:

Ti, V>>Cr > Fe > Co

The temperature at which the acetylene polymerization is performed has a large impact on its course, and more particularly on the structure of the resulting polyacetylene. Table 10.30 shows that at low temperatures in the overwhelming amount the "cis" polyacetylene is formed. With increasing polymerization temperature the amount of "trans" form rises too. At the temperature of 150°C only trans-polyacetylene is formed.

The mechanism of acetylene polymerization on coordination catalysts can be illustrated by the following scheme:

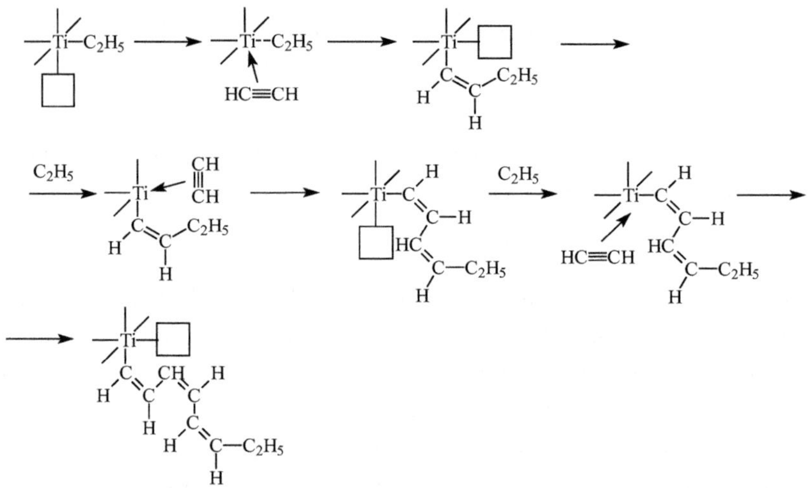

TABLE 10.30 Effect of Temperature on the Structure of Polyacetylene Obtained in the Presence of Shirakawa* Catalyst with Concentration of 10 mol/dm^3

Polymerization temperature[0°C]	Isomer content	
	cis [%]	trans [%]
–78	98.1	1.9
–18	95.4	4.6
0	78.6	21.4
18	59.3	40.7
50	32.4	67.6
100	7.5	92.5
150	0	100

*Catalyst: Ti(OC4H9)4–Al(C2H5)3, Al/Ti = 4.

* T. Ito, H. Shirakawa, S. Ikeda: J. Polym. Sci. Polym. Chem. Ed. 12, 11 (1974).

According to this scheme always formed is polyacetylene with cis – transoidal structure only. It is obtained exclusively from the polymerization of acetylene at low temperatures.

Polyacetylene can be also obtained by applying the Diels-Alder synthesis reaction. One example of the polyacetylene synthesis using this method is the addition of cyclobutadiene to aromatic compounds. The polymerization of the obtained in this way adduct and the thermal elimination of aromatic compounds from the resulting polymer gives polyacetylene. The reaction can be presented as follows:

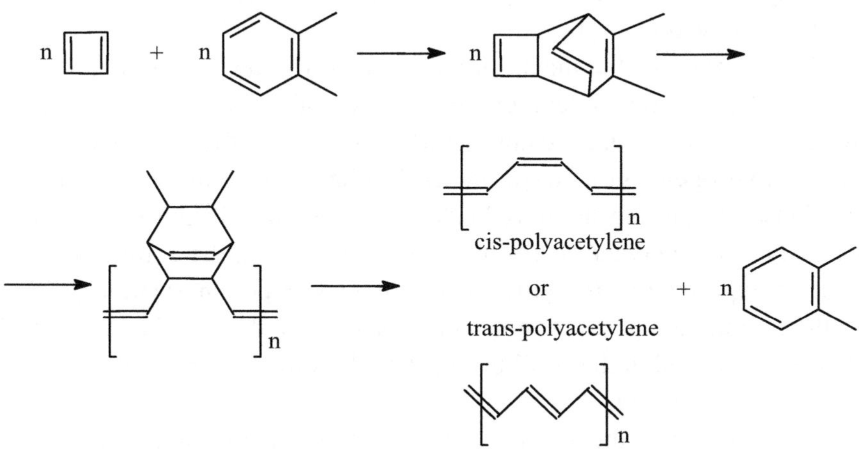

10.5 CONDENSATION POLYMERIZATION (POLYCONDENSATION)

10.5.1 PROCESS OVERVIEW

The condensation polymerization, often called as polycondensation, is a condensation reaction of a large number of monomer molecules or comonomers to the polycondensate macromolecules, during which released are: water, hydrogen chloride, ammonia and other simple compounds as byproducts.

In the polycondensate macromolecules the main chain is formed not only of carbon atoms but it includes also atoms of other elements such as, for example, oxygen, nitrogen, phosphorus, boron or silicon.

Polycondensation is a special case of substitution reaction, in which multifunctional molecules react with each other and the equilibrium conditions do not interfere with the formation of long-chain molecules.

In contrast to the addition polymerization the condensation polymerization is a gradual reaction. At all stages of this process formed are the transitional and the permanent products, which can be well distinguished. If in the reaction are involved two types of molecules having at least two functional groups, the reaction is called heteropolycondensation, whereas when react among themselves molecules of the same type (e.g., hydroxyacid molecules of type $HO\text{-}(CH_2)_x\text{-}COOH$) the process is called the homopolycondensation.

The increase of the macromolecule chain is slow. The kinetics of this process depends on the temperature, the rate of removal of low molecular weight byproducts and the amount and the nature of the catalyst (usually on the hydrogen ion concentration). Similarly as in the case of normal condensation, during the polycondensation an equilibrium is reached for each stage of the reaction. As it is generally known, the equilibrium constant is expressed as the ratio of the product of polemerization products to the product of substrates concentrations. Since the reaction of polycondensation is fixed in the equilibrium state therefore one takes the averaged equilibrium constant.

Assuming that the value of the equilibrium constant (K) does not change during the process duration the equilibrium constant is given by the following equation:

$$K = \frac{(N_0 - N)N_a}{N^2}$$

where: N_a – number of small molecular mass molecules of byproduct at the equilibrium state; N_0 – initial number of difunctional monomer molecules particles equal to the initial number of functional groups X and Y; N – number of formed macromolecules, corresponding to the number of functional groups X or Y in equilibrium state; (N_0–N) – number of new bonds created and simultaneously the number of bonds extinct.

$$\frac{N_0 - N}{N_0} = n_z$$

- is the mole fraction of bonds per the basic polymer segment, which at the same time is also the degree of conversion

$$n_a = \frac{N_a}{N_0}$$

- mole fraction of small molecule compound per the polymer mer in the equilibrium state

$$\frac{N_0}{N} = \overline{P}$$

- average degree of polymerization.

Dividing the numerator and the denominator in the above equation for the equilibrium constant by N_0^2 one gets:

$$K = \frac{\left(\frac{N_0 - N}{N_0}\right)\frac{N_a}{N_0}}{\left(\frac{N}{N_0}\right)^2} = \frac{n_z n_a}{\left(\frac{1}{\overline{P}}\right)^2}$$

and after its transformation:

$$\overline{P} = \sqrt{\frac{K}{n_a}\frac{1}{n_z}}$$

With the high degree of polycondensation the ratio (N/N_0) is so small that it can be neglected, and then:

$$\overline{P} = \sqrt{\frac{K}{n_a}}$$

This means that the average degree of condensation polymerization is directly proportional to the square root of the equilibrium constant and inversely proportional to the square root of the molar fraction of the secreting small molecule compound. Therefore, the removal of byproduct during the reaction increases always the average degree of polycondensation.

An important parameter in the polycondensation process is the degree of conversion, given by the following formula:

$$\alpha = \frac{2(N_0 - N)}{N_0 f} = \frac{2}{f} - \frac{2}{\overline{P} f}$$

where: f – functionality of monomer.

When the value of P is large then $\alpha = 2/f$.

In the case when one of the molecules of substances involved in the polycondensation reactions contain more than two functional groups, the resulting products may undergo cross-linking under the influence of temperature. It means they become to be thermosetting.

The value of the equilibrium constant K allows to determine the course of the polycondensation reaction. The increase in the value of equilibrium constant K results in a higher polymer yield. This issue can be illustrated by the synthesis of polyesters and urea-formaldehyde resins. The polyesterification reaction is a reaction typically reversible and the establishing state of equilibrium is unfavorable for the emerging polyester.

As the application in excess of one of the substrates leads to obtaining a compound of small molecular weight, the only way to shift the balance in favor of the emerging polyester, according to the Le Chatelier rule, is the removal of water as a small molecule by product. In that case it is possible to obtain polyester of high molecular weight.

Other examples of polycondensation are the reactions of hydroxymethyl derivatives of urea or phenol. These reactions are characterized by a relatively high value of equilibrium constant allowing even the synthesis of polymer in the aquatic environment. In this case the water removal from the reaction medium serves the synthesis of a product with higher molecular weight.

Overall, it was assumed that the polycondensation reactions, in which the equilibrium constant is less than 10^3, belong to the typical equilibrium reactions. If the value of the equilibrium constant is larger than 10^3, then such reactions are considered as unidirectional and the process is called the non-equilibrium polycondensation. An example of such a reaction is the synthesis of polyesters from glycols and dicarboxylic acid chlorides carried out in an alkaline environment:

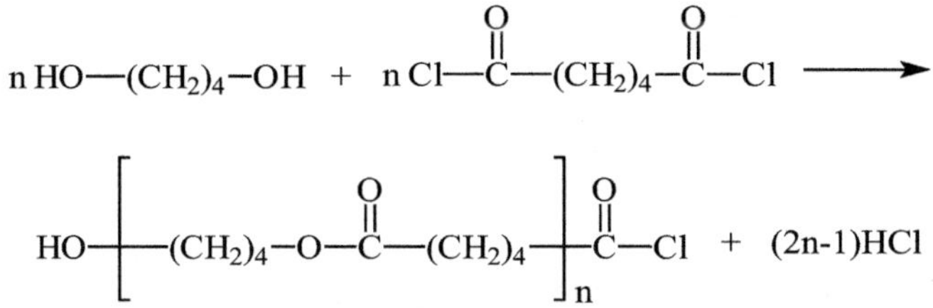

10.5.2 KINETICS OF POLYCONDENSATION

The polycondensation kinetics calculations are based on the assumption that the reactivity of the functional groups does not depend on the molecular weight of molecules in which such groups exist. This assumption was experimentally verified and confirmed.

In a classical polycondensation process the monomers are bifunctional, used in 1:1 molar ratio of reactants.

The measure of determining the reaction rate may be the disappearance of functional groups of one of the reactants.

The general scheme of the polycondensation reaction is as follows:

When using a catalyst, the process is a second-order reaction, and its rate V can be described by the following formula:

$$v = -\frac{d[X]}{dt} = k[X][Y]$$

assuming that:

$$[X] = [Y] = c$$

$$v = -\frac{dc}{dt} = kc^2$$

After integration the equation becomes:

$$kt = \frac{1}{c} + const$$

Because the degree of conversion *s* at time *t* is given by:

$$s = \frac{c_0 - c}{c_0} \text{ then } c = c_0(1-s)$$

where: c_0 – initial concentration of functional groups; c – concentration of functional groups at the time *t*.

Therefore, the equation of the polycondensation reaction rate takes the following form:

$$kt = \frac{1}{c_0(1-s)} + const$$

$$ktc_0 = \frac{1}{1-s} + const$$

If the polycondensation reaction proceeds without a catalyst and the functional groups X have catalytic properties, as it is the case in poly-esterification reaction, the process may run as a second order with one respect to X

Then:

$$\frac{d[X]}{dt} = k[X]^2[Y]$$

namely:

$$\frac{dc}{dt} = kc^3$$

after integration:

$$2kt = \frac{1}{c^2} + const$$

$$2c_0^2kt = \frac{1}{(1-s)^2} + const$$

The value of the constant k is read from the graph in the (y, x) coordinate system. Depending on the order of reaction the y coordinate is

$$\frac{1}{1-s} \text{ or } \frac{1}{(1-s)^2},$$

while the x-axis is the reaction time *t* in minutes. The reaction rate constant k is tan of the angle between the y-axis and the reaction line.

The average degree of polycondensation P is determined by the formula derived from its definition:

$$\overline{P} = \frac{N_0}{N} = \frac{c_0}{c} = \frac{1}{1-s}$$

10.5.2.1 Effect of Temperature on the Course of Polycondensation

Temperature is an important parameter used in the polycondensation processes. Its impact on the course of a particular reaction is various depending on the type of monomers used and the method of conducting the process. There are examples of polymer synthesis by the condensation method at low, around −40°C, as well as at high temperatures, of the order of 350°C.

An example of the polycondensation reaction at low temperatures is the synthesis of polyamides from dichlorides of carboxylic acids and diamines.

Other polycondensation reactions require higher temperature, whose value depends on the reactivity of certain functional groups and on an eventual use of catalysts, which reduce the activation energy. A typical example of such a reaction is the synthesis of polyesters. For example during the reaction of glycerol with phthalic acid anhydride it was found that the primary hydroxyl groups are more reactive than the secondary. For this reason this reaction, conducted at the temperature of 180°C, proceeds with the formation of a linear polyester, as if glycerol was a bifunctional monomer:

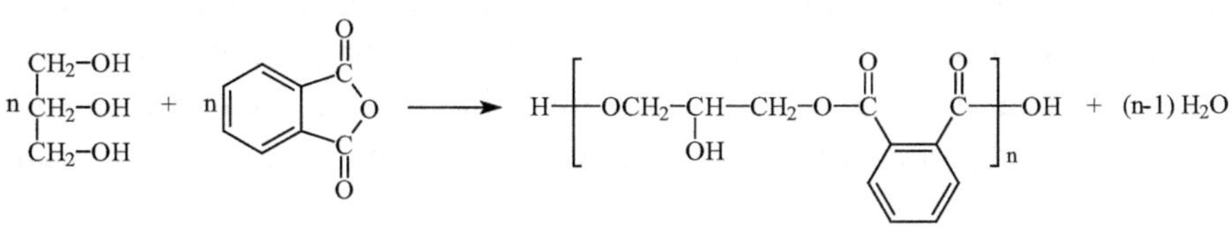

The increase of temperature to 220°C activates the secondary hydroxyl groups, which may under these conditions react with molecules of phthalic anhydride with the formation of a cross-linked polymer with spatial structure. Under these conditions glycerol acts as a trifunctional monomer.

The polycondensation reactions are often accompanied by the cyclization reactions, which can run as secondary or side reactions.

An example of the influence of temperature on the course of synthesis of polymers with heterocyclic structure in the main chain is the synthesis of polyimides. In the first stage of the synthesis, carried out at 20–70°C, formed are polyamid acids. These compounds heated under a reduced pressure at temperatures above 200°C cyclizes to the corresponding polyimides:

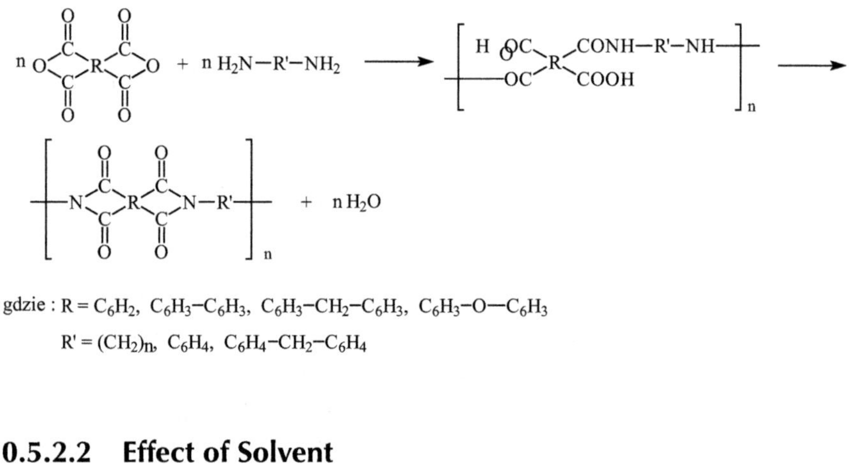

gdzie : R = C_6H_2, $C_6H_3{-}C_6H_3$, $C_6H_3{-}CH_2{-}C_6H_3$, $C_6H_3{-}O{-}C_6H_3$

R' = $(CH_2)_n$, C_6H_4, $C_6H_4{-}CH_2{-}C_6H_4$

10.5.2.2 Effect of Solvent

The condensation polymerization reaction may be carried out also in solution.

Solvent plays an important role in this process. Due to the ionic nature of polycondensation one uses mostly the polar solvents, which affect the kinetics of the reaction.

An example of such a reaction is the non-equilibrium polycondensation of dicarboxylic acid chlorides with amines conducted in amide solvents.

It passes always through a stage of a reactive complex of chloroanhydride with amine. The process runs with a simultaneous dehydrochlorination by dimethylacetamide molecules with the formation of an appropriate hydrochloride:

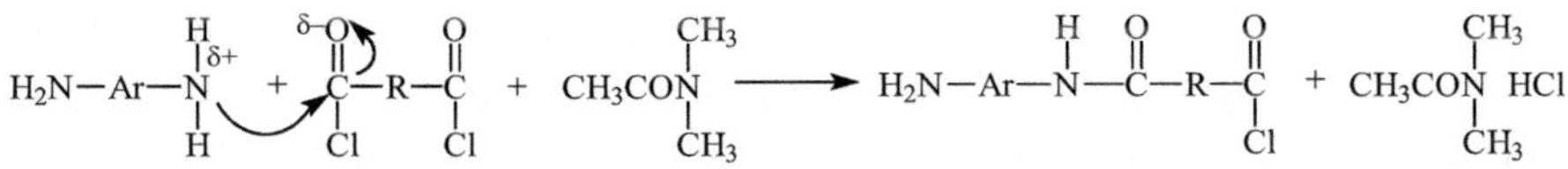

It was also found that during this reaction formed are also intermediate adducts between the chloroanhydride and the solvent molecules. The formed adducts react with amines. The polycondensation process depends thus on the constant of the equilibrium balance of the following reaction:

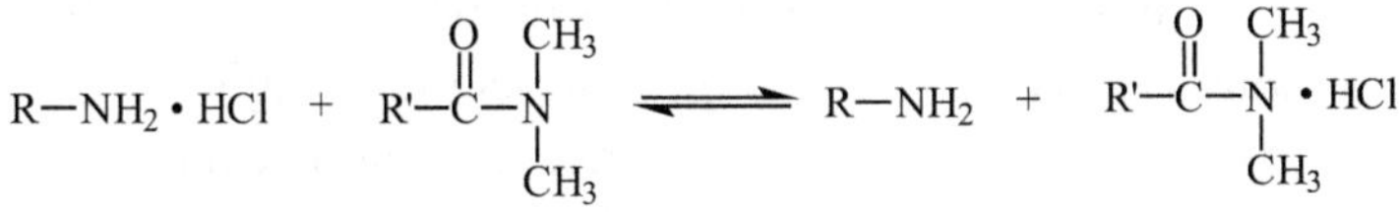

In addition, the transamidation reaction may take place too, as illustrated by the following example:

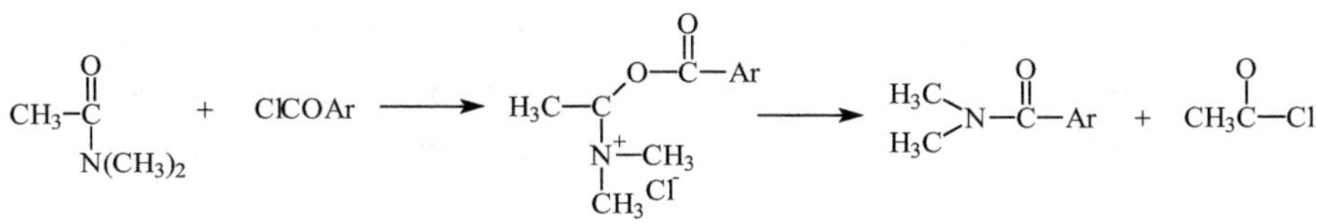

Application of the labeled ^{14}C atoms technique showed that the growth rate of the polymer chain is 280 times larger than its termination. The activation energy of chain growth is of 6.27 kJ/mol.

The kinetic studies of the polyamide acid chlorides synthesis showed that it could be carried out in temperatures below 80°C. At higher temperatures the competitive reactions of transamidation and chain growth termination begin to outweigh.

10.5.2.3 Effect of Monomer Concentration

The process of polycondensation reaction takes place mostly as a second or third order, therefore the concentration of reactants has a significant impact on the speed of its course.

Changing the concentration of monomers in the balanced polycondensation reaction affects not only its course and rate but also the molecular weight of the resulting polymer.

In dilute solutions predominates often the competitive cyclization reaction.

As an example one can cite the polycondensation reaction of sebacic acid (didecane acid) with hexamethylenediamine at equimolecular ratio and, in m-cresol environment. When the concentration of reactants is changed from 5% to 50% the global reaction rate increases by about 5 to 6 times. The rate constants for this reaction, depending on concentration, are practically unchanged at a given temperature. The obtained experimental results are presented in Table 10.31.

With the increasing monomer concentration, up to a certain optimal value, one obtains polymers with higher molecular weight.

Further increase of the monomer concentration affects the molecular weight of the product, decreasing it. This is explained by the increase of viscosity of the medium. During the synthesis of polyamides the best results are obtained with monomers at a concentration of 50%. In the case of polyesterification it is preferable to use 20% solutions.

During the low temperature polycondensation of chloroanhydride dicarboxylic acids with diamines, conducted at the phase boundary, the best solution is to use solutions with concentration equal to or slightly greater than 10%.

10.5.3 EFFECT OF CATALYSTS ON THE COURSE OF POLYCONDENSATION

The polycondensation reactions, depending on the starting monomers and the type of reaction, can be accelerated with cationic, anionic or ion-coordination catalysts.

TABLE 10.31 Effect of Concentration and Temperature on the Course of Polycondensation of Sebacic Acid with Hexamethylenediamine in m-Cresol

No.	Monomer concentration [mol/kg]	Reaction rate constant [kg/(mol•min•10^2)]				Activation energy [kJ/mol]
		433 K	443 K	456 K	463 K	
1	0.57	1.1	3.1	3.5	6.5	94.5
2	0.15	1.2	2.7	4.1	7.7	100.3
3	0.71	1.2	2.7	3.3	7.5	101.6
4	0.3	1.2	2.4	3.7	7.6	99.6
5	1.57	1.3	3.8	5.7	7.8	99.5

The action mechanism of various types of catalysts can be summarized as follows:

- mechanism of cationic catalysis

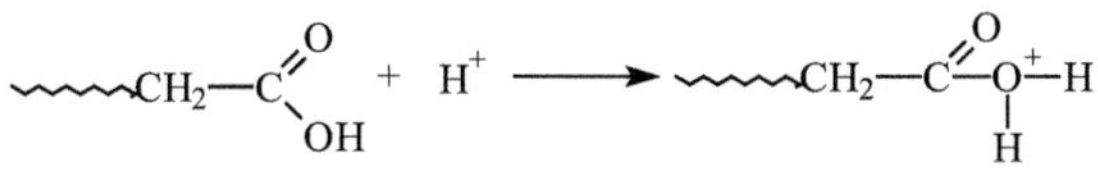

- mechanism of anionic catalysis

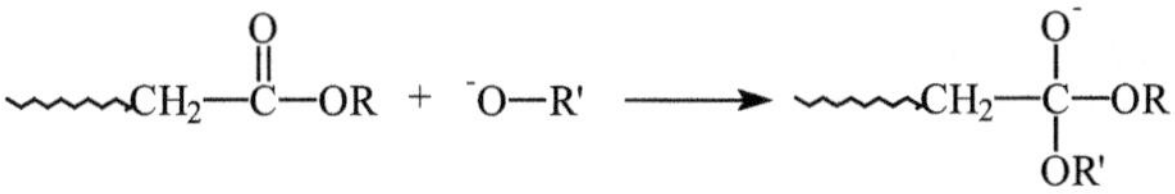

- mechanism of ion-coordination catalysis

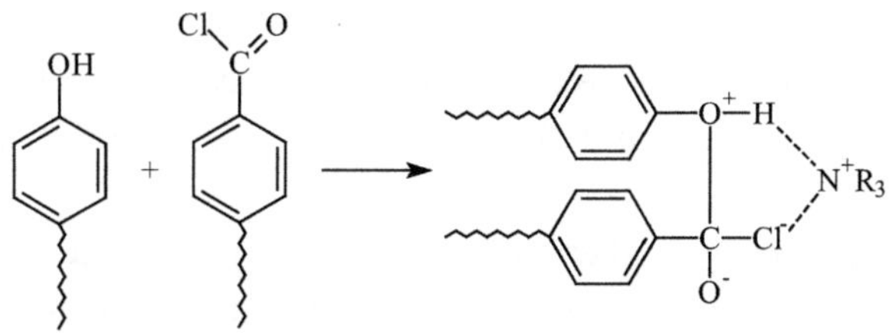

There exists now a number of polycondensation catalysts, derived from elements of practically all groups of the periodic table. Their catalytic activity depends on the particular type of polycondensation reaction.

Catalysts used in the synthesis of polyamides are generally tertiary amines or phosphines. The effect of basicity of amines on the course of synthesis of poly(hexamethylenediamine sebacate) is shown in Table 10.32.

Recently, more and more important are the interfacial catalysts used in the polycondensation reactions carried out at the interface. The most commonly used for this purpose compounds are the quaternary ammonium salts and the crown ethers.

As a result of the polycondensation reaction of α,α'-dichloro-p-xylene in a mixture of tetrahydrofuran and dimethyl sulfoxide at the interface of aqueous solution of sodium hydroxide a polymer with the following structure is formed:

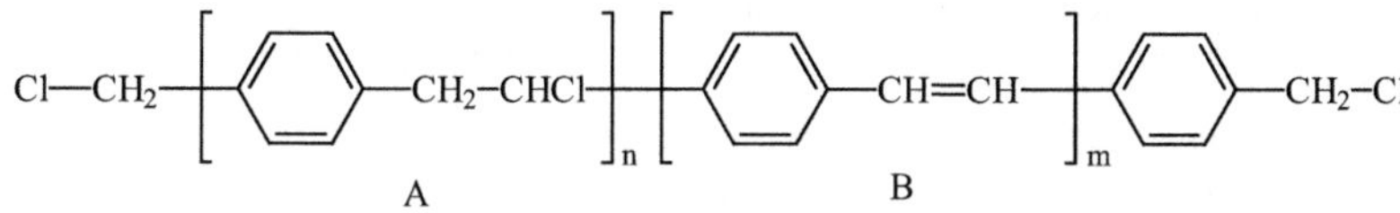

TABLE 10.32 Effect of the Basicity of Anionic Catalyst on the Course of Polycondensation of Sebacic Acid with Hexamethylenediamine in the Presence of Triphenylphosphine and Carbon Tetrabromide

No.	Base	pKa	Yield [%]	Intrinsic viscosity in H2SO4 at concentration 10 g/dm^3 and at 303 K
1	Quinoline	4.81	58	0.17
2	Pyridine	5.25	62	0.27
3	3-picoline	5.63	79	0.42
4	2-picoline	5.94	21	0.13
5	4-picoline	6.03	64	0.3
6	2,6-lutidine	6.6	24	Insoluble
7	Morphine	8.33	0	—

*After N. Ogata et al., Polym. J. 16, 569 (1984).

The influence of interphase ammonium catalysts on the course of this reaction, taking into account the emerging elements of structure A:B is presented in Table 10.33.

Application of crown ethers as catalysts of the interface polycondensation made possible the development of a new method of synthesis of polycarbonates from alkylaryl halides and carbon dioxide or potassium carbonate.

TABLE 10.33 Effect of Catalysts on the Course of Interphase Polycondensation of α,α'-Dichloro-p-Xylene in the Reaction with Sodium Hydroxide

No.	Catalyst	Concentration NaOH [%]	Yield [%]	
			A	B
1	—	50	10	6
2	benzyltriethylammonium chloride	50	21	12
3	benzyltriethylammonium bromide	50	29	17
4	benzyltriethylammonium iodide	50	54	31
5	tetrabutylammonium chloride	50	24	14
6	Tetrabutylammonium bromide	50	51	30
7	cetyltrimethylammonium chloride	50	20	12
8	benzyltrimethylammonium chloride	40	12	7
9	benzyltrimethylammonium chloride	30	8	5

A – based on the xylidine structure.

B – based on the amount of reacted monomer.

*After N. Yamazaki, Y. Imai, Polym. J. 15, 905 (1983).

The reactions proceeds according to the following schemes:

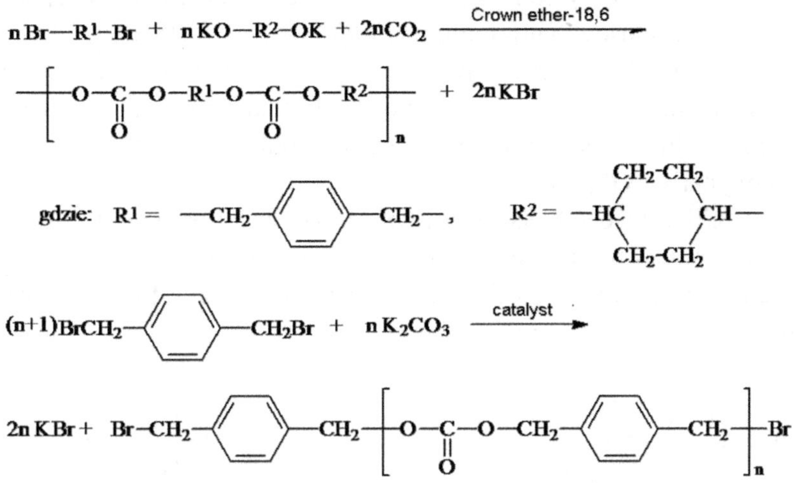

The obtained experimental results for this reaction are listed in Table 10.34.

In a similar way one can obtain polithiocarbonates. It is done by the catalyzed interface polycondensation of diphenols with thiophosgene (thiocarbonyl chloride) as shown on the following scheme:

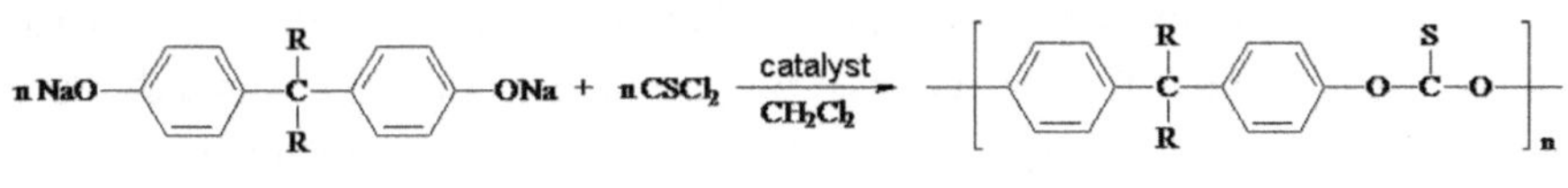

TABLE 10.34 Characteristics of Polycondensation of α,α'-Dibromo-p-Xylene with Potassium Carbonate in the Presence of Crown Ether 18–6

No.	α,α'-dibromo-p-xylene [mmol]	K_2CO_3 [mm]	Crown ether [mmol]	Solvent	Temp. [°C]	Time [hr]	Yield [%]
1	1.89	3.17	—	benzene	60	45	0
2	3.03	5.55	0.38	benzene	75	40	49.3
3	3.41	4.75	1.97	benzene	60	44	66.8
4	3.79	4.04	1.89	benzene	85	70	51.5
5	2.27	3.8	3.03	dioxane	70	44	67.1
6	3.03	4.75	3.03	dioxane	100	42	24.3

*After K. Soga, S. Hosoda, S. Ikeda: J. Polym. Sci. Polym. Lett. Ed., 15, 611 (1977).

As catalysts of this reaction one can use tetrabutyl ammonium bromide, methyl cetyltrimethyloammonium bromide, tributylcetylphosphonium bromide, methyltrioctylammonium chloride or crown ethers.

Known are also are the ways to catalyze the polycondensation by formation of charge transfer complexes with some polymers such as, for example, nitrated polystyrene during the polycondensation of 1,4-bis (methoxycarbonyl) piperazine.

In addition to catalysts in order to carry out effectively the polycondensation reaction one can use also the condensing means. An example of such a variant of synthesis is the polycondensation of p-aminobenzoic acid using the silicon tetrachloride:

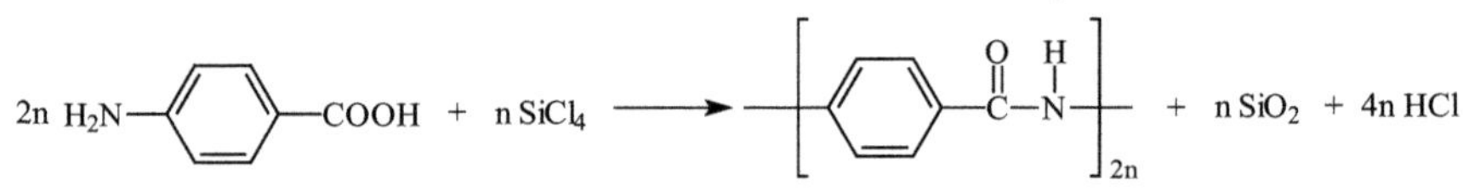

10.5.4 EFFECT OF MONOMER STRUCTURE ON THE POLYCONDENSATION PROCESS

The course of polycondensation is affected by both the chemical structure of the reagents used as well as their quantitative ratio. As already mentioned, in the different types of polycondensation reactions a dynamic equilibrium state is established which is more or less shifted in favor of the nascent polymer. The use of a multifunctional monomer can lead to obtaining a cross-linked product, while the use of a single function monomer leads to the chain growth termination and the formation of low molecular weight compounds. A very important factor influencing the molecular weight of the resulting linear polymer is the quantitative ratio of used monomers. Excess of one of them causes a mutual blocking of functional groups and thus the formation of a product with low molecular weight. In order to obtain a polymer with high molecular weight in the equilibrium polycondensation process it is essential to preserve the stoichiometry of reacting monomers. A striking example of this may be the synthesis of Nylon (polyamide-6, 6). For the polycondensation reaction one uses the salt formed from adipic acid and hexamethylenediamine, keeping always the molar ratio to be exactly 1:1.

In the non-equilibrium polycondensation process, conducted at the interface, it is not necessary to use reagents in the molar ratio of 1:1, as a possible excess of one of them remains in a separate phase and does not take part in the reaction.

The dependence of the polymer molecular weight resulting from the excess of one of the monomers in equilibrium and non-equilibrium polycondensation reactions is presented in Figure 10.8.

10.5.5 *COPOLYCONDENSATION*

The copolycondensation, known also as condensation copolimerization, proceeds when into the reaction environment introduced are different reagents with a similar chemical structure. For example, in the polyesterification process of dicarboxylic acid with diol various other dicarboxylic acids or glycols can be used.

In general, the copolycondensation process can be represented schematically by:

HOOC—R—COOH + HO—R''—OH ⇌ HOOC—R—COO—R''—OH + H_2O

HOOC—R'—COOH + HO—R''—OH ⇌ HOOC—R'—COO—R''—OH + H_2O

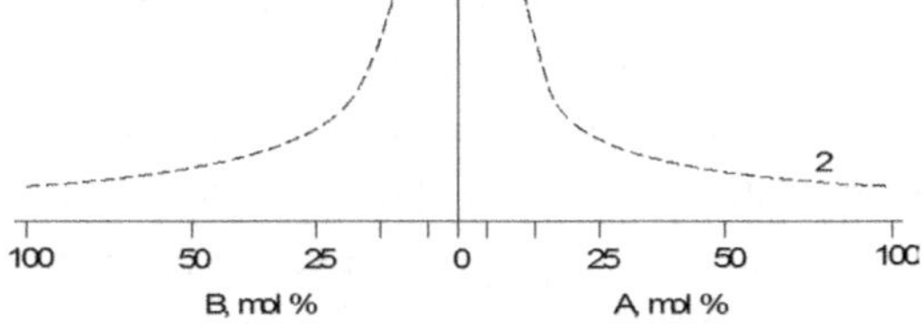

FIGURE 10.8 Dependence of the average molecular weight M of polyhexamethylene diamine on the excess of the components used: 1 – non-equilibrium process (at interphase), 2 – equilibrium process (in blend).

At the end a copolymer with the following structure is obtained:

$$HO{-}\left[OC{-}R{-}COO{-}R''{-}O\right]_n{-}\left[OC{-}R'{-}COO{-}R''{-}O\right]_m{-}H$$

where: $R = {-}(CH_2)_x{-}$, ${-}C_6H_4{-}$ $R' = {-}(CH_2)_y{-}$, ${-}C_6H_4{-}$ $R'' = {-}(CH_2)_z{-}$

The rate of these reactions is proportional to the concentration of the groups R – COOH, R” – OH, R’ – COOH and can be described by the following equations:

$$\frac{-d[R-COOH]}{dt} = k_1[R-COOH][R''-OH]$$

$$\frac{-d[R'-COOH]}{dt} = k_2[R'-COOH][R''-OH]$$

After dividing these equations by themselves one gets:

$$\frac{-d[R-COOH]}{-d[R'-COOH]} = \frac{k_1[R-COOH]}{k_2[R'-COOH]} = k\frac{[R-COOH]}{[R'-COOH]}$$

The last equation shows that the composition of the reaction mixture will correspond to the copolymer composition only when $k = 1$.

During the test of this process it appeared that some secondary reactions can run too, such as acidolisis, alcoholysis or transesterification. As result of these reactions the previously formed homopolymer is converted into a copolymer and the composition of polycondensation product corresponds to the composition of reactants used. The condensation copolymers may be also formed by heating a mixture of polymers with similar structure, e.g., two polyamides. Similarly, by heating the block copolymer one can obtain a normal statistical copolymer. This is possible by establishing a series of equilibrium states in the polycondensation process. The choice of appropriate reaction parameters makes possible the synthesis of copolycondensates with a more uniform structure, such as, for example, block copolymers.

10.5.6 BASIC TYPES OF POLYCONDENSATION REACTIONS

The most important examples of polycondensation reactions include:

- polyesterification:

$$n\,HO{-}R{-}COOH \rightleftharpoons H{-}[ORCO]_n{-}OH + (n-1)H_2O$$

$$n\,HO{-}R{-}OH + n\,HOOC{-}R'{-}COOH \rightleftharpoons {-}[OROOCR'CO]_n{-}OH + (2n-1)H_2O$$

$$n\,HO{-}R{-}OH + n\,ClOC{-}R'{-}COCl \longrightarrow H{-}[OROOCR'CO]_n{-}Cl + (2n-1)\,HCl$$

gdzie : $R = {-}(CH_2)_x{-}$ $R' = {-}(CH_2)_y{-}\ ,\ {-}C_6H_4{-}$

- polyamidation:

$$n\,H_2N{-}R{-}COOH \rightleftharpoons H{-}[HNRCO]_n{-}OH + (n-1)H_2O$$

$$n\,H_2N{-}R{-}NH_2 + n\,HOOC{-}R'{-}COOH \rightleftharpoons H{-}[NHRNHCOR'CO]_n{-}OH + (2n-1)H_2O$$

$$n\,H_2N{-}R{-}NH_2 + n\,ClOC{-}R'{-}COOH \longrightarrow H{-}[NHRNHCOR'CO]_n{-}Cl + (2n-1)\,HCl$$

- polyanhydrization:

$$2n\,HOOC{-}R{-}COOH \rightleftharpoons HO{-}[\underset{\underset{O}{\|}}{C}{-}R{-}\underset{\underset{O}{\|}}{C}{-}O{-}\underset{\underset{O}{\|}}{C}{-}R{-}\underset{\underset{O}{\|}}{C}{-}O]_n{-}H + (2n-1)H_2O$$

- polyamination:

$$2n\,H_2N{-}R{-}NH_2 \rightleftharpoons H{-}[NH{-}R{-}NH{-}R]_n{-}NH_2 + (2n-1)NH_3$$

- polyacetylation:

$$2n\,HO{-}R{-}OH + n\,R'{-}C\overset{\displaystyle O}{\underset{\displaystyle H}{\lessgtr}} \rightleftharpoons H{-}[O{-}R{-}O{-}\overset{\overset{H}{|}}{\underset{\underset{R'}{|}}{C}}{-}OR]_n{-}OH + (2n-1)H_2O$$

- polysulfuration:

$$2n\,Cl{-}R{-}Cl + n\,Na_2S \longrightarrow Cl{-}[R{-}S]_n{-}Na + (2n-1)NaCl$$

- polyoxymethylation:

a) phenols:

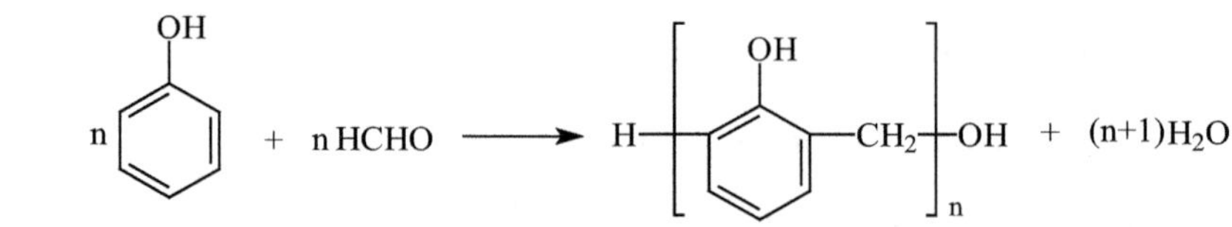

b) urea:

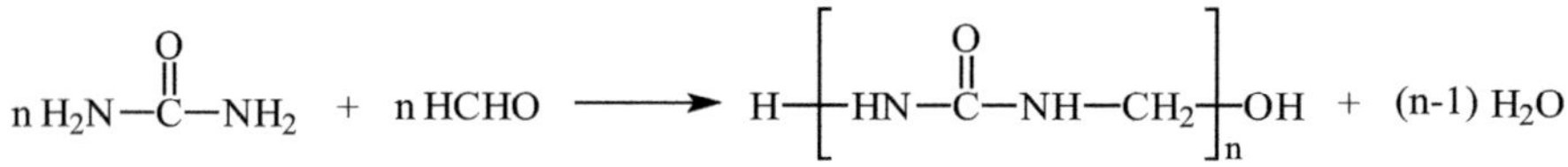

With an excess of formaldehyde the polyoxymethylation of phenol or urea continues and causes a cross-linking of the product.

- polybenzimidazolation:

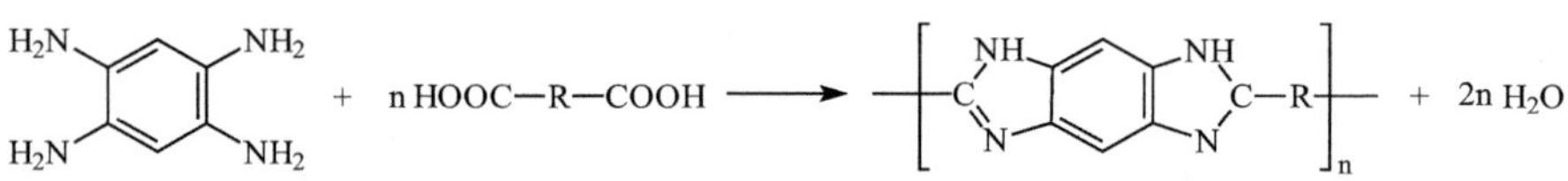

- polyimidation:

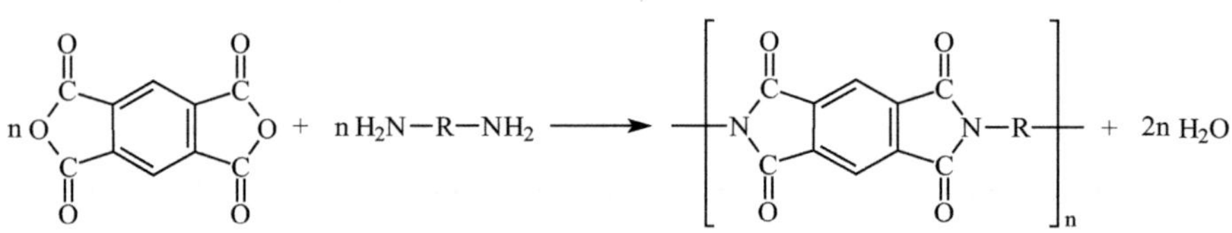

- polysiloxanation:

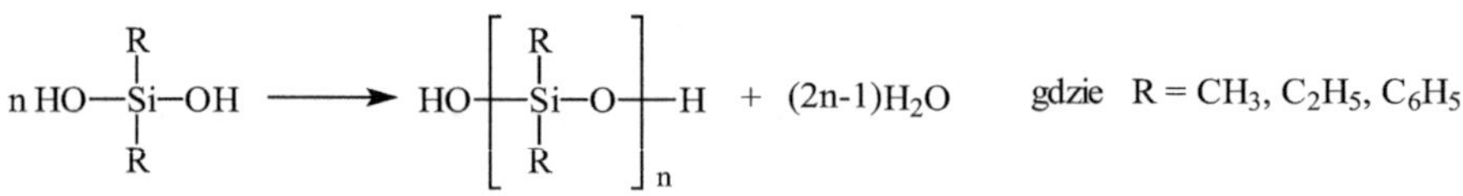

- polycomplexation:

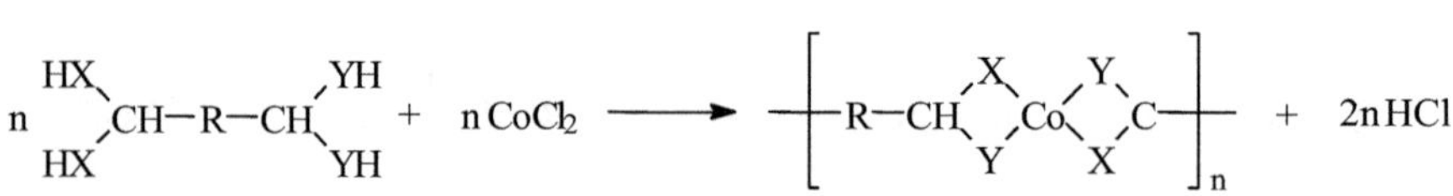

- polyhydrazidation:

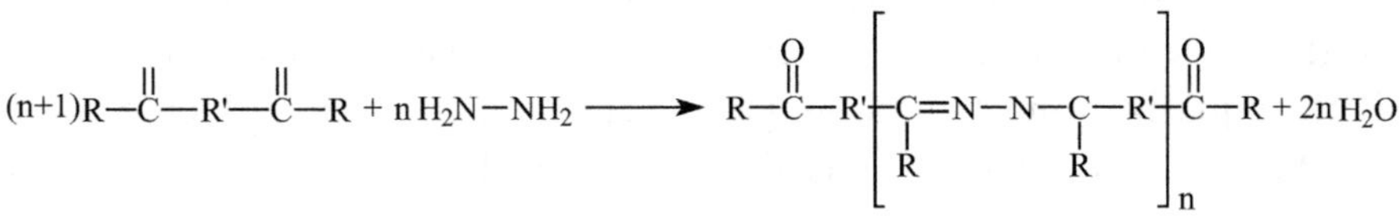

where: R = CH3, (CH2)xCH3, R' = –(CH2)y-,

- dehydropolycondensation:

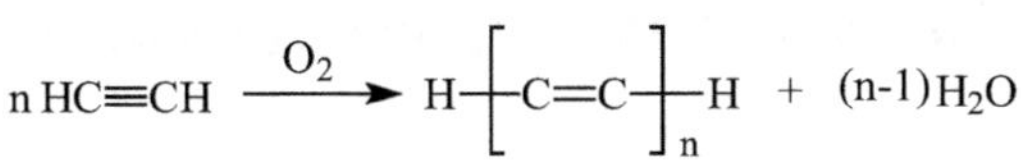

The side reactions accompanying the polycondensation reaction:

The cyclization reaction is an important byproduct of the process, competing with the homopolycondensation. This reaction takes place particularly easily when it is possible to form a permanent ring of five, six or seven members. This reaction has an intramolecular character:

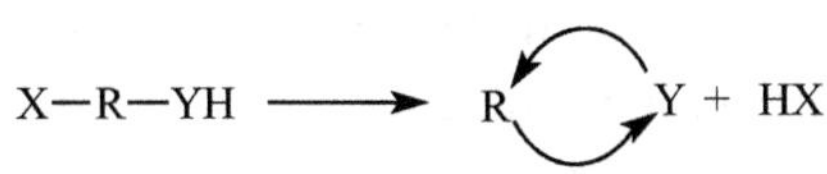

Running the polycondensation reaction at low substrate concentration in the reaction mixture favors the cyclization process. Moreover, increasing the process temperature favors also the cyclization reaction. Despite the use of optimal parameters a balance between both competitive reactions establishes and in the obtained product, in addition to the polymer, an amount of a low molecular mass cyclic compound is present. This compound is removed by extraction method, in a similar way as, for example, removal of caprolactam from polyamide-6 or by distillation as it the case during the synthesis of polysiloxanes.

Another important group of processes accompanying the polycondensation reactions are the degradation reactions caused by the presence of bonds in polymer macromolecules, disintegrating under the influence of compounds present in the reaction mixture. Examples of such reactions may be the rupture of ester bond in the polyester molecule under the influence of alcohol (alcoholysis) or acid (acidolysis).

The decrease in molecular weight of polycondensation products caused by the degradation reactions is the greater, the greater is the initial molecular weight of degraded polymer. The consequence of this is the equalization of chain lengths of individual macromolecules, thus by the same, decrease of polydispersity. For this reason one introduces monofunctional organic acids during the synthesis of polyamide as well as conducted is also the linearization process during the synthesis of polyorganosiloxanes.

10.5.7 CROSS-LINKING IN THE POLYCONDENSATION PROCESS

The use of tri-or more functional monomers in the process of polycondensation leads, in consequence, to cross-linking the formed polymer. The point at which the cross-linking reaction takes place is called gelation.

Flory has developed a statistical method for determining the gelation point, depending on the degree of conversion. He assumed that the probability of gelation depends on the branching factor α giving the probability of connecting two branching molecules.

The gelation point determines the value of critical branching factor α_k, which depends on the functionality of the reacting monomers.

$$\alpha_k = \frac{1}{f-1}$$

where f is the monomer functionality.

This formula shows that for trifunctional monomers (f = 3) $\alpha_k = ½$. It means that gelation will occur at 50% of reacted substrates.

One can also provide the dependence of α_k coefficient on the degree of conversion p.

If in the reaction mixture only trifunctional monomer molecules are present, such as

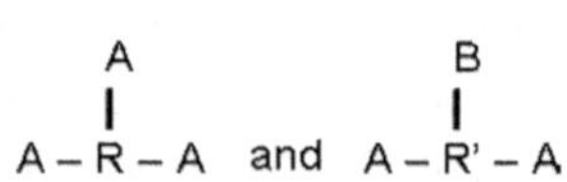

then the probability of addition of one of them to the second one is equal to the probability of conversion of so many groups, namely:

$$\alpha_k = p$$

If however, in the reaction mixture are present difunctional and trifunctional monomer molecules:

```
  A          B
  |          |
A – R – A  and  R' – B
```

then, assuming that the number of groups A is equivalent to the number of groups B, the probability of reaction for a molecule with group A terminated macromolecule will depend both on whether the group A will react with group B, as well as whether the other end of the monomer B – R' – B will react with the second group A of the macromolecular compound.

In this case:

$$\alpha_k = p$$
$$p^2 = \frac{1}{f-1} = \frac{1}{3-1} = 0.5$$
$$p = \sqrt{0.5} \approx 0.7$$

It follows from this that the point of gelation of equivalent mixture of difunctional and trifunctional monomers will take place at the conversion rate of 70%.

If the reaction mixture contains difunctional and trifunctional monomer molecules containing identical functional groups A and difunctional monomers reacting with groups A, then the coefficient α_k is given by the following expression:

$$\alpha_k = \frac{rp^2\rho}{1 - rp^2(1-p)}$$

where

```
  A
  |
A—R—A   A—R'—A   B—R'—B  - used monomers
```

$$r = \frac{N_0^A}{N_0^B} = \frac{\text{number of functional groups A}}{\text{number of functional groups B}}$$

$$\rho = \frac{N_{0\,rozg}^A}{N_0^A} = \frac{\text{number of functional, groups A in a trifunctional monomer}}{\text{number of functional groups A}}$$

10.5.8 TECHNICAL METHODS OF POLYCONDENSATION PILOTING

Depending on the type of reaction the polycondensation process can be carried out in a single or a heterogeneous phase system.

The reactions carried out in a single-phase system include the polycondensation processes in (a) solution and (b) melt.

The process of polycondensation in solution can be carried out for both the equilibrium and non – equilibrium reactions. A characteristic feature of this process is that it takes place in a homogenous environment of relatively low temperature and the use of homogeneous catalysts is possible. The emitted low molecular weight byproducts are most frequently removed by distillation as exemplified by the azeotropic removal of water during the polyesterification. Another way to remove the low molecular weight byproducts is the possibility of binding them by solvent molecules. A typical example of such a reaction is the synthesis of polycarbonate. The emitted in this reaction hydrogen chloride reacts with pyridine, used as a solvent, with formation of pyridine hydrochloride,.

The course of polycondensation reaction in solution is influenced by the type and the amount of used solvent, catalyst and the reaction temperature.

The advantages of this method are:

- easy evacuation of reaction heat;
- possibility of obtaining polymers of different molecular weight in the same reactor;
- possibility of influencing the reaction rate by changing the viscosity of the solution;
- relatively small number of side reactions; and
- possibility of a direct use of the product in the form of a glue.

The primary disadvantage of this process, from an environment protection perspective, is the necessity to operate often with toxic solvents and the need to extract polymer from solution.

Polycondensation in the melt is performed at relatively high temperatures, usually above 200°C.

The possibility of conducting polycondensation in the melt is, however, limited by the thermal resistance of both the monomers used, as well as the resulting product.

In order to inhibit undesirable side reactions such as, for example, the oxidation, the process is carried out in an inert atmosphere or under reduced pressure. The use of both: the high temperature and the low pressure makes easy removal of low molecular mass byproducts. The advantage of this method is the possibility of producing polymers with a relatively high molecular weight and fabrication of synthetic fibers directly from the melt.

Examples of such processes is the production of poly(ethylene terephthalate) and of polyamide 6.6.

Polycondensation in a heterogeneous system is performed:

- at phase boundary;
- in emulsion; and
- in solid phase.

The polycondensation at the interface takes place when different substrates are soluble in various solvents, not miscible with each other. A typical example of such a reaction is the non-equilibrium polycondensation of dicarboxylic acid chlorides with diamines or glycols.

As mentioned previously, in this case the interfacial catalysts can be used and the addition of a binder to the aqueous phase makes it easy to remove the byproduct.

The advantages of polycondensation at the interface are:

- high reaction rate;
- possibility of conducting the process at room temperature;
- no need for a strict control of the monomers purity;
- possibility of using in excess one of the substrates; and
- obtaining of polymers with high molecular weight.

The disadvantages of the method are the need to use solvents and the evacuation of large quantities of sewage.

The polycondensation in emulsion is a variant of the method carried out at the interface. It allows a much larger contact area of reactants, than it did at the interface as well as a partial or complete elimination of solvents. This process is performed in similar cases as the condensation on the phase boundary. As a result of the polycondensation of dicarboxylic acid dichlorides with diamines a competitive reaction of hydrolysis takes place:

$$2\,H_2N{-}R{-}NH_2 + ClOC{-}R'{-}COCl \xrightarrow{k_a} H_2N{-}R{-}NHCOR'CONHR{-}NH_2$$

$$ClOC{-}R'{-}COCl + 2H_2O \xrightarrow{k_r} HOOC{-}R'{-}COOH + 2HCl$$

The prevalence of the amidation reaction rate over the hydrolysis reaction determines the following expression:

$$\frac{k_a}{k_r} = \frac{\lg\left(\frac{C_{A0}}{C_{A\infty}}\right)}{\lg\left(\frac{C_{(H_2O)0}}{C_{(H_2O)\infty}}\right)}$$

where: C_{A0}, $C_{(H_2O)\infty}$ – initial concentration of amine groups (or - COCl) and water; $C_{A\infty}$ – concentration of amine groups (or - COCl) after the time $\tau = \infty$; $C_{(H_2O)\infty}$ – concentration of water after the time $\tau = \infty$.

The degree of conversion in the amidation reaction is given by:

$$\alpha = \frac{C_{A0} - C_{A\infty}}{C_{A0}}$$

The obtained experimental results are presented in Table 10.35.

The method of polycondensation in the solid phase is practically applied in the process of cross-linking of polymers. Depending on the type of polymer and eventual catalyst used it may be carried out at different temperatures.

This method is used mainly in the processing of plastics.

TABLE 10.35 Reaction Rate Constant for Amidation and Hydrolysis in the System: Tetrahydrofuran – Water – Na_2CO_3 – NaOH by Benzyl Chloride (1–6) or Aniline (7–11)

No.	Monomer	α	*ka/kr*
1	hexamethylenediamine	0.946	3170
2	p-phenylenediamine	0.954	4460
3	m-phenylenediamine	0.887	1050
4	o-phenylenediamine	0.832	646
5	benzidine	0.877	1030
6	2,4-diaminophenol	0.714	260
7	2,6-naphthalene dicarboxylic acid dichloride	0.964	4800
8	Sebacylic acid dichloride	0.959	3470
9	Terephtalic acid dichloride	0.93	1920
10	Isophtalic acid dichloride	0.928	1630
11	oxalyl dichloride	0.11	6.2

After L.B. Sokołov, W.I. Kogulova, Wysokomol. Soied. A. (1979) 21, 1050.

10.6 CONDENSATION POLYMERS

10.6.1 PHENOPLASTS

Phenoplasts (phenolics) are synthetic resins formed by the condensation polymerization reaction of phenol or its derivatives with aldehydes. The most commonly used are phenol formaldehyde resins (PF).

The polycondensation reaction of phenol with aldehydes is catalyzed by both the hydrogen and the hydroxide ions. Because of the favorable equilibrium states the reaction can be carried out in an aqueous solution.

In the first stage of the reaction takes place an electrophilic binding of aldehyde molecule in ortho or para position to the phenolic group with a simultaneous rearrangement of the hydrogen atom to the oxygen atom.

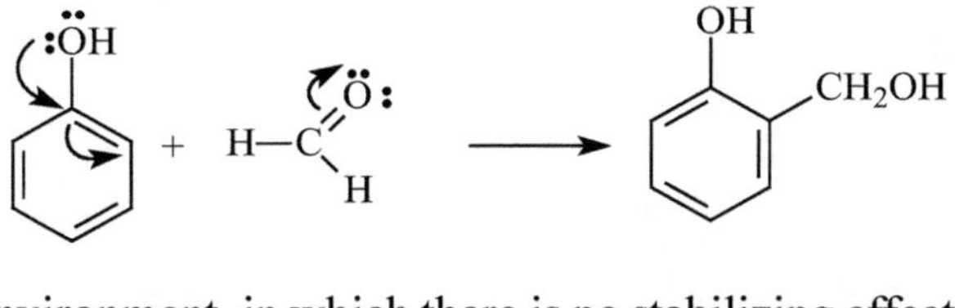

The acidic environment, in which there is no stabilizing effect of hydrogen bonds, the rate of condensation reaction of hydroxymethylene groups is very large. In the case of trifunctional phenol it results in an immediate cross-linking. To avoid this the polycondensation reaction in an acidic environment is performed in excess of phenol. In practice the maximum acceptable molar ratio of formaldehyde to phenol is 10:12. The optimum ratio giving the possibility of safe operation of the process and obtaining the best properties of the resin is the ratio of 26.5–27.5 g of formaldehyde per 100 g of phenol. It corresponds approximately to the molar ratio 6:7. The resins obtained in these conditions, called novolacs, are thermoplastic and are soluble in organic solvents, what indicates their linear structure:

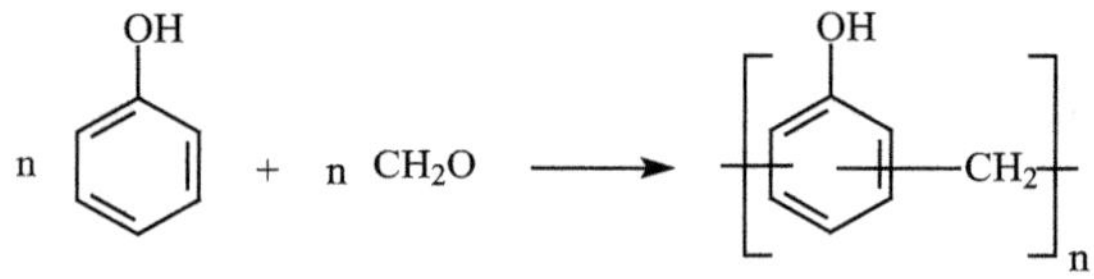

Novolacs can be cross-linked by heating with polyoxymethylene or urotropin.

During the polycondensation of phenol with formaldehyde in alkaline environment an excess of formaldehyde is applied. Since the rate of this reaction is much lower than in an acidic environment a soluble resin with linear structure can be isolated. This resin is called resol resin.

During the polycondensation of phenol with formaldehyde the following reactions take place:

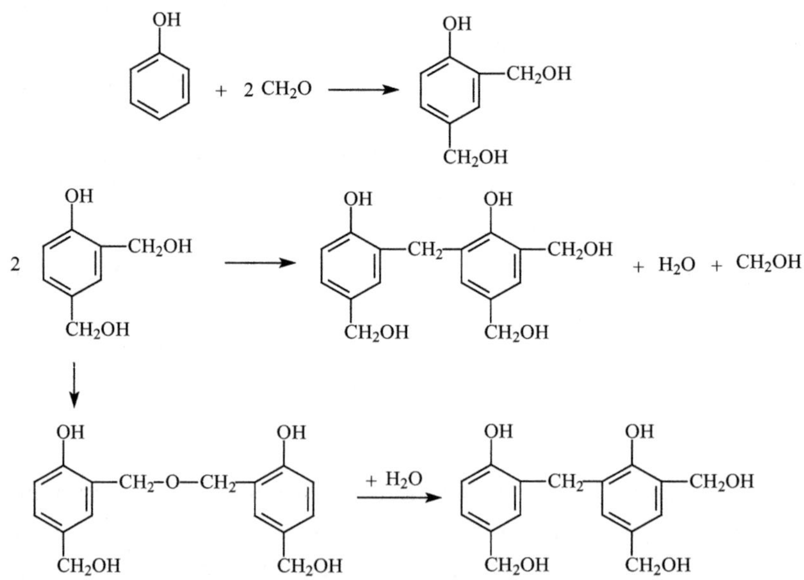

Repeated many times the above processes lead to the formation of macromolecules, in which the aromatic rings are connected by methylene and dimethylenether bridges.

Under heating the resols transform into partially cross-linked resitols, and then into the insoluble and infusible resites.

The reactive hydroxymethyl groups allow a chemical modification of these resins. The phenol formaldehyde resins, etherified by butanol or esterified by fatty acids, are more soluble and more flexible, allowing their use as lakhs for painting purposes.

Known are also phenol formaldehyde resins (PF – phenoplasts) produced during the condensation of phenol with other aldehydes such as acetaldehyde and furfural.

10.6.2 AMINOPLASTS (CARBAMIDE RESIN)

Aminoplasts mean generally the macromolecular reaction products of compounds containing amino groups in the molecule or amide groups with aldehydes. Main types of aminoplasts are products of polycondensation reaction of formaldehyde with urea, melamine, dicyanodiamide, guanidine, thiourea, etc.

As a result of the reaction of amine compounds with aldehydes the aldehyde molecule is attached to the nitrogen atom of amino group with a simultaneous rearrangement of the hydrogen atom to the oxygen atom of aldehyde group. When using formaldehyde formed are compounds containing hydroxymethylamino groups, which are easy to separate. With an excess of formaldehyde it is possible to replace all the hydrogen atoms of amine groups by methylol groups.

$$R{-}NH_2 + CH_2O \longrightarrow R{-}NH{-}CH_2OH$$

$$R{-}NH{-}CH_2OH + CH_2O \longrightarrow R{-}N(CH_2OH)_2$$

The reaction for obtaining methylol (hydroxymethyl) derivatives runs in the aquatic environment and is catalyzed both by hydrogen and hydroxyl ions. The pH value determines also the rate of reaction of urea with formaldehyde. The process is the slowest one in the pH range of 4–8.

Increasing pH above 8 or a reducing it below 4 speeds up significantly the process. The resulting methylol compounds undergo polycondensation with the formation of macromolecules linked by methylene and dimetylene ether bridges.

$$—NH—CH_2OH + H_2N— \longrightarrow —NH—CH_2\text{–}NH— + H_2O$$

$$—NH—CH_2OH + HOCH_2\text{–}NH— \longrightarrow —NH—CH_2\text{–}O—CH_2\text{–}NH— + H_2O$$

The dimetylene ether bridges arise when the polycondensation reactions are carried out in a neutral or alkaline environment. They may also occur in an environment slightly acidic with a large molar excess of formaldehyde.

The polycondensation reaction of multifunctional amine compounds with formaldehyde is conducted in an inert environment in order to avoid a premature cross-linking of product. The obtained urea or melamine-formaldehyde derived polycondensates are brought to pH = 7 and in this state they remain stable for about 6 months. Cross-linking of these resins is accelerated by hydrogen ions and by elevated temperature. The resulting products are used as adhesives, binders for molding and laminates and as impregnating materials.

The urea or melamine-formaldehyde modified with butyl alcohol resins are used as chemically setting varnishes. In these products a part of methylol groups undergoes an etherification by butyl alcohol. The butyl chains cause inner plastification of the formed resin.

10.6.3 POLYESTERS

Polyesters are formed by polycondensation of multifunctional acids with alcohols with evolution of water molecules. In this reaction the state of equilibrium establishes rapidly and the resulting water acts as a hydrolyzing factor. Therefore, in order to shift the equilibrium state, it is necessary to remove water from the reaction environment by carrying out the process at high temperature (220–250°C) or by azeotropic distillation of water from toluene or xylene. The reaction proceeds according to the scheme:

$$n\,HO{-}R{-}OH + n\,HOOC{-}R'{-}COOH \rightleftharpoons H{-}\left[O{-}R{-}O{-}\overset{\overset{\displaystyle O}{\|}}{C}{-}R'{-}\overset{\overset{\displaystyle O}{\|}}{C}\right]_n{-}OH + (2n\text{-}1)\,H_2O$$

Catalysts for this reaction are acids, such as, for example, sulfuric acid and p-toluene sulfonic acid. The use of elevated temperatures leads to a series of side reactions and to eliminate them the process is conducted in nitrogen atmosphere or in a solvent whose use allows to lower the reaction temperature.

Among the other methods of obtaining polyesters are:

a) homopolycondensation of hydroxyacids in which the number of methylene groups exceeds 5. It ensures a complete elimination of cyclization:

$$n\,HO{-}R{-}COOH \rightleftharpoons H{-}\left[O{-}R{-}\overset{\overset{\displaystyle O}{\|}}{C}\right]_n{-}OH + (n\text{-}1)\,H_2O$$

b) alcoholysis reaction of dimethyl esters of dicarboxylic acid with glycols, catalyzed by acid or acetates of cadmium, zinc and lead

$$n\,HO{-}R{-}OH + n\,H_3CO{-}\overset{\overset{\displaystyle O}{\|}}{C}{-}R{-}\overset{\overset{\displaystyle O}{\|}}{C}{-}OCH_3 \rightleftharpoons H{-}\left[O{-}R{-}O{-}\overset{\overset{\displaystyle O}{\|}}{C}{-}R{-}\overset{\overset{\displaystyle O}{\|}}{C}\right]_n{-}OCH_3$$

$$+\ (2n\text{-}1)\,CH_3OH$$

c) polycondensation reaction at the interface of glycols with carboxylic acid dichlorides:

$$n{+}1\ HO{-}R{-}OH + n\ Cl{-}\overset{\overset{\displaystyle O}{\|}}{C}{-}R'{-}\overset{\overset{\displaystyle O}{\|}}{C}{-}Cl \longrightarrow H{-}\left[O{-}R{-}O{-}\overset{\overset{\displaystyle O}{\|}}{C}{-}R'{-}\overset{\overset{\displaystyle O}{\|}}{C}\right]_n{-}O{-}R{-}OH$$

Polyesters, formed as result of the above-described reaction can be divided into linear polyesters, alkyd resins, unsaturated polyesters and polyarylates.

Linear polyesters are obtained from dicarboxylic acids and glycols. Aliphatic polyesters are characterized by a low melting temperature (below 80°C) and are well soluble in many organic solvents, such as, for example, chloroform, benzene, acetone, tetrahydrofuran and ethyl acetate. The ester bindings of aliphatic polyester are easily hydrolyzed at room temperature under the influence of the aqueous solution of base. The resistance of aliphatic polyesters to hydrolysis increases with the increase of the number of methylene groups separating ester bonds. The linear polyesters derived from aromatic dicarboxylic acids are characterized by a greater degree of crystallization and higher melting temperatures. The biggest importance in this group of polymers has poly(ethylene terephthalate) (PET), which is obtained by the reaction of dimethyl terephthalate with ethylene glycol. This polymer is used for packing purposes (bottles for beverages) and also to produce foils and synthetic fibers (Dacron, Polyester, Terylene, Elana).

The alkyd resins are polyesters obtained from aliphatic polyols and phthalic acids. In the first stage of reaction obtained are linear polyesters, which can be then cross-linked. These resins are often modified with fatty acids in order to improve their flexibility, solubility and resistance to water.

The unsaturated polyester resins are obtained by polycondensation of glycols with fumaric acid or maleic acid anhydride. Here it is possible to use as a comonomer other carboxylic acids, such as adipic, phthalic and hexachloroendomethylenetetrahydrophthalic (so called HET acid).

The introduction of chlorine, bromine, phosphorus or boron atoms to the polyester molecule reduces the flammability of the product. The obtained linear polyesters, containing double bonds, can be cross-linked by radical copolymerization with vinyl or allyl compounds. The most frequently used cross-linking agent is styrene.

Poliarylates are products of polycondensation of diphenols with aromatic dicarboxylic acids. The characteristic chemical structure of these polymers, whose main chains are composed almost exclusively of aromatic rings linked together by chemical bonds, gives them some specific properties. Above all they are characterized by a high softening temperature, thermal resistance and resistance to chemical agents.

10.6.4 POLYCARBONATES

Polycarbonates are linear polyesters of carbonic acid. Due to special methods of their obtaining and to specific physico chemical properties and thermoplasticity they constitute a distinct group of polymers.

Depending on the structure of the polymer chain polycarbonates are divided into aliphatic, aliphatic-aromatic and aromatic. The greatest practical importance represents the aromatic polycarbonates.

Polycarbonates are obtained by condensation polymerization reaction of bifunctional alcohols or phenols with phosgene or carbonic acid esters. The reaction may be carried out at the interface or in a solvent.

The reaction at the interface is carried out in two-phase system composed of alkaline aqueous solution of biphenolate and phosgene in an organic solvent (usually in methylene chloride). The process to obtain a polycarbonate is a two-step process. In the first stage the reaction proceeds in the presence of excess of phosgene and leads to obtaining polycarbonates with molecular weight of 5,000–10,000 and terminal chloromethyl groups:

$$n\ NaO{-}Ar{-}ONa + (n+1)\ COCl_2 \longrightarrow Cl{-}\overset{\overset{\displaystyle O}{\|}}{C}{-}\left[O{-}Ar{-}O{-}\overset{\overset{\displaystyle O}{\|}}{C}\right]_n{-}Cl + n\ NaCl$$

An excess of phosgene is necessary as a part of it is hydrolyzed by an undesirable side reaction

$$COCl_2 + 4\ NaOH \longrightarrow Na_2CO_3 + 2\ NaCl + H_2O$$

The second phase of the reaction is carried out with the excess of sodium hydroxide. Then follows the hydrolysis of terminal chloroformate groups to phenolic and a further polycondensation by the reaction of chloroformate groups with phenolic ones:

$$Cl{-}\overset{\overset{\displaystyle O}{\|}}{C}{-}\left[O{-}Ar{-}O{-}\overset{\overset{\displaystyle O}{\|}}{C}\right]_n{-}Cl + 5\ NaOH \longrightarrow Na{-}\left[O{-}Ar{-}O{-}\overset{\overset{\displaystyle O}{\|}}{C}\right]_n{-}OH + 2\ NaCl + Na_2CO_3 + 2\ H_2O$$

$$Na{-}\left[O{-}Ar{-}O{-}\overset{\overset{\displaystyle O}{\|}}{C}\right]_n{-}OH + Cl{-}\overset{\overset{\displaystyle O}{\|}}{C}{-}\left[O{-}Ar{-}O{-}\overset{\overset{\displaystyle O}{\|}}{C}\right]_n{-}Cl \xrightarrow[-\ NaCl]{NaOH} Na{-}\left[O{-}Ar{-}O{-}\overset{\overset{\displaystyle O}{\|}}{C}\right]_{n+m}{-}OH$$

The reaction is catalyzed by a small amount of tertiary amines and quaternary ammonium bases.

The synthesis of polycarbonates can be also carried out in a solution in presence of a substance binding the emitted hydrochloride. Most commonly used for this purpose are tertiary amines and among them pyridine.

Another method for obtaining polycarbonates is the ester exchange reaction of biphenols with the alkyl or aryl carbonates.

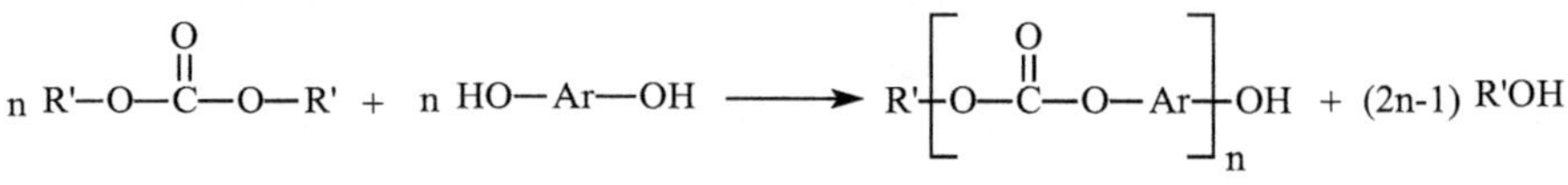

In this reaction an equilibrium state establishes and it is necessary to remove from the reaction environment its byproducts: alcohol or phenol. Therefore, the process is carried out in the second stage at 553–573 K (280–300°C) and under a reduced pressure. As products one obtains polycarbonates with molecular weight not exceeding 50,000.

The ester exchange catalysts are oxides, hydroxides, hydrides and alkoxides of alkali metals and alkaline earths (oxides of Ca, Sr, Ba and Ra) as well as oxides, hydroxides and salts of heavy metals.

In this process the side reactions may occur, associated with the rearrangement and cross-linking of the product.

The greatest practical importance exhibit polycarbonates obtained by reacting phosgene with 4,4'-d dihydroxy diphenylpropane, called Dian (bisphenol A).

The reaction proceeds according to the scheme:

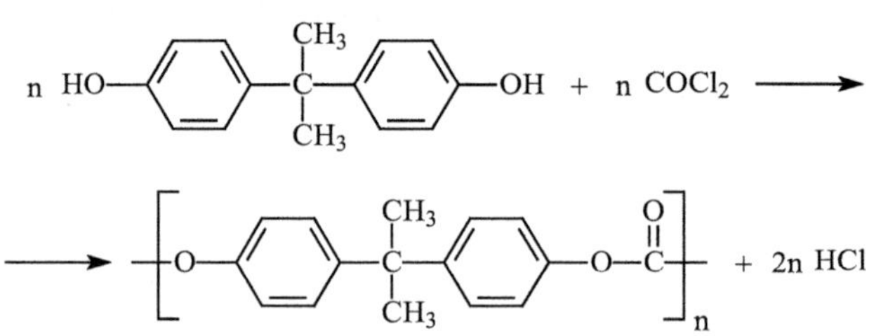

Poly(dian carbonate) is characterized by transparency, excellent hardness and resistance to heat in a very wide range. It is used in the manufacture of bulletproof glass in cars, in diplomatic representations and as windows in banks, to cover the shielding's of precious stained glass windows, screens for the audience at the hockey rinks, cabs in supersonic jets, diving masks and shields used by the anti-terrorist brigade. The possibility of sterilizing the polycarbonate plastic by steam made it also an interesting material to manufacture parts for medical equipment, safety glasses, infant feeding bottles, containers and packaging.

10.6.5 POLY(ETHER-KETONES) (PEEK)

Poly(ether-ketones) are obtained by polycondensation of diphenols with 4,4'-difluoracetophenon.

The reaction of hydroquinone with 4,4'-difluoracetophenon is carried out in diphenylsulfon and in the presence of anhydrous potassium carbonate at temperature of 300–350°C.

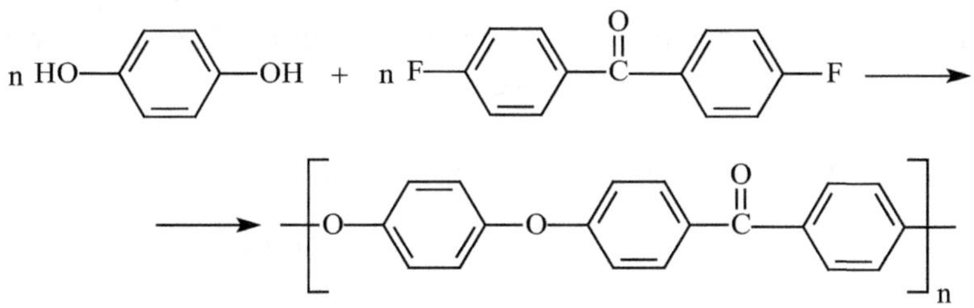

The obtained in this way poly(ether-ether-ketone) (PEEK) contains 30–40% of crystallites with low flammability (oxygen index = 35). It can be used up to the temperature of 280°C. The polymer is produced since 1982 by an English company ICI under the trade name of Victrex.

In 1990 an US company DuPont Co. began to produce poly(ether-ketone-ketone (PEKKA) by the reaction of diphenyl oxide with terephthalic acid dichloride:

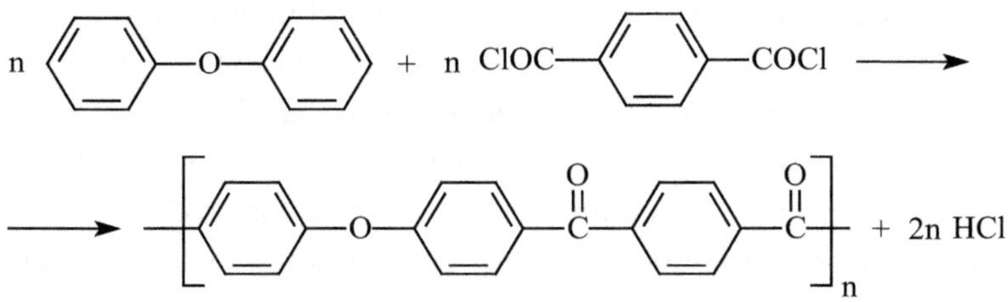

The emitted hydrogen chloride is neutralized by the addition of a weak base. Replacing a part terephthalic acid dichloride by an equivalent amount of izoftaloil dichloride reduces the degree of crystallinity and improves the processing properties of polymer.

10.6.6 POLYAMIDES

Polyamides are polymers with a chain structure in which the individual hydrocarbon segments are connected by amide -CO-NH- bonds.

These connections can be obtained by three methods:

- polycondensation of dicarboxylic acids with diaminami;
- polycondensation of ω-amino acids; and
- polymerization of lactams.

In laboratory of W. H. Carothers at Du Pont Co. the synthesis of poly(adipate hexamethylendimine), called polyamide 6.6, was elaborated in 1936. It proceeds following the scheme:

$$n\ H_2N(CH_2)_6NH_2 + n\ (CH_2)_4(COOH)_2 \longrightarrow H{-}\left[HN(CH_2)_6NHCO(CH_2)_4CO\right]_n{-}OH + (2n\text{-}1)\ H_2O$$

The industrial production of this polymer known as Nylon (NY – New York, LON – London) started in 1940. At the beginning it was used for production of thin fibers designed for stockings. The outbreak of war caused the use of all production of nylon in the manufacture of parachutes. For this reason polyamides influenced the course of the war and contributed significantly to the increase of research on polymers.

Another important polyamide is polycaprolactam, called polyamide 6. It is obtained from caprolactam (caproate acid lactam), which is subjected initially to hydrolysis into aminocaproic acid and then maintained at temperature of 240–250°C with a simultaneous distillation off the water. The polycondensation reaction of caprolactam is an equilibrium reaction and 8–12% of unreacted caprolactam remains in the product. For this reason the obtained polyamide 6 cannot be used directly for production of fibers as it is the case of polyamide 6.6 but must be fragmented and subjected to extraction of unreacted caprolactam with water or matanol. After extraction the product is re-melted and processed into fibers or granulates.

The production and the processing of polyamides at high temperatures must take place in a neutral atmosphere like nitrogen, because the product in melt phase oxidizes easily and darkens.

Polycaprolactam can be also obtained by anionic polymerization of caprolactam. The obtained polymer is characterized by a very high molecular weight and a low liquidity, which prevents its thermal processing. The product is suitable for machining only.

Polyamides found application in the manufacture of synthetic fibers and as construction materials processed by injection molding.

10.6.7 AROMATIC POLYAMIDES

Introduction to the polyamide chain of alicyclic or aromatic rings results in improving the thermal resistance of the product.

Among the various aliphatic-aromatic polyamides noteworthy is the polyhexamethylene terephtalamide, called polyamide 6T. It is obtained by polycondensation of salts of terephthalic acid with hexamethylenediamine. The resulting salt is easy to clean crystalline compound. Its use in polycondensation, instead of the mixture of diamine with dicarboxylic acid, ensures the maintenance of equimolar ratio of functional groups during the process.

The fiber from polyamide 6T is formed from sulfuric acid solutions in a wet way, or from 15–25% solution of polymer in trifluoroacetic acid by a method involving evaporation of solvent. The newly formed fibers are stretched by 200–400% at the temperature of 160–180°C. It provides them the strength of 0.2–0.5 N/tex(?) and the elongation at break of 10–40%. An important advantage of products made from these fibers is to maintain the shape at elevated temperatures. They have also a good resistance to the action of bases. Dyeing of his type of fibers difficult. A good coloring provides only the dispersed dyes.

A growing interest attracts aromatic polyamides derived from dicarboxylic acids and aromatic diamines. The highest importance from this polyamide group has the m – phenylene isophtalamide, produced under the trade name "Nomex" in US and "Feniton" in Russia, respectively.

The aromatic polyamides are obtained by reacting diamines with dicarboxylic acid chlorides:

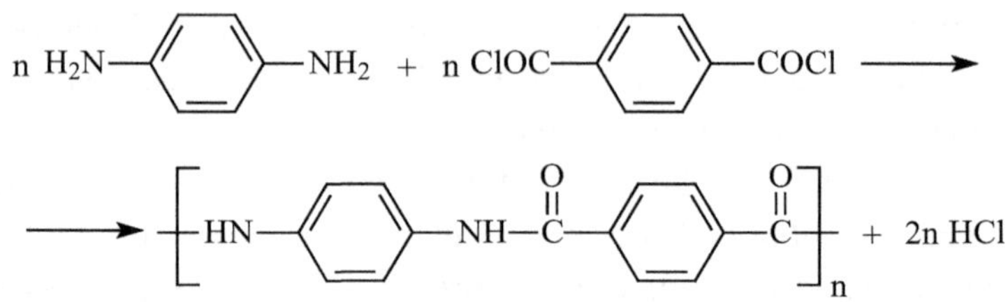

This reaction can be carried out by polycondensation at the interface, emulsion or in solution. In industrial practice the solution method is the mainly used. It has the advantage that with the formation of polymer one obtains the spinning solution with sufficient concentration from which directly fibers can be produced. Dimethylacetamide is used as solvent. It has the advantage to be acceptor of hydrogen chloride, which is secreted during the reaction. Nevertheless, the calcium hydroxide is added to the polymer solution, which neutralizes the reaction medium. The formed during the reaction calcium chloride increases the stability of the spinning solution. The addition of alkali metal salts to the solution, in which the polycondensation takes place, affects positively the course of the reaction ensuring the obtaining of a polymer with higher molecular weight.

The fibers made from aromatic polyamides can be formed dry by evaporation of the solvent or wet by precipitation of polymer in form of fibers in a coagulating solution. As the coagulation bath the aqueous solvent (dimethylformamide or alcohol) solutions are used.

10.6.8 POLYIMIDES

Polyimides are products of polycondensation of tetracarboxylic acids or their dianhydrades with diamines. In the synthesis most frequently the pyromellitic acid anhydride or its derivatives and various diamines are used. The polycondensation reaction is carried out in solution. Dimethylformamide, dimethylacetamide, dimethylsulfoxide and N-methylpyrrolidone are used as solvents.

In the first stage of the synthesis, carried out at 253–343 K (–20–70°C) formed are polyamide acids. Heated at temperatures above 473 K (200°C), under reduced pressure or in the presence of dehydrating factors such acetic anhydride and ketenes, the polyamide acids undergo a cyclization process to polyimides.

The reaction proceeds according to the scheme:

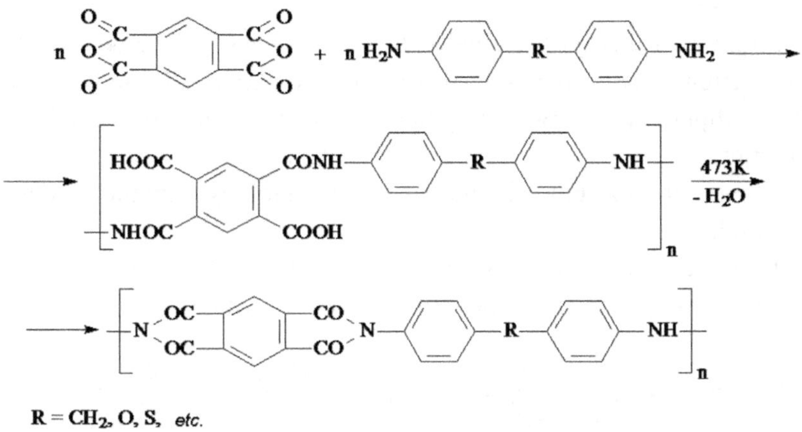

Polyimides are thermostable polymers because they do not melt and do not soften and do not ignite, but undergo a slow pyrolysis. Through the thick polyimide fabric flame penetrates after 15 minutes only.

Polyimides are not soluble and do not swell in most known organic solvents. The prolonged action of steam causes a slight decrease in fiber strength of polyimide only. However, the action of concentrated acids and bases causes a harm to the material through its hydrolysis.

Because of their good properties polyimide are applied to produce heat and flame –resistant fibers.

Polyimide fibers are obtained from the solution polyamidoacid through its coagulation, or by evaporation of the solvent.

In the case of wet processing the content of polymer in the spinning solution should not exceed 10%. The solution is introduced into a coagulation bath consisting of a mixture of solvent and aqueous solution of inorganic salts (e.g., calcium thiocyanate).

The formed fiber is dried and subjected to a staged heat treatment with the task of providing the fibers with adequate physico – mechanical properties. During the heat treatment the polymer undergoes cyclization and the fiber is stretched by 170–200%.

Polyimide fibers are manufactured under the brand names: Arimid (Russia) and Kapton (USA).

10.6.9 POLYSULFONES

Polysulfones are polymers containing in the main chain the sulfone groups $-SO_{2-}$.

Bindings of this type can be obtained as a result of the polycondensation reaction of chlorinated derivatives of diphenylo sulfone with sodium salts of diphenyls or by copolymerization of sulfur dioxide with vinyl monomers.

A practical importance in the polymer technology only the first method has found.

On industrial scale polysulfones are obtained by the reaction of 4,4'-dichlorodiphenylosulfone with bisphenol A (dian):

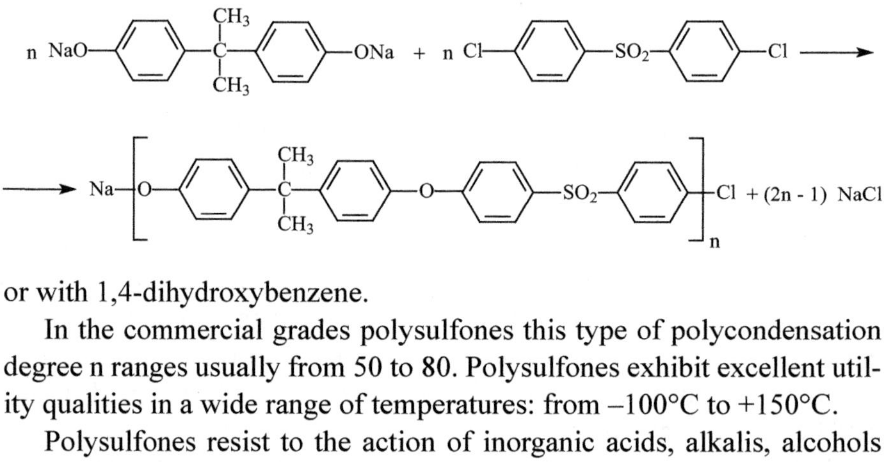

or with 1,4-dihydroxybenzene.

In the commercial grades polysulfones this type of polycondensation degree n ranges usually from 50 to 80. Polysulfones exhibit excellent utility qualities in a wide range of temperatures: from –100°C to +150°C.

Polysulfones resist to the action of inorganic acids, alkalis, alcohols and aliphatic hydrocarbons. They are soluble in chlorinated hydrocarbons and are partially soluble in esters, ketones and aromatic hydrocarbons.

The commercial products are obtained from polysulfones by the injection method. Polysulfones can be also processed into panels characterized by a good transparency.

10.6.10 POLYSULFIDES

The earliest known polymers containing sulfur in the main chain are thioplasts (polysulfide rubbers), called thiocols (polysulfide resins). They are obtained by the reaction of alkyl dichloride s with sodium polysulfide. An example of such a reaction is the synthesis of thiocol A with 1,2-dichloroethane:

$$n\ ClCH_2CH_2Cl + n\ Na_2S_4 \longrightarrow \left(CH_2{-}CH_2{-}\underset{\underset{S}{\|}}{S}{-}\underset{\underset{S}{\|}}{S} \right)_n + 2n\ NaCl$$

The polycondensation process proceeds easily at 80°C in an aqueous medium and in the presence of magnesium hydroxide suspension. The resulting latex is subjected to the coagulation process with acids and the isolated polymer is separated from the residue, washed with water, dried and pressed into blocks.

As a result of the action of bases on thiocols one can remove the lateral sulfur atoms with formation of a disulfide polymer with chemical formula:

$$\left(CH_2{-}CH_2{-}S{-}S \right)_n$$

Thioplasts are obtained also from the dichlorodiethyl ether (Thiocol D) and from glycerol dichlorohydrin (Vulcaplas).

Thioplasts have found application as synthetic rubbers and can be vulcanized using the zinc oxide. During the vulcanization process a slight tear gas is emitted. It hinders the technology process and necessitates cooling of forms in the press before removing the vulcanizates from them. Thioplasts can be also cross-linked by mixing them with polyester, epoxy and phenolic resins. The mechanical properties of thioplast vulcanizates are worse than those of rubber. Moreover, thiocols have an unpleasant smell and tends to hardening during storage, what limits their application.

An important advantage of thioplasts is their very low solubility in oils and in many other organic solvents, low gas permeability and resistance to ozone. Therefore, thiocols are used in production of gaskets and hoses resistant to gasoline and oils, as well as a guard for cables instead of lead. They are also used as binders in the manufacture of solid rocket propellants due to their excellent combustibility.

Polysulfides are obtained also during the opening of three-, four-and five member cyclic sulfides, in the presence of cationic, anionic and coordination catalysts. An example of this type of reaction is the polymerization of the ethylene sulfide.

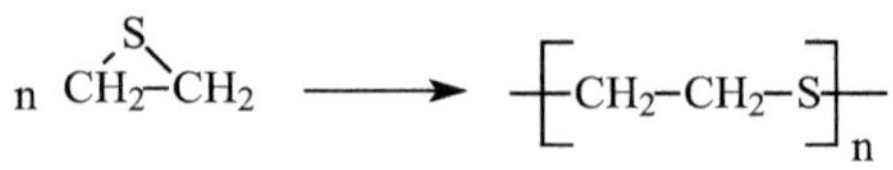

Good results are also obtained during the reaction of dithiols with dialkenes. If this reaction is carried out by the radical mechanism then it runs in disagreement with the Markovnikov's rule and the formed polymer does not contain side groups.

$$n\ HS-R-SH + n\ CH_2=CH-R'-CH=CH_2 \longrightarrow \left[S-R-S-CH_2CH_2R'CH_2CH_2 \right]_n$$

During the ionic reaction catalyzed by hydrogen or hydroxyl ions the resulting polymer contains the side methyl groups:

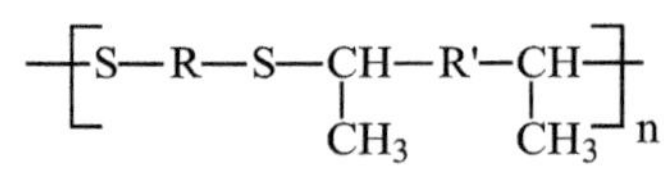

Polysulfides with a high molecular weight (20,000–60,000) are obtained in reaction of dithiols with alkynes through the stage of formation of a reactive vinyl sulfide. This reaction is usually initiated by γ rays.

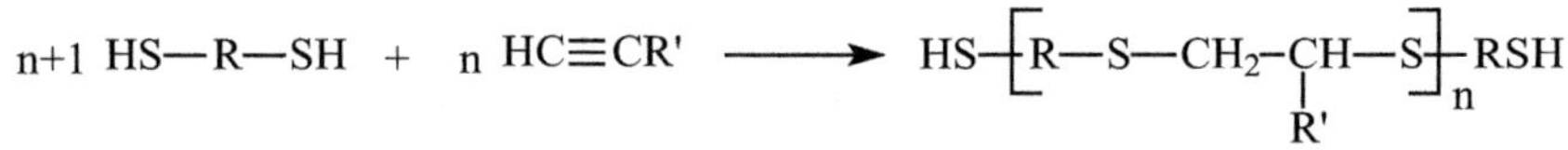

In this way the polymers obtained are used as additives for adhesives and putty.

Dithiols react with aldehydes and ketones forming the polythioacetals, which are more resistant to hydrolysis:

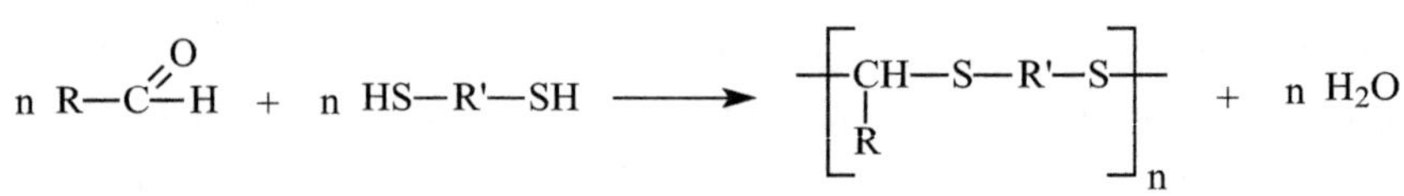

Known are also aromatic polysulfides, characterized by a good thermal and chemical resistance as well as the property of self-extinguishing.

They are formed by the reaction of p-dichlorobenzene with sodium sulfide or by the alkaline hydrolysis of p-chlorothiophenol:

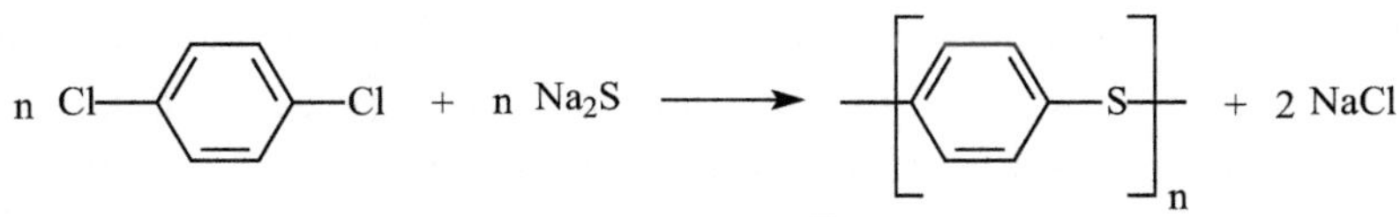

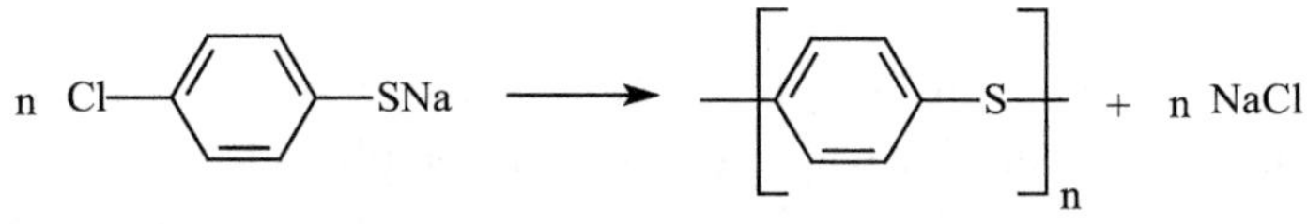

During the condensation of diphenylether with sulfur chlorides a polymer containing ether and sulfide groups is formed.

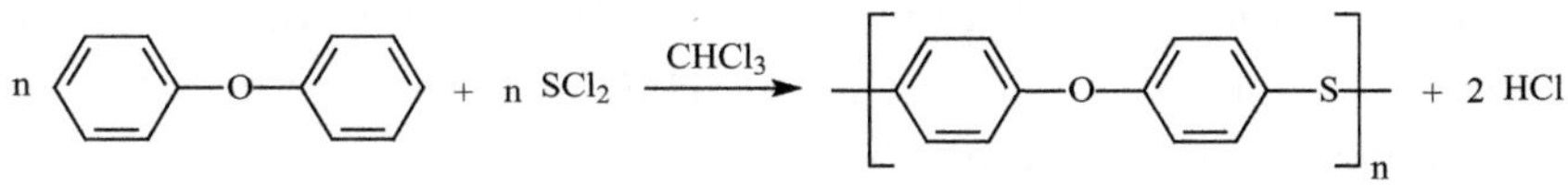

The formed polytetrafluoro-p-phenylensulfide is characterized by a good thermal resistance and is used as a semiconducting material with conductivity 0–10^{-11} $\Omega^{-1}cm^{-1}$.

10.6.11 POLYSELENIDES

Recently a method was developed for the synthesis of polyselenides by a reaction of aromatic chlorine-containing derivatives with sodium selenide (metallic sodium and black selenium) in dimethylformamide or in pyridine at the temperature of 120°C.

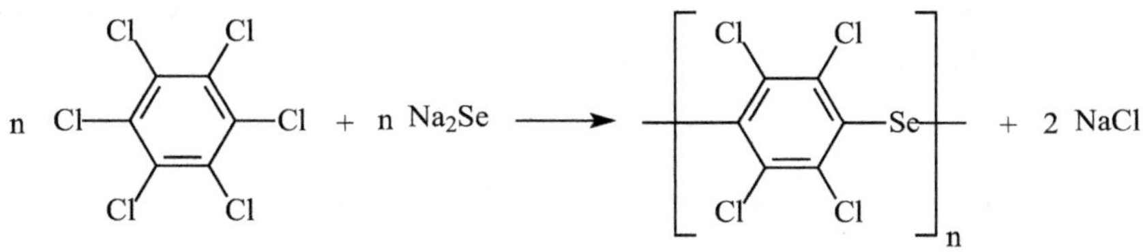

The obtained polymer has a glass transition temperature of 104°C, thermal resistance up to 320°C and conductivity of 1.5 x 10^{-11} Ω^{-1} cm^{-1}.

10.6.12 SILICONES

Silicones are macromolecular organosilicon compounds, owing to their excellent properties, which find increasing industrial applications. They are obtained from alkyl- or arylochloro silanes by hydrolysis, followed by polycondensation of multifunctional monomers. At present the arylochlorosilanes are obtained by reaction alkyl halogens with silicon alloy with copper.

The reaction mixture of synthesis products of is separated on distillation columns. In a similar way are obtained the phenylchloro silanes.

By hydrolysis of dichlorosilanes and neutralization with soda of formed siloxanes one obtains products with low molecular weights, applicable as silicone oils. In order to obtain a specific grade of the oil the polymerization process is led in an appropriate proportion mixture of single – and bifunctional monomers with a catalyst (H_2SO_4), at room or elevated temperature with a simultaneous control of viscosity increase. The silicone oils can be converted into emulsions, pastes, sealants and lubricants.

Of practical importance are silicone rubbers, produced from bifunctional monomers with a high degree of purity. One obtains polysiloxanes with consistency of a syrup and molecular weight of the order of 300–800 thousands. These polymers are mixed in a crusher with fillers, pigments and stabilizers. The obtained premix can be stored for several months until the vulcanization. The vulcanization may be carried out with organic peroxides or sulfur, together with accelerators. The vulcanization of silicone rubber with peroxides is carried out in the ordinary and transfer molding press for 20 minutes at temperature of 100–180°C and under pressure of 50–100 atm. The silicone rubbers after vulcanization require still an annealing for 2–10 hours at elevated temperatures (200–250°C). Silicones have found also application in paint industry.

10.7 CHEMICAL REACTIONS OF POLYMERS

10.7.1 GENERAL CHARACTERISTICS

During the storage or the use of polymers and derived from them plastics they may be subject of a number of different chemical reactions, both intentional and incidental as a result of actions by various physical or chemical agents.

The intended chemical reactions carried out on polymers change their properties and allow obtaining from one polymer a large number of new varieties. The latter are quite different from the original, broadening in this way the assortment of plastics and possibilities of their applications. The research area dealing with this issue is called the chemical modification of polymers. It is now widely expanded and further developed. The knowledge of routes of these reactions allows the synthesis of new polymers with specific chemical structure and intended for pre-planned applications.

The chemical reactions carried out on polymers can be divided into several types, depending on whether they run on individual polymer molecules with or without the participation of other small molecule reagents, or as an intermolecular reaction occurring between the macromolecules of the same or different polymers.

The most important types of reactions of chemical modification of polymers include:

- introduction of functional groups to polymer molecules;
- transformation of functional groups in polymer macromolecules;
- intramolecular cyclization;
- oxidation and reduction;
- polymer grafting;
- cross-linking;
- degradation of polymers.

The chemical reactions carried out on polymer macromolecules are characterized by a defined specificity. In his kinetic studies Flory has shown that the reactivity of carboxyl and hydroxyl groups during the synthesis of polyester does not dependent on chain length of the nascent polymer. On this basis he concluded that the reactivity of functional groups in polymer molecules do not differ from the reactivity of the same functional groups in small molecule compounds. This suggestion was confirmed when testing a variety of reactions in dilute solutions of polymers and small molecule compounds. However, it is important to keep in mind the complex structure of polymers, whose chains are tangled, and in the solid state they are often combined with other chains by the Van der Waals forces. A lot of them form crystalline areas and all of these factors make the access of a reactive particle (e.g., a radical) to a specific location in the chain of macromolecules may be difficult or even impossible. For this reason the chemical modification of polymers with long chains runs in accessible places and the resulting product has usually the structure of a statistical copolymer. It means that it is not substituted in each mer.

The course of chemical reactions in polymers, enabling the introduction of new functional groups or transformation of functional groups in the macromolecules, is affected by the following factors:

- availability of functional groups;
- chain length (degree of entanglement of macromolecules);
- effect of neighboring groups;
- effect of configuration;
- electrostatic effects;
- cooperative effects;
- effect of inhomogeneous activity;
- solution concentration;
- supermolecular effects in heterogeneous reactions;
- effect of mechanical stress.

10.7.1.1 Availability of Functional Groups

For an effective course of chemical modification of polymer it is important that its functional groups exhibit the same reactivity. Large macromolecules, contained in solution, are partially tangled. In the solid they are associated with other macromolecules via intermolecular forces.

This macrostructure of the polymer molecules makes that in the reaction environment the functional groups in polymer are not equally sensitive to the reactive agents. Especially in a heterogeneous system an important role plays the diffusion factor. Functional groups present in the outer layers of polymer ball react relatively easily, while inside the reaction depends on the possibility of reactive particles to penetrate in the part of polymer.

Functional groups inaccessible for steric reasons remain unchanged in the reaction product.

An example of the impact of the availability of functional groups on the reaction rate can be the amidation reactions with aniline of various types of copolymers of maleic anhydride in dimethylformamide. The rate of amidation of the maleic anhydride copolymer with ethylene is, for example, two orders of magnitude higher than the rate of amidation of maleic anhydride copolymer with norbornene.

10.7.1.2 Influence of the Chain Length

Typical macromolecular reactions, not observed in the chemistry of small molecule compounds, are the degradation processes (chain decay into smaller fragments) and the intramolecular cyclization.

As the polymer chain length grows the viscosity of its solutions or blends increases. It hampers the mobility of molecules and thus reduces their reactivity. Also entanglement of polymer molecules increases with the chain length growth.

The course of reactions of functional groups in polymers of different molecular weights depends on the macrostructure of molecules, what is associated with the availability of functional groups.

On the basis of experimental studies it was shown that the effect of the chain length of macromolecules on the reactivity of functional groups may vary, depending on the nature and structure of the polymer. For example, it was found that during the reaction of phenylisocyanate with polyesters of different molecular weight the reaction rate decreases with the increasing chain length of polyester.

Similarly, during the esterification of maleic anhydride copolymers with (poly)propylene, with molecular weights of 42,000, 51,000, 91,000, and 103,000, and the reaction rate decreases with increasing average molecular weight.

Quite different results were obtained during the amidation of styrene maleic anhydride copolymers with average molecular weights of 24,000 and 46,000. In this case the reaction rate practically does not depend on the chain length.

The results of these reactions are presented in Table 10.36.

10.7.1.3 Effect of Neighboring Groups

The neighboring groups, both the nearest and the further lying away from the functional group formed in the reaction center, impact its electronic structure and thus its reactivity. The effect of neighboring groups may lead both to a reduction and to an increase of the reaction rate.

The hydrolysis of poly(vinyl acetate) to poly(vinyl alcohol) is a characteristic example of a reaction which reflects the influence of the neighboring

TABLE 10.36 Influence of Molecular Weight of Maleic Anhydride Copolymer with Styrene on the Amidation Reaction

No.	Temperature [°C]	Amine used	k x 102 [dm^3/mol s]	
			M = 24,000	M = 46,000
1	60	aniline	1.45	1.38
2	45	p-toluidine	2.37	2.1
3	30	p-anisidine (p-methoxyaniline)	1.75	1.8
4	45	m-chloroaniline	0.68	0.71

After M. Rätzsch, V. Phien, Faserforsch. u. Textiltech. (1975) 26, 165.

groups on the reactivity of the substituent in the polymer molecule. With the progressing process of hydrolysis increases the reaction rate, because the hydroxyl groups activate the neighboring acetoxyl groups. Therefore, the reactivity of various groups increases in the following order:

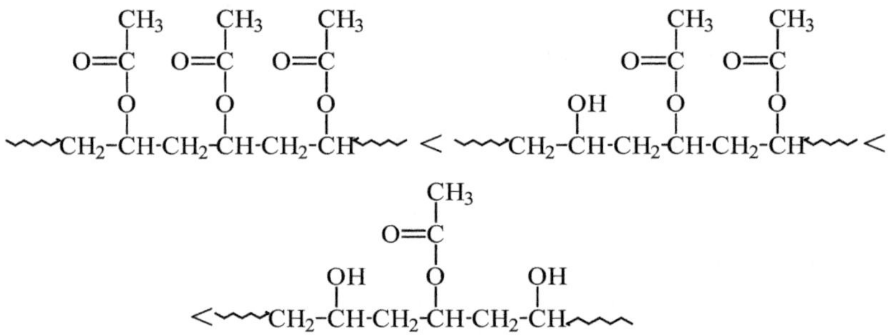

In some cases the active groups can react with neighboring groups. In that case the reactivity of intramolecular groups influences also the other substituents contained in the polymer molecule.

An example of it is the imidation reaction of copolymers of acid derivatives of phenyl malamate with styrene or ethylene, which runs according to the following scheme:

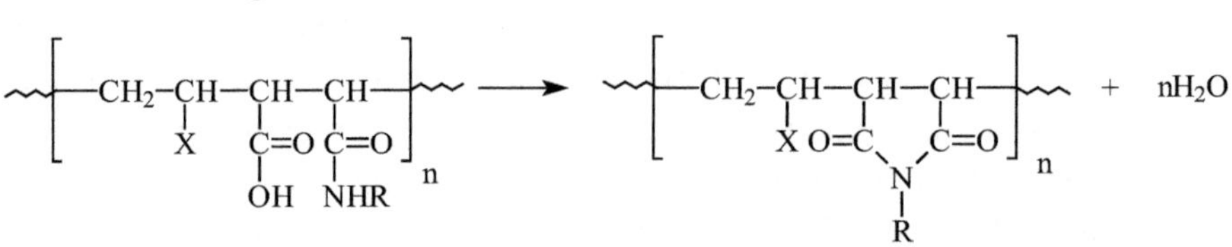

where X = H, C6H5

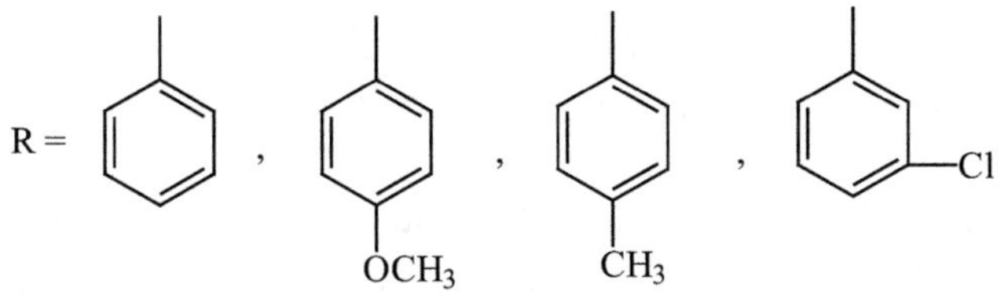

With a decrease of the basicity of amide substituent decreases the cyclization reaction rate. The imidation rate of copolymers with styrene is much higher than that with ethylene. It shows the influence of phenyl substituent on the course of this reaction.

10.7.1.4 Configuration Effects

Configuration is called the distribution of atoms in molecule, which is characteristic for a particular stereoisomer.

The substituents present in polymer chains can be arranged either statistically (atactic polymers) or regularly (syndiotactic and isotactic polymers).

The reactivity of functional groups in polymers change with a change of spatial configuration of macromolecules.

The configuration effects in the chemical reactivity of polymers depend on the kind of functional groups present in macromolecules and on the type of chemical reaction.

During the study of the hydrolysis of poly(methyl methacrylate) it was found that the isotactic polymer reacts with a high rate whereas the atactic and the syndiotactic p react slowly. This phenomenon is explained by the fact that the ester groups in the isotactic polymer form complexes with the neighboring carboxyl groups, so that the hydrolysis reaction proceeds more easily.

During the similar study done with the labeled carbon atoms it was observed that the isotactic poly(methyl methacrylate) reacts 16 times faster than the atactic and 43 times faster than the syndiotactic polymer, respectively.

Some tactic polymers such as polyethylene and polypropylene, are characterized by a much larger degree of crystallinity than the atactic polymers. During a chemical reaction in the solid state the reaction takes place in the amorphous areas of polymer, while the crystalline areas are weakly reactive.

An example of such a reaction is the chlorination. Atactic polypropylene is much easier to chlorinate than the isotactic polypropylene with a high degree of crystallinity. Dissolving polymer in a solvent or conducting the process above the melting point of polymer crystallites enables its more uniform substitution.

10.7.1.5 Conformational Effects

Conformations are the possibilities of mutual transformation of different spatial arrangements of atoms in molecules, as a result of the rotation around single bonds. There are two basic types of conformational structures: opposite conformation (called also eclipsed conformation) and alternate conformation (called also staggered conformation). Several intermediate conformations occurring between the two are called skew conformations.

Conformational effects, which take place during the reaction of functional groups in macromolecules overlap often with the other, discussed effects.

The phenomenon occurs primarily during the reactions carried out in solutions. The polymer molecules during the dissolution pass into solution in the form of coils, which depending on the impact on them of suitable solvent molecules are more or less developed.

In good solvents and in dilute solutions the availability of functional groups is much greater, what affects the reaction rate. However, the reaction rate changes clearly when the polarity of the solvent differs significantly from the polarity of the resulting modified polymer. For this reason it is advisable to carry out the reaction in a solvent (and not in a mixture) to facilitate the attainability of polymer functional groups at least at the beginning of the reaction. If a mixture of solvents is used, such as, for example, a solvent for the output polymer and for the product after the modification, and possibly for the small molecule compound then one obtains a compound with various degrees of substitution.

The conformational effects play a particularly important role in the intramolecular cyclization reaction, carried out in dilute solutions. The defined conformation and the distance between the chain ends influence the course of formation of stable rings with 5–12 members.

10.7.1.6 Electrostatic Effects

The electrostatic effects are observed during the reactions with the use of ionic polymers, in comparison with similar reactions carried out on small molecule compounds.

As an example, the use of protonated poly(4-vinylpyridine) as a catalyst for the hydrolysis of sodium 4-acetoxy-3-nitrobenzenesulfonate is shown below. Poly(4-vinylpyridine) shows in this reaction a much higher catalytic activity than 4-methylpyridine.

The mechanism of this reaction is shown in the following scheme:

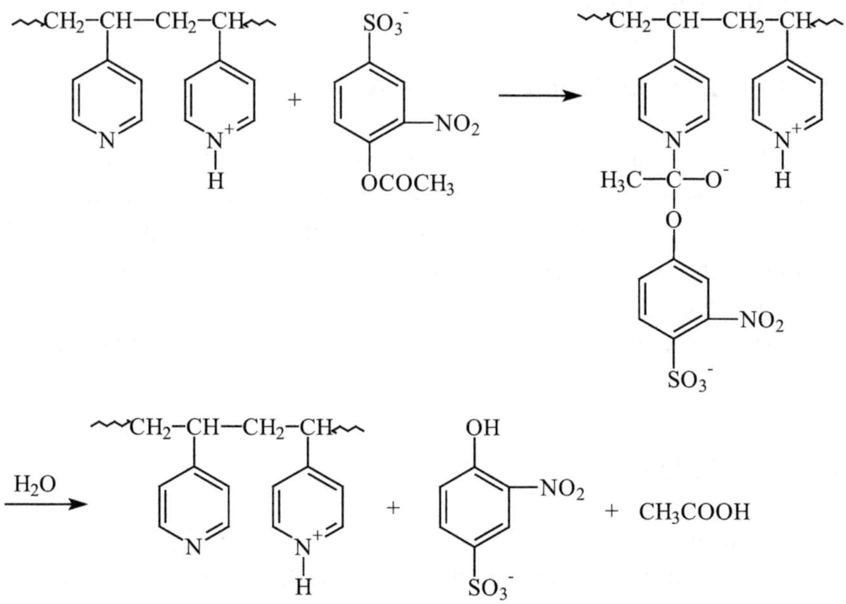

The electrostatic interactions depend also on the solvent polarity, pH and on the type of neighboring groups.

10.7.1.7 Cooperative Effects

The cooperative effects in chemical reactivity of polymers may be subject to the influence of neighboring groups, association processes and catalysts presence. They play an important role in many biochemical processes involving proteins and nucleic acids, and in particular enzymes. The cooperative action in course of chemical reactions involving macromolecules can be presented schematically as follows:

The functional group A in the polymer chain becomes an active site during a reaction with functional group M of a small molecule compound. At the same time a mutual interaction between the group A' and another functional group N of the reagent takes place. It facilitates the main reaction by stabilizing the transition compound. Groups A and A' in the polymer chain may be the same or different.

10.7.1.8 Effect of Inhomogeneous Activity

This effect is related to the possible formation of hydrogen bonds in polymer as well as to the hydrophobic interactions, which influence the reactivity of different functional groups. During the amidation reaction of maleic anhydride copolymers with aliphatic amines at a stoichiometric molar ratio the substitution at the polymer chain does not take place completely, but only to half when the reaction carried out on small molecule compounds is running to the end.

10.7.1.9 Influence of Concentration

The course of the reaction of chemical modification of polymer in solution depends on the concentration of polymer, the actual concentration of active groups, and the concentration of small molecule reagent. During the substitution or replacement of functional groups one uses usually a 2.5-fold excess of small molecule reagent.

In very dilute solutions possible are the cyclization reactions. A significant increase in the concentration of polymer solution leads to an increase of viscosity and to a series of side reactions, thereby increasing the heterogeneity of the product.

During the chemical modification of polymers, carried out in a homogenous environment, the processes typically run as second-order reactions. The rate of this reaction depends on both solvent polarity and on solvation effects.

10.7.1.10 Supramolecular Effects in Heterogeneous Reactions

The influence of supramolecular polymer structures plays an important role during their chemical modification. On the reactions running in solutions impose at the same time conformational effects, which are difficult to distinguish from supramolecular.

A typical example of the impact of supramolecular effects on the process takes place during the acetylation of cellulose. The course of reaction depends on the supramolecular structure of cellulose. During the acetylation of natural cellulose fibers for 24 hours one obtains 26.3–27.7% substitution degree of acetyl groups in cellulose. Acetylation in the same conditions of viscose fibers leads to the substitution degree of order of 14–16% only.

Similarly, the thermal dehydration of poly(vinyl alcohol) depends on the ordering degree of polymer chains. Samples of poly(vinyl alcohol), containing oriented molecules, are more resistant to the action of temperature.

Chemical reactions of polymers in the solid phase can run easily in amorphous areas or on the surface of amorphous or crystalline areas. Exploiting this phenomenon one can obtain polyethylene with a given degree of crystallinity by the destruction of the amorphous areas with fuming nitric acid. In the reaction conditions the crystalline polyethylene does not react with nitric acid.

10.7.1.11 Effect of Mechanical Stress

The mechanical stress in polymer molecules can affect the activation of chemical bonds in macromolecule, something that plays a particularly important role in such reactions as the degradation of polymers, cross-linking and grafting. The mechanochemistry of polymers is dealing with these issues.

The most important external factors that cause the effect of mechanical strain are the forcing processes, such as: extrusion, calendering, injection molding rolling, grinding, hammering, crushing, osmotic pressure, phase transitions and the impact of electro-hydraulic as well as ultrasound actions.

Another example of the effect of mechanical stress is the process of *deacetylation* of poly(vinyl) acetyl imidazol in the presence of 3-nitroterephthalic monoester. Depending on the type of esters used different results are obtained. The rate of this *deacetylation* reaction in the presence of octadecyl monoester is 100 to 400 times larger than when using an analogous ethyl ester. The reason for this is the interpenetration of molecules and the mutual interactions of aliphatic chains and acid groups in polymer molecule.

10.7.2 MODIFICATION REACTIONS OF POLYMERS

10.7.2.1 Introduction of New Functional Groups

Introduction of new functional groups into the polymer molecules may be carried out by the substitution reaction of hydrogen atom or other functional groups, as well as by the addition reaction to double bond when using the unsaturated polymers. A model for this type of process is the chlorination reaction. It may follow different mechanisms, depending on the type of used chlorinating agent and the process conditions. Chlorination of polymers by free radical mechanism may take place by using chlorine or sulfuryl chloride in the presence of initiators such as organic peroxides, dinitryl azoisobutyric acid or by exposure to ultraviolet light. The reaction can be carried out in solution, suspension and in fluidized bed.

During the chlorination of polymers in solution the substitution of polymer chains by chlorine proceeds initially statistically, and then observed is the influence of substituted chlorine atoms on the kinetics of the further chlorination reaction.

The advantage of the method of chlorination in solution is the destruction of the crystalline structure of polymer as a result of its dissolution and the solvation of individual polymeric ball (more or less coiled chains). Thus the chlorinating factor has an easier access to various areas of polymer chain. This allows a relatively uniform chlorine substitution. The mechanism of the free radical chlorination reaction is the following:

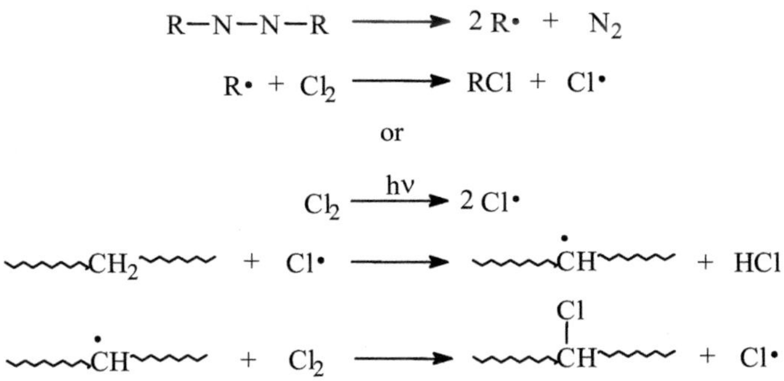

A similar reaction takes place when carried out using the sulfuryl chloride:

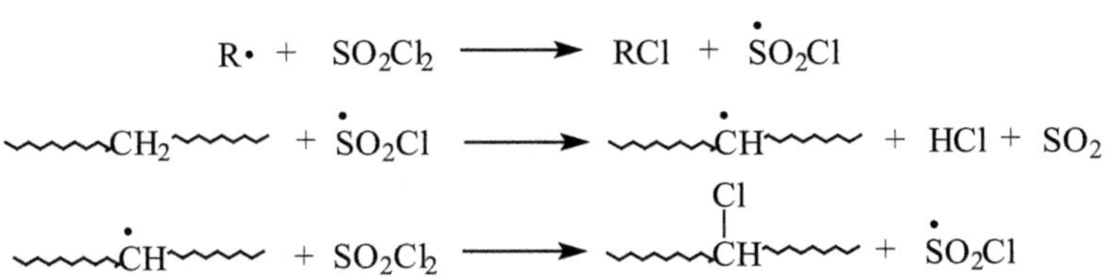

The disadvantage of chlorination reactions in solution is the necessity of using often the flammable and toxic solvents and the difficulty with polymer extraction.

Unlikely runs the process of chlorination carried out in a fluidized bed or in suspension. Polymers containing crystalline areas chlorinate then only in the amorphous areas. Only the destruction of the crystalline structure, by carrying out the reaction at a temperature above the melting point of crystallites, allows a more uniform process of chlorination. There are also difficulties with penetration of chlorinating agent in the internal layers of polymer. It involves the influence of diffusion factors on the course of this reaction. It follows from the above considerations that the chlorination products of polymers in solution and in solid phase differ in both the micro-and the macrostructure. The products chlorinated in the solid phase are characterized by substitution in certain areas of the molecules only and retain the original crystalline structure. They are also characterized by a poorer solubility than polymers chlorinated in solution.

An important parameter influencing the coarse of chlorination reaction is temperature at which the chlorination process is conducted. Figure 10.9 shows the influence of temperature on the coarse of chlorination process

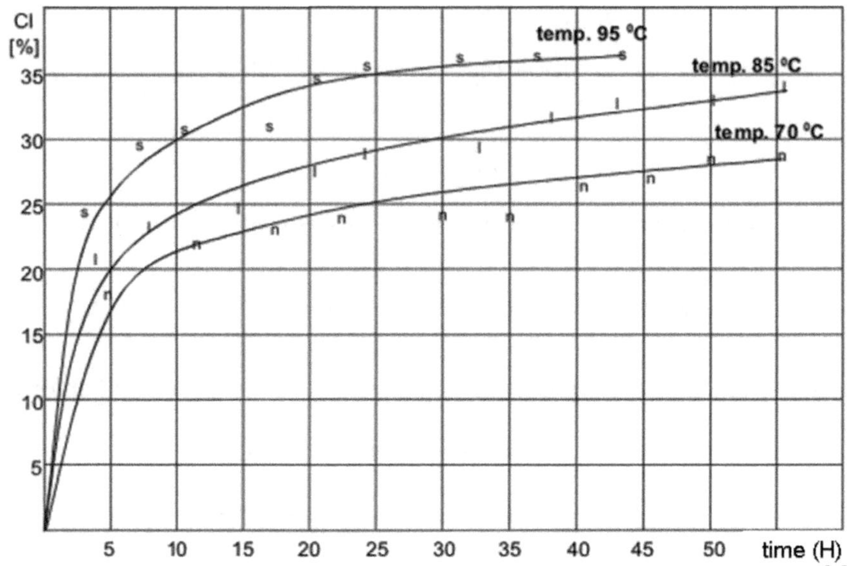

FIGURE 10.9 Influence of temperature on the course of chlorination of polyethylene in suspension (after A. Puszyński; Pol. J. Chem. 1981, 55, 2143).

of a low-density polyethylene, in suspension, by chlorine produced in the reaction environment by dispensing a solution of potassium chlorate to the solution of hydrochloric acid. The increase of reaction temperature increases the rate in the initial period of the course, what is associated with better accessibility of chlorinating agent to outer layers.

In the next stage of the reaction the diffusion processes dominate and the reaction rate decreases.

During the chlorination of polymers in suspension also the reaction medium affects the process. The ions in solution, in which the polymer powder was dispersed, act on the polymer matrix, causing acceleration or hinder the chlorination reaction. This applies, first of al, to the polymers containing crystalline groups.

Figure 10.10 presents, as example, the impact of salt solutions at concentration of 0.5 mol/dm^3 on the course of the chlorination process of isotactic polypropylene dispersed in an aqueous solution of hydrochloric acid.

It follows from the shown graph that most salts increase the effectiveness of chlorination reactions in the first period, while the copper cations show a clear inhibitive effect.

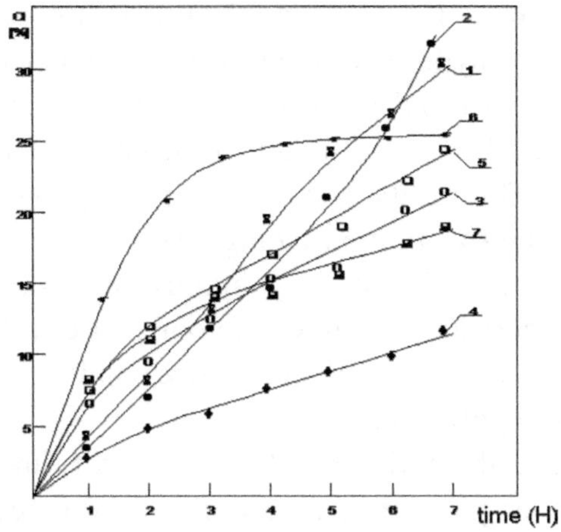

FIGURE 10.10 Influence of the reaction environment on the course of the chlorination process of polypropylene in an aqueous solution of hydrochloric acid and in the presence of salts at a concentration of 0.5 mol/dm^3: 1 – without salt, 2 – $CoCl_2$, 3 – $(NH_4)_2MoO_4$, 4 – $CuCl_2$, 5 – KCl, 6 – NaCl, 7 – NaCl with concentration of 1.5 mol/dm^3 (after A. Puszyński, J. Dwornicka: Angew. Makromol. Chem. 1986, 139, 123).

Known are also chlorination methods of polymers through the cationic mechanism by using Lewis acid type catalysts, for example, the aluminum chloride.

In the case of chlorination in these conditions of aliphatic polymers one obtains often cross-linked products.

Impact of the mechanism on the type of products formed during the reaction can be illustrated by the example of polystyrene chlorination. During this process, running according to the free radical mechanism, the substitution takes place to the polymer chain, while in the course of ionic polymerization the substitution reaction takes place in the ring.

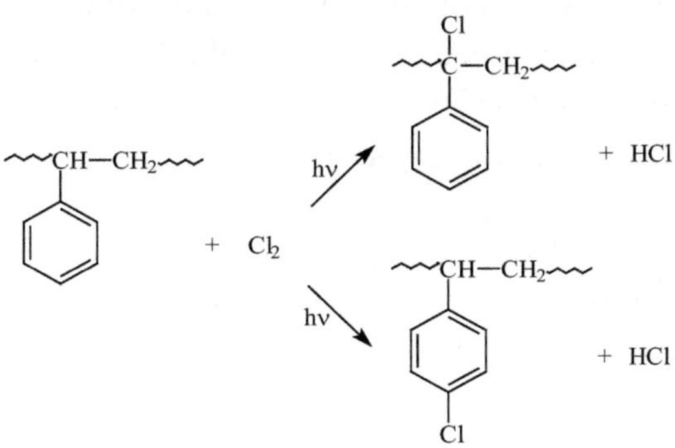

Similar results are obtained during the bromination of polymers.

The exchange of hydrogen atoms in the polymer molecule to other functional groups is carried out using conventional methods of organic preparation. Especially easy to run are electrophilic substitution reactions in aromatic rings. During the nitration of polystyrene one obtains poly (p-nitrostyrene). The substitution-taking place almost exclusively in para position is due to the steric factors. Similarly, during the sulphonation of polystyrene or its derivatives formed are polymeric sulfonic acids, which found application as soluble polyelectrolytes and ion exchangers in the case of cross-linked polymers.

The method of substitution can be also used to modify the polymer molecules in finished products. An example of this process is the fluorination of polyethylene bottles. It is possible to obtain a fluorinated polymer coating with a thickness of 0.1 mm.

10.7.2.2 Transformation of Functional Groups

Conversion of one functional group into another one is mainly used to obtain polymers, which are difficult to synthesize by other methods or cannot be obtained by a direct polymerization.

A classic example of such a reaction is obtaining of poly(vinyl alcohol). Vinyl alcohol does not exist, because it rearranges immediately to acetaldehyde. For this reason, poly(vinyl alcohol) is obtained by hydrolysis of poly(vinyl acetate) or poly(vinyl formate). The hydrolysis reaction can be carried out in an acidic or alkaline environment. However, the formed product contains always a certain amount of acetyl or formyl groups and actually is a copolymer of vinyl alcohol with initial ester. Very good results can be obtained by running the hydrolysis reaction in the presence of crown ethers as its catalysts.

During the hydrolysis of polyacrylonitrile formed is poly(acrylic acid):

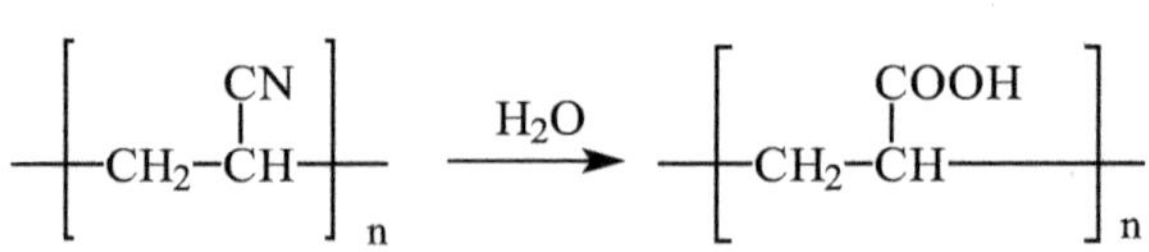

Syndiotactic and isotactic poly(methacrylic acid) is obtained by the hydrolysis of poly(trimethylsilyl methacrylate):

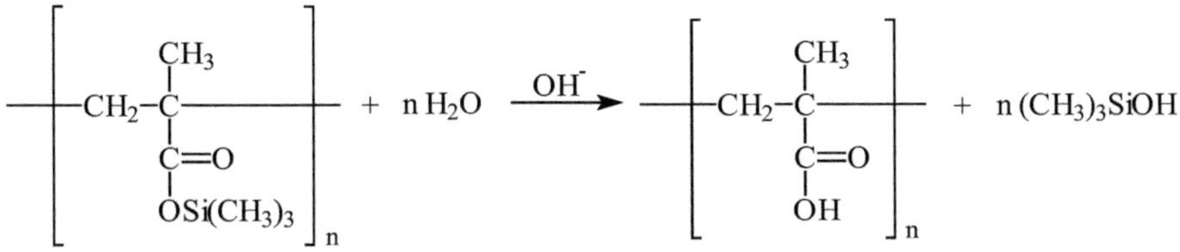

Syndiotactic poly(trimethylsilyl methacrylate) is obtained by anionic polymerization carried out in tetrahydrofuran, while the isotactic polymer is formed in toluene.

Another example of substitution reaction used to synthesize new polymers is the use of Hofmann method to produce polyamines from an easily accessible polyacrylamide.

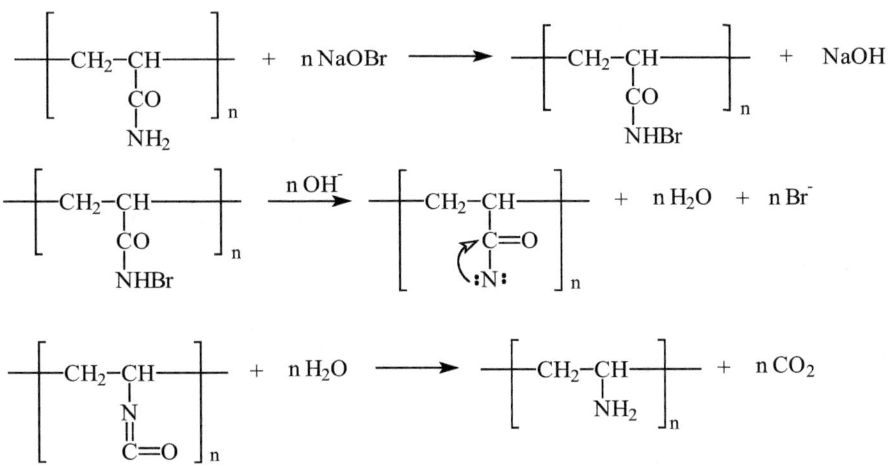

The chloromethylated polystyrene, formed by the reaction of polystyrene with chloromethyl ether, is an important raw material for further modification. It can be used to obtain a series of derivatives with the following structures:

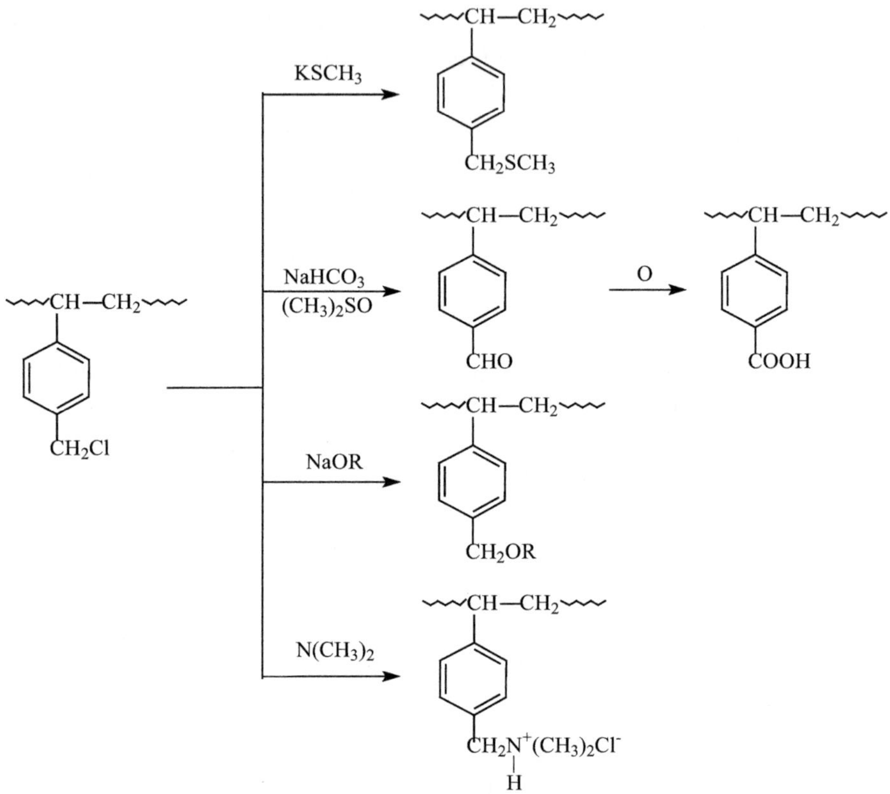

A series of polystyrene derivatives can be also obtained through the stage of lithium polystyrene:

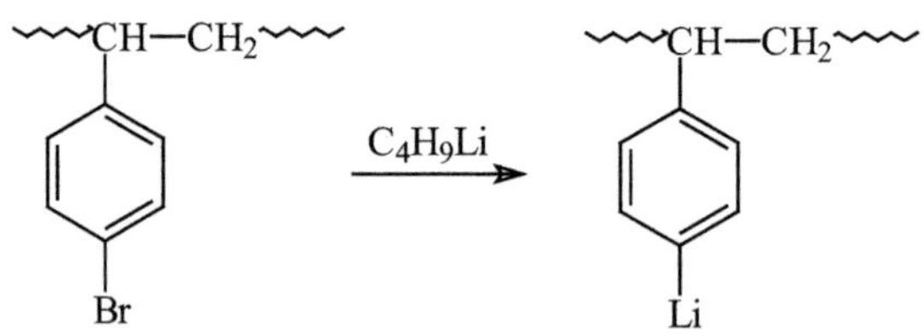

Lithium polystyrene is obtained by the reaction of n-polybromostyrene with n-butyllithium. It can react with such compounds like chlorodialkyl phosphine, boric acid ester, carbon dioxide or dimethyldisulfide with formation of products with the following structure:

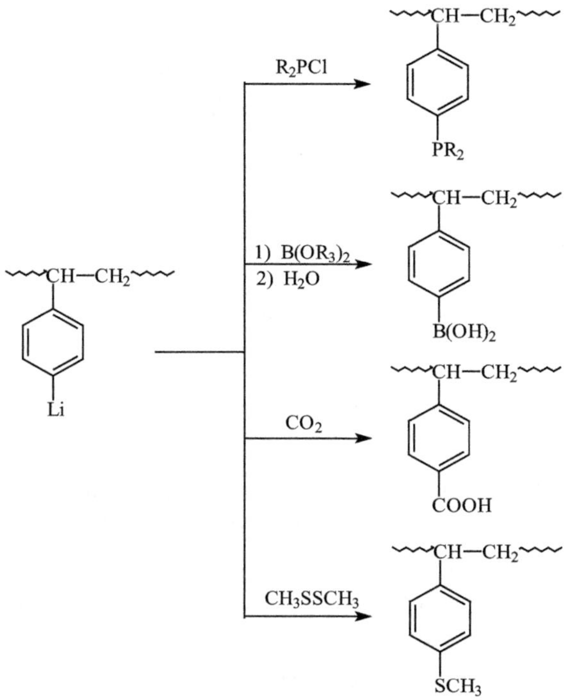

Introduction to the polymer molecule of segments containing primary aromatic amines or phenolic groups enables the synthesis of polymeric dyes. For this purpose one runs the diazotization reaction on polymer containing primary amino groups at the aromatic carbon atom and then the coupling reaction with phenols or aromatic amines. It is also possible to run the coupling reaction of diazonium compounds with polymers containing hydroxyl or amine groups attached to the benzene ring. The course of these processes provide the following reactions:

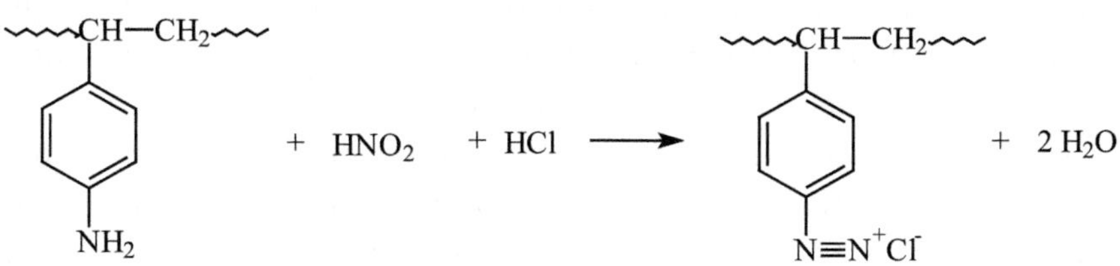

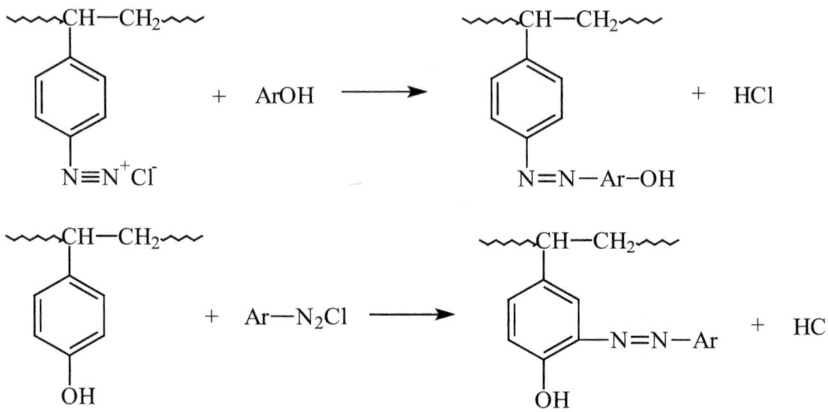

The synthesis of copolymers in which one component contains reactive groups provides further opportunities for chemical modification. An example of this can be the modification of copolymers of maleic anhydride, which can be carried out into suitable esters or imides.

The reaction with aniline proceeds through the stage of phenyl malamate acid derivative and leads to the formation of phenylmaleimide copolymer.

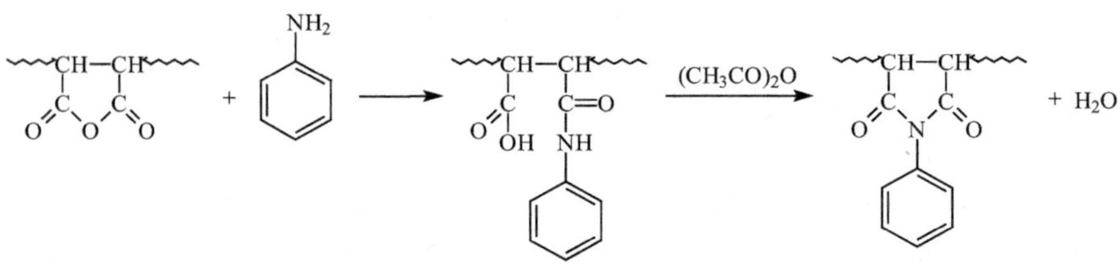

The chemical modification combined with the transformation or splitting off of functional groups can sometimes take place in the plastics processing industry as an adverse reaction. An example of such a reaction is the dehydrochlorination of poly(vinyl chloride).

The above examples show that due to the high reactivity of polymers the possibilities their chemical modification are very large. Some of these reactions have already found practical application on industrial scale.

As an example of the synthesis of chemically modified polymers may serve the previously discussed synthesis of poly(vinyl alcohol), polyvinylo-acetals, copolymers of ethylene with vinyl alcohol, cellulose nitrate (known also as nitrocellulose), cellulose acetate, ethyl cellulose, carboxymethyl cellulose, chitosan, chlorinated rubber, butoxylated melamine-formaldehyde resin, ion exchangers and many other economically useful products.

10.7.2.3 Intramolecular Cyclization

The cyclic systems are introduced into the polymer chains mainly in order to increase their rigidity by inhibiting the free rotation of individual segments of the chain or to provide them the ladder structure, which increases the heat resistance of polymer

The cyclization of an intramolecular reaction takes place under the influence of catalysts or elevated temperatures.

The cyclization of polydienes takes place under the influence of protonic acids, tin tetrachloride and titanium tetrachloride. The cyclization reaction of rubber is an ionic reaction and runs through the stage of formation of carbonion ion, the last arising as a result of the proton or titanium tetrachloride attachment to the double bond:

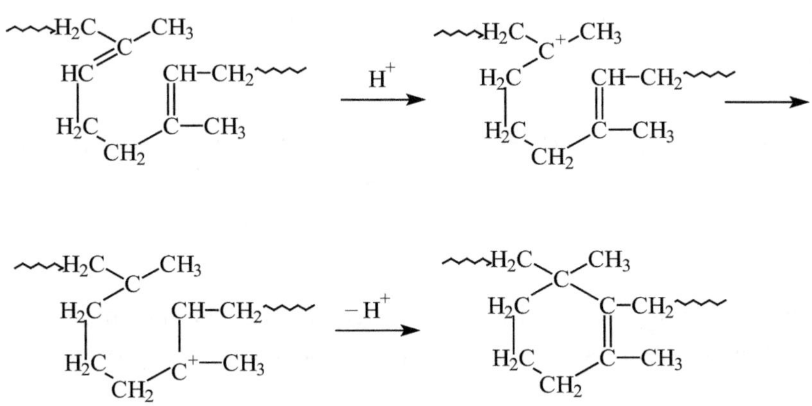

The cyclization of rubber is accompanied by a decrease of the unsaturation degree of polymer. Catalysts are not part of the cyclization products, but affect significantly the structure of the cyclized rubber.

An example of the synthesis of a ladder polymer is the cyclization of polyacrylonitrile at temperature of 200°C with the formation of a heterocyclic heat-resistant polymer with semiconductor properties:

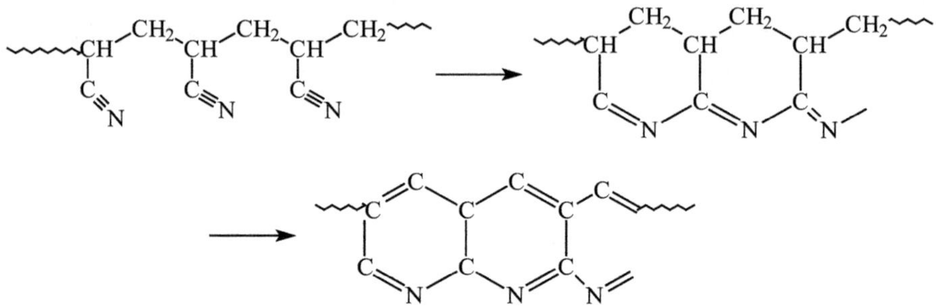

The ladder structures are formed also during the intramolecular aldol condensation of poly(methyl vinyl ketone) in the presence of an alkaline catalyst.

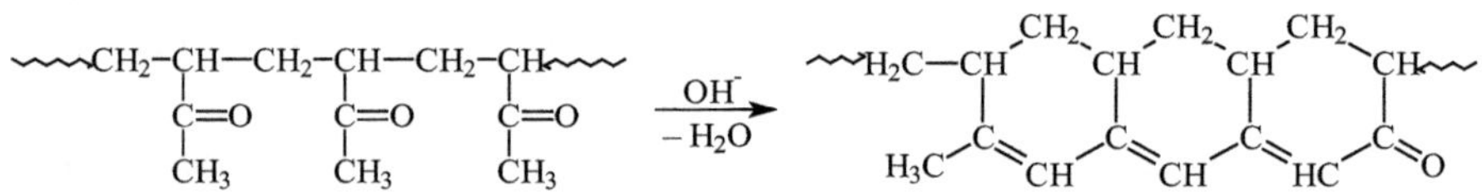

During the dehydration of polyacrylic acid a corresponding polyanhydride is formed:

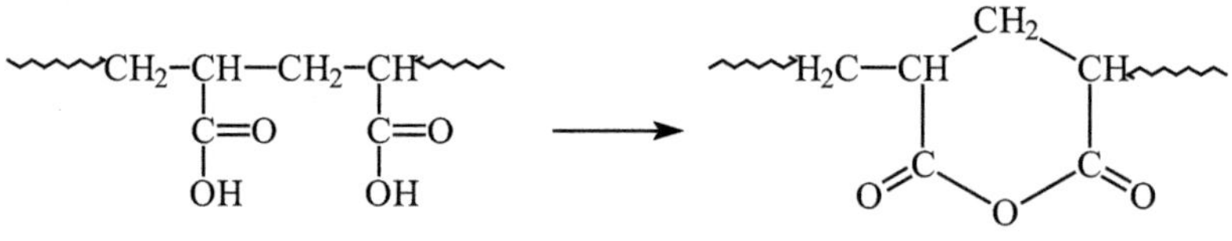

By attaching the aldehyde and phenylhydroxylamine molecules to unsaturated polymers formed are polymers with isooxazine rings:

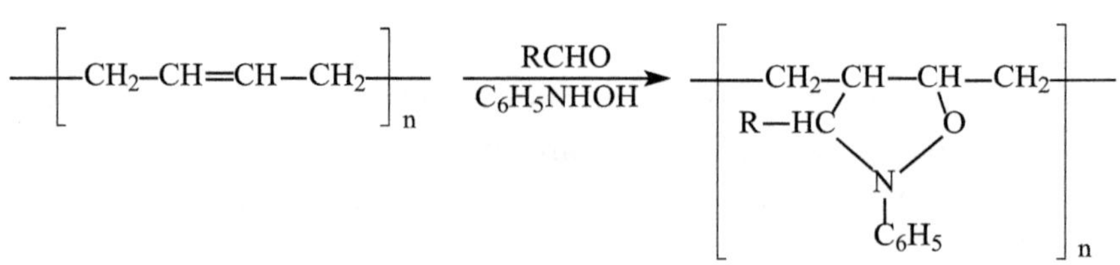

10.7.2.4 Oxidation of Polymers

The total oxidation of polymer leads to its decay with liberation of carbon dioxide, water and possibly other degradation products. In less dramatic terms this process is often linked to the degradation of macromolecules.

By selecting an appropriate oxidizing agent and using proper reaction parameters it is possible to run a selective oxidation enabling the chemical modification of the polymer.

One of the most important reactions of this type is the formation in unsaturated polymers of oxirane (epoxy) units.

Selective oxidizing agents are in this case the carboxylic peracids such as, for example, the peracetic acid.

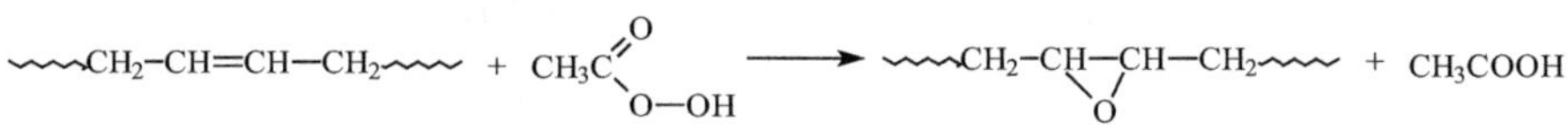

It is possible, in a similar way, to oxidize poly(2-vinylpyridine) to a corresponding N-oxide, which found application as a biologically active polymer in the treatment of silicosis:

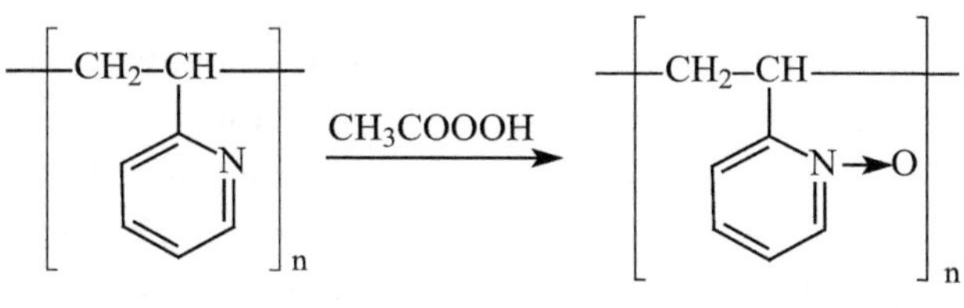

Application of a very mild oxidizing agent, such as, for example, pyridine dichromate, allows the oxidation of primary alcohol groups in poly(methylallyl alcohol) to poly(methyl acroleine):

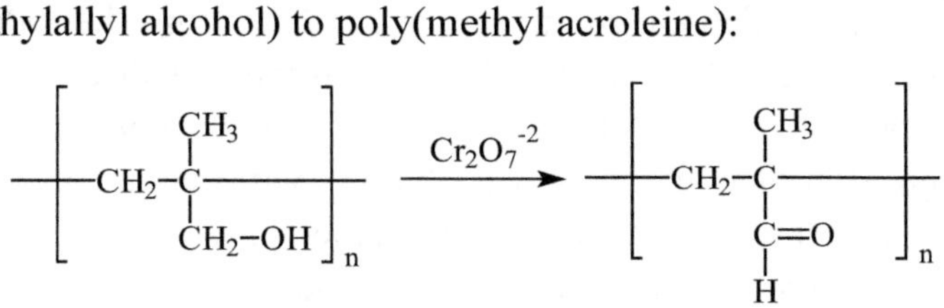

The reaction of oxidation of polymers is also often connected with the cross-linking process. For example, by mixing a solution of poly (vinyl alcohol) with a solution of potassium dichromate a photosensitive mixture is formed.

Covering the polyamide grid with this mixture and evaporating the solvent in dark permits to obtain a photosensitive coating, which in exposed places becomes to be insoluble. This procedure allows to obtain matrices for screen printing of fabrics.

10.7.2.5 Reduction of Polymers

The reduction method can also be used to modify polymers. It enables the formation of a polymer molecule of amine groups by reducing the nitro groups:

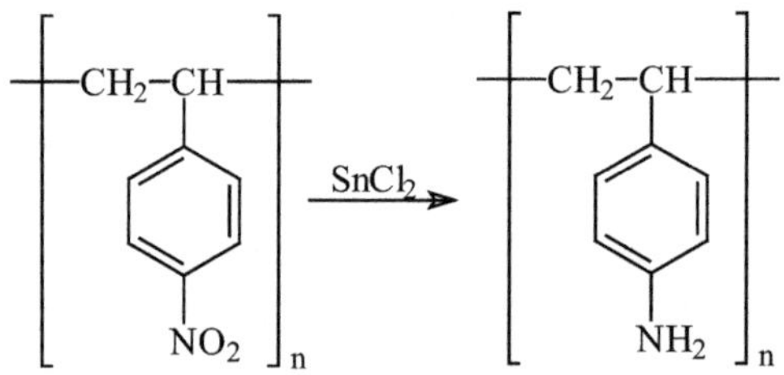

A highly selective reducing agent used in polymer chemistry is the lithium aluminum hydride, as it permits the removal of halogen atoms contained in the polymer molecule. In reaction of lithium aluminum hydride with poly(vinyl chloride) a partial removal of chlorine atoms from molecules takes place and the resulting product has a structure similar to ethylene copolymer with vinyl chloride or chlorinated polyethylene.

The lithium aluminum hydride allows also a selective reduction of the ester group to alcohol group. For example, during the reduction of poly(methyl methacrylate) with lithium aluminum hydride in dioxane, depending on the polymer concentration in solution, one can obtain the poly(methyl allyl alcohol) or its copolymers with methyl methacrylate. Some results of this process are listed in Table 10.37.

The reaction proceeds according to the scheme:

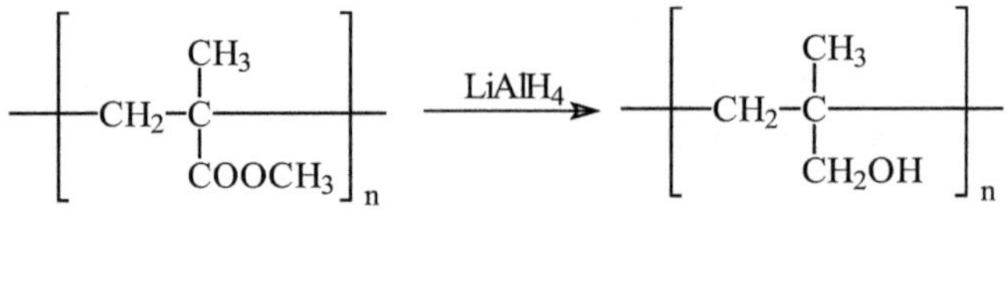

TABLE 10.37 Effect of concentration of poly(methyl methacrylate) (PMMA) in dioxane on the course of its reduction with lithium aluminum hydride (after R. T. Sikorski, Z. Rykowski, A. Puszyński: Plast u. Kautschuk, 31, 250 (1984)).

No.	PMMA concentration [g/dm3]	Yield [%]	Saponification number [mgKOH/gw]	Conversion [% mol]
1	10	89.4	349	37.5
2	6.9	77.2	372	33.4
3	4	84	0	100
4	2	78	46	91.8

In the same way one can get poly(allyl alcohol) by reduction of poly(ethyl acrylate)

10.7.2.6 Grafting of Polymers

Grafting of polymers is an important method of synthesis of copolymers by chemical modification of monomers.

The principle of this process consists on formation of an active center along the polymer chain, which starts polymerization of another monomer. The formed active centers could be either radical or ionic. This process can be presented by the following scheme:

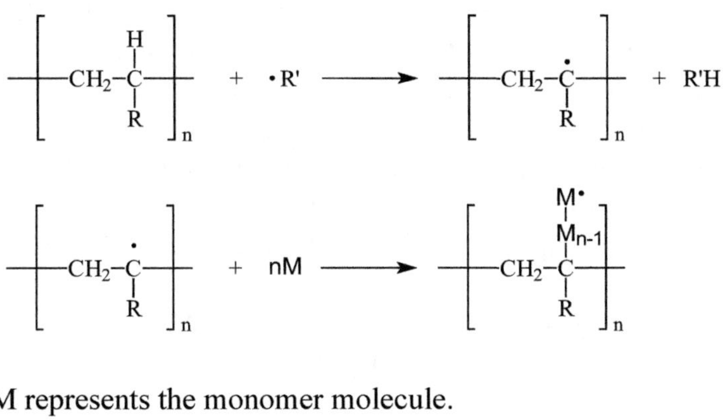

where M represents the monomer molecule.

Initiation of a grafting reaction can be carried out by the ozonization or oxidation method:

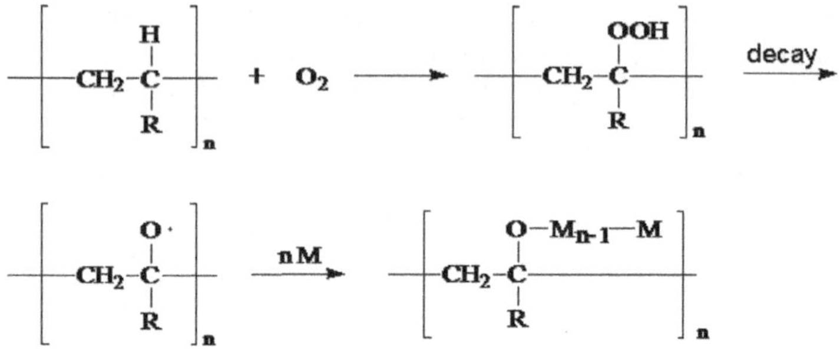

The free radical grafting reactions of polymers are often initiated by cerium (IV) ions, which are reduced under the influence of sunlight with formation of free radicals:

$$Ce^{+4} + OH^{-} \longrightarrow Ce^{+3} + \cdot OH$$

These ions can also initiate a competitive homopolymerization reaction, because they generate free radicals of M• type:

$$M + Ce^{+4} \longrightarrow Ce^{+3} + \cdot M + H^{+}$$

Similar results are obtained with Mn^{+3} ions.

During the grafting reaction of acrylonitrile on starch one obtains 95% yield of the grafted product and the molecular weight of grafted polyacrylonitrile chain is equal to 86,000. The content of the incidentally formed homopolymer is 4.2–4.9%. Replacing acrylonitrile by methyl methacrylate diminishes the grafting efficiency to 87% and the amount of incidentally formed poly(methyl methacrylate) is equal to 13.3%.

In order to characterize the grafted copolymers one introduced the concept of grafting efficiency (G), total conversion (C_t), grafting efficiency (GE) and grafting frequency (GF), which are defined as follows:

$$G = \frac{\text{mass of grafted copolymer - mass of polymer}}{\text{mass of polymer}} x100\%$$

$$C_t = \frac{\text{mass of grafted polymer - mass of homopolymer}}{\text{mass of polymer}} x100\%$$

$$GE = \frac{\text{mass of grafted polymer}}{\text{mass of copolymer} + \text{mass of grafted polymer}} x100\%$$

$$GF = \frac{\text{mass of grafted polymer}}{\text{mass of polymer}} \cdot \frac{\overline{M}_{polymer}}{\overline{M}_{grafted\ polymer}} x100\%$$

In the case of grafting of acrylonitrile on starch by the mass of polymer one understands the mass of starch and by mass of grafted polymer – the mass of polyacrylonitrile attached to starch.

Known are also ionic methods of grafting. Using the cationic method it is possible to graft tetrahydrofuran on polyethylene or polypropylene.

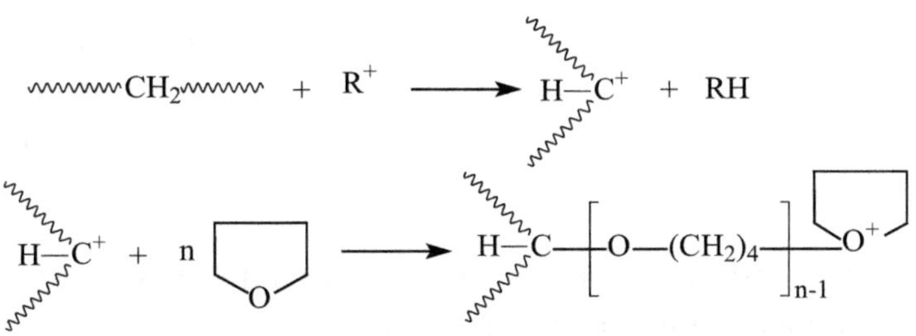

Polymers containing reactive side groups can be also transformed ionic processes into grafted copolymers. For example, the chloromethylated polystyrene dissolved in carbon disulfide in the presence of aluminum tribromide reacts with 2-methylpropene, forming mainly polystyrene grafted with poly(2-methylpropene). Similarly, poly(vinyl chloride) undergoes a cationic grafting reaction by styrene in the presence of aluminum trichloride or by butadiene in the presence of the diethylaluminum chloride complex with cobalt.

There were performed also the anionic grafting processes. An example of such a reaction is the grafting of polyacrylonitrile on poly(p-chlorostyrene) in tetrahydrofuran in the presence of sodium naphthalene:

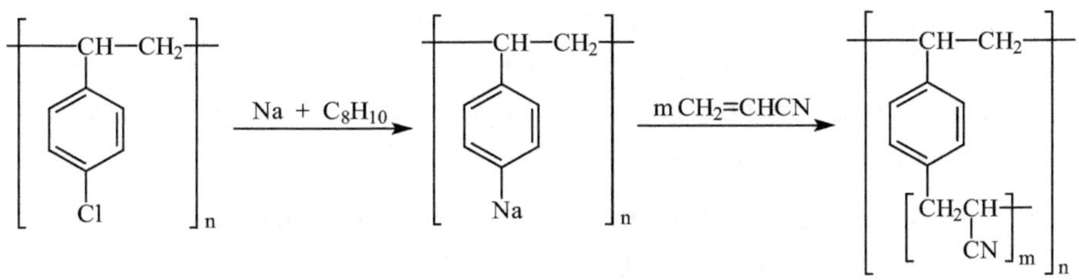

The unsaturated polymers can be grafted with vinyl monomers using the Ziegler-Natta type catalysts. In the first stage they are subjected to react with diethylaluminum hydride. One obtains in this way the alkyl-aluminum polymeric compounds. Then, by the reaction of halogens with metals such as titanium trichloride one gets active centers initiating the grafting reaction. When using propylene the method leads to the formation of side chains being an isotactic polypropylene:

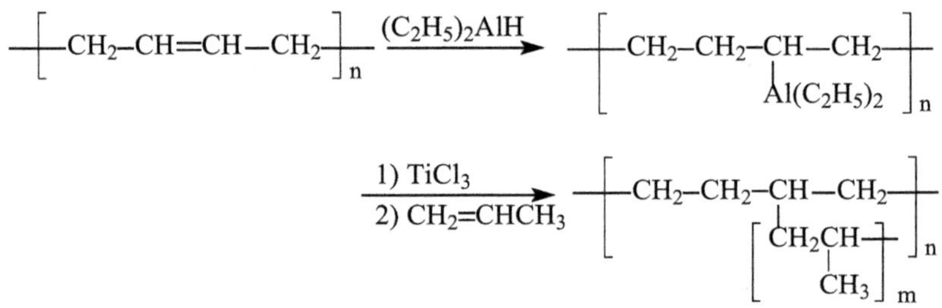

The above-described polymer grafting examples do not cover all possibilities for conducting this important process.

10.7.2.7 Cross-Linking of Saturated Polymers

Cross-linking is an important reaction of chemical modification of polymers making possible obtaining fusible and insoluble products, often with better properties than linear polymers. A small degree of cross-linking reduces the mobility of polymer chains and prevents their relative motion to each other, providing them properties of an elastomer.

During the tension or compression of partly cross-linked polymers, the extended lateral bonds cause return to the previous shape after stopping the deforming stress. The increase of cross-linking density of polymers reduces their flexibility and too dense cross-linking causes fragility of the product. The cross-linked polymers are often called duroplasts.

Depending on chemical structure of polymers there are several ways of their cross-linking. In particular one distinguishes:

- cross-linking of polymers containing no functional groups;
- cross-linking of unsaturated polymers;

- cross-linking of polymers containing reactive functional groups reacting with each other;
- cross-linking of polymers containing functional groups, able to react with other small or large molecule compounds, known as hardeners.

10.7.2.8 Cross-Linking of Polymers Without Functional Groups

Polymers without functional groups can cross-link using the free radical reactions, initiated by the decay of organic peroxides as well as the UV radiation. The cross-linking reaction of such compounds can be also induced by radiation with streams of photons, electrons, neutrons or protons. Sometimes the reaction may take place under the influence of ionic catalysts of Friedel – Crafts type.

In order to cross-link a polymer with organic peroxides one prepares the so – called “preblends”. The polymer is mixed with an organic peroxide and fillers (addition of inorganic or organic substances modifying material properties). The mixture is then annealed in forms at temperature allowing a decomposition of peroxide to free radicals. This method has found practical application in cross-linking of ethylene-propylene, polyester and silicone rubbers.

The radiolysis of polymers under the influence of ionizing radiation can lead to both the cross-linking reaction and the degradation. In the case of high doses of radiation used dominates the degradation process. In other case (low dose) the reaction course depends on the polymer structure.

Geminally bisubstituted polymers, that is, containing two identical substituents at the same carbon atom and the chlorinated polymers tend to split the chain, thus its degradation. The other types of vinyl polymers are easier to cure.

The mechanism of radiation induced cross-linking is a free radical process and proceeds through the stage of the extraction of hydrogen atom with the formation of a macroradical.

The reaction proceeds according to the scheme:

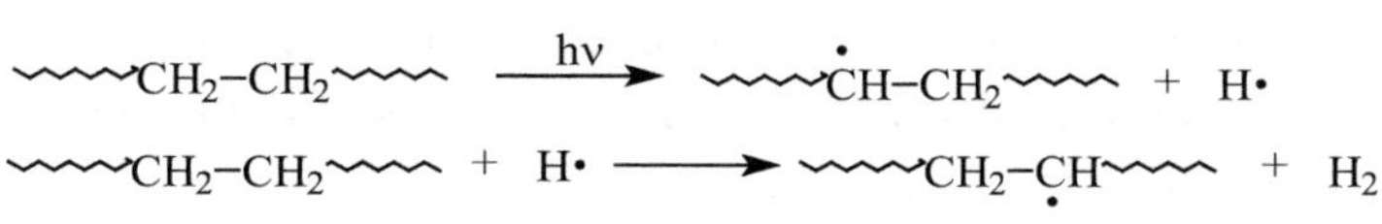

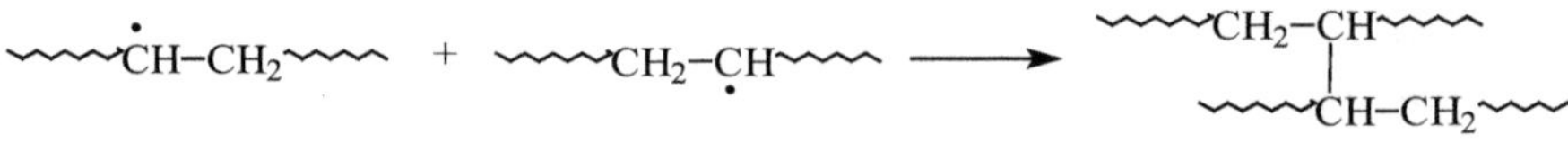

In a similar way runs the photolysis reaction of polymers under the influence of ultraviolet or visible light. But the process is more specific than the radiolysis. In this case to proceed necessary are well-defined polymer structure elements or additives, enabling absorption of light radiation in an appropriate wavelength range.

If one adds to polymer some sensitizers, such as benzophenone, the absorption in ultraviolet causes excitation of sensitizer molecule and then splitting off the hydrogen atom from the polymer molecule. It leads to the formation of a macroradical capable of cross-linking by recombination with other macromolecules.

If the chromophore group is present in polymer molecule then the absorption of an energy quantum can lead to both the cross-linking as well as the degradation of polymer.

An example of such a reaction is the process-taking place during the irradiation of vinyl polymers containing ester groups as side groups:

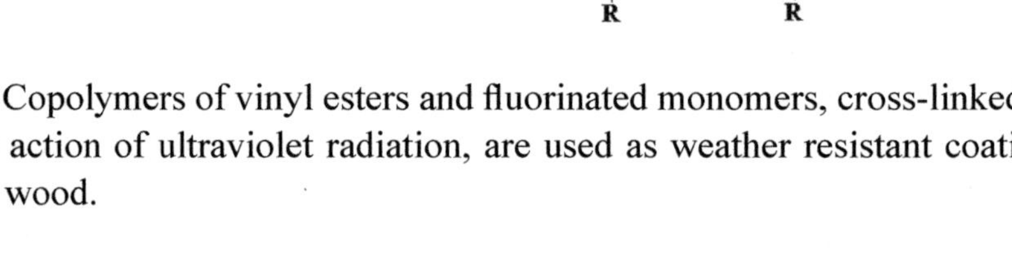

Copolymers of vinyl esters and fluorinated monomers, cross-linked by the action of ultraviolet radiation, are used as weather resistant coatings for wood.

10.7.2.9 Cross-Linking of Unsaturated Polymers

The cross-linking process of unsaturated polymers is often called vulcanization. The vulcanization may be carried out under the influence of organic peroxides, sulfur, disulfur dichloride (S_2Cl_2) and other cross-linking agents such as the 1,3-phenylene disulfone acid.

In the case of cross-linking with organic peroxides the split off of hydrogen occurs most frequently at the allyl carbon atom, for example, at the atom that is adjacent to the carbon atom forming a double bond. A macroradical is then formed and the cross-linking reaction takes place as a result of the recombination of radicals or through binding processes connected with charge transfer:

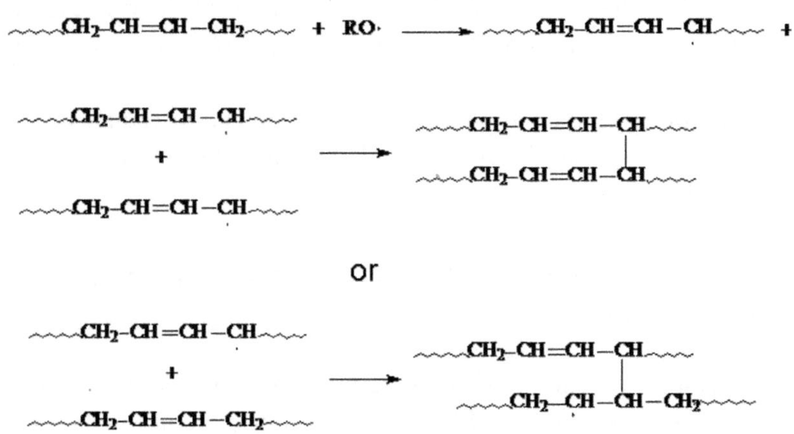

The initiation of vulcanization using sulfur is carried out by attaching the active hydrogen atom to the sulfur molecules.

$$\sim\sim CH_2{-}CH{=}CH{-}CH_2\sim\sim + S_x \longrightarrow \sim\sim CH_2{-}CH{=}CH{-}\dot{C}H\sim\sim + S_xH$$

The formed radical reacts with the next sulfur molecule and the resulting addition product with radical on the sulfur atom joins the unsaturated carbon atom of another chain of polydiene.

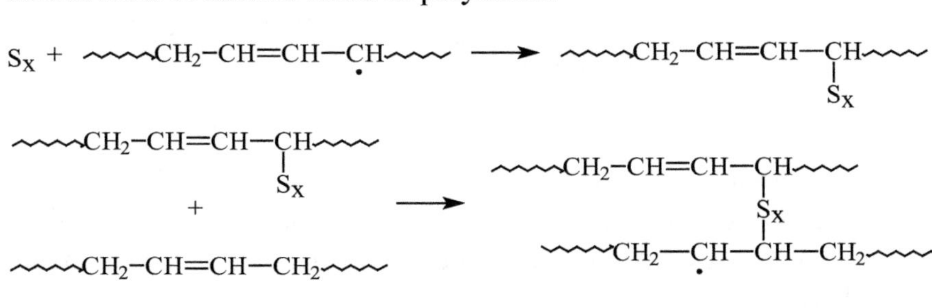

Deactivation of the formed macroradical takes place by transfer to the next polydiene molecule. The whole process runs on the principle of a chain reaction, up to a total exhaustion of the unbounded sulfur contained in the system.

The cross-linking of polydienes by using the sulfur is very slow. Therefore, one uses for this purpose accelerators of this process, such as tetramethylthiuram disulfide, dithiocarbamine acid derivatives, thiazole derivatives, diphenylguanidine and the others.

In order to achieve the full effect of speeding up the process of vulcanization one uses additionally other activators such as zinc oxide and stearic acid.

Acceleration of the cross-linking process with sulfur reduces the number of cyclic monosulfide groups and increases the number of lateral mono- and disulfide bonds.

10.7.2.10 Cross-Linking of Polymers Containing Reactive Functional Groups

If the polymer molecule contains reactive functional groups, capable of condensing with each other, than under appropriate conditions, possible are intermolecular reactions combined with the cross-linking of polymer.

Examples of this type of reactive polymers are phenol-formaldehyde resins of rezol type as well as urea – or melamine-formaldehyde resins. A characteristic feature of such compounds is the presence of hydroxymethyl groups, also called methyl groups.

These groups react with each other with evolution of water or formaldehyde molecules. At the same time a cross-linked polymer is formed, which contains methylene-ether or methylene bridges.

$$—CH_2OH + HOH_2C— \nearrow —CH_2–O—CH_2 + H_2O$$

$$—CH_2OH + HOH_2C— \searrow —CH_2— + HCOH$$

This reaction can be catalyzed by hydrogen or hydroxidel ions, or initiated at elevated temperatures. Since these reactions proceed readily at higher temperatures therefore such type of polymers are called thermosetting polymers.

Polymers containing reactive groups in the polymer chain can react with other suitable multifunctional compounds with the formation of cross-linked products. This type of polymers is called chemoreactive compounds. A striking example of this type of connections are polymers with epoxy groups, which can be cross-linked in a polymerization reaction, combined with opening of the oxirane ring as in the reactions with multifunctional amines or carboxylic acids.

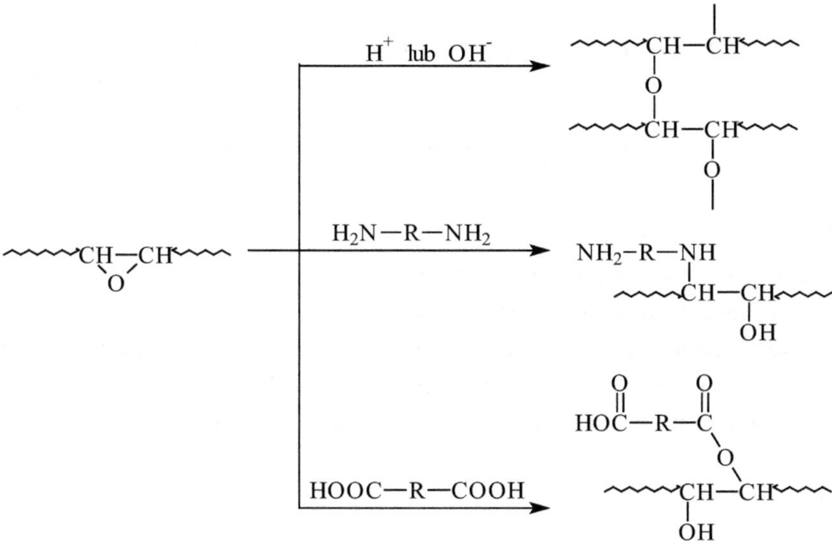

The others, after attachment to the epoxy group, free amino or carboxyl groups, react with epoxy groups of other polymer chains, forming cross-linked products.

Very good cross-linking agents, reacting with almost all reactive groups containing mobile hydrogen atoms, are multifunctional isocyanates, resulting in the formation of urethane bridges.

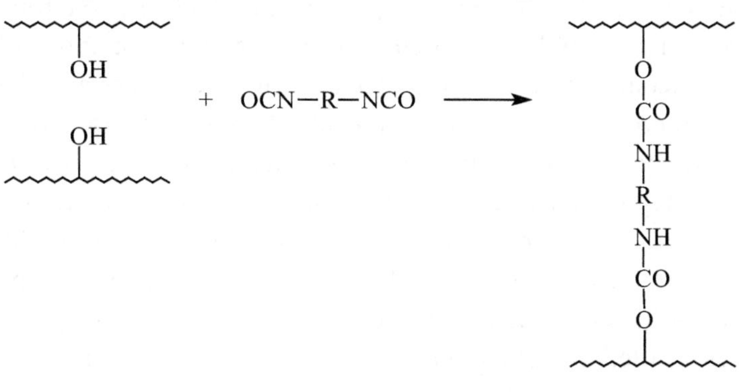

Highly reactive groups, enabling cross-linking, are chlorosulfonated groups or other groups containing mobile chlorine atom.

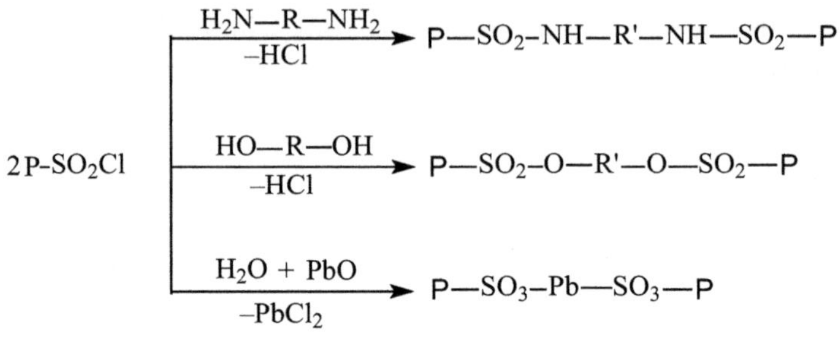

where P – polymer chain.

The possibilities are also known as cross-linking the linear polymers through the formation of coordination bonds.

The Friedel-Crafts reaction enables a catalytic cross-linking of polymers by ionic mechanism. A typical its example is the cross-linking of polystyrene with p-bis (chloromethyl)benzene in the presence of $SnCl_4$.

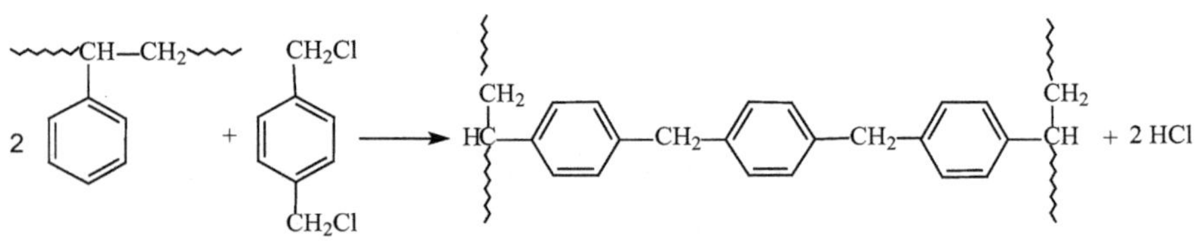

10.7.3 DEGRADATION OF POLYMERS

The degradation of polymers is a process in which the macromolecules decompose into smaller fragments. The latter may be molecules with smaller molecular weight or the products of partial decomposition with changed chemical composition as a result of split off or transformation of certain substituents. A special case of the degradation of polymers is the reaction of depolymerization leading to the monomer formation.

The process of degradation of polymers, run in a controlled way, is of a great practical importance. It allows for a reduction in molecular weight of polymer molecules in order to standardize the length of chains (a process called linearization) and to facilitate processing operations.

Examples of practical use of controlled degradation is the mercerization of cellulose in the manufacture of cellulose fibers, mastication of rubber before vulcanization obtaining of peptides and amino acids by hydrolysis of proteins and the recovery of methyl methacrylate by depolymerization of poly(methyl methacrylate) waste and also the linearization of polyorganosiloxanes.

The uncontrolled polymer degradation process is often harmful, limiting their practical applications.

Factors affecting the degradation of polymers can be both physical and chemical.

10.7.3.1 Degradation of Polymers by Physical Agents

Various physical factors affect natural polymers, such as heat, light illumination and radiation, ultrasounds and mechanical forces during grinding, rolling and other mechanical processes and cause their degradation.

The polymer degradation caused by the physical agents is often referred to as the polymer aging process.

The persistence of individual bonds in macromolecule depends on their dissociation energy, which causes disintegration of a particular bond. According to the Boltzmann distribution at any temperature T [K], there is a finite probability of "W" that a given bond has reached the critical level of binding and dissociation energy E_d.

$$W = e^{-\frac{E_d}{kT}}$$

where k is the Boltzmann constant.

As a result of dissociation of a specific bond two macroradicals are formed, which undergo the disproportionation reaction. A new macroradical is then formed and a polymer with shorter chain and with double bond.

If the macroradical is located at the end of the chain, then a polymer rupture reaction can take place, during which a monomer molecule is released:

The polymer degradation process, running in the solid state, is accompanied by a parallelly progressing cross-linking process, which was discussed previously.

At sufficiently high temperatures the polymer degradation occurs in the absence of oxygen.

The photodegradation of polymers occurs under the influence of ultraviolet radiation and sometimes visible.

The energy of photon radiation (quantum) for a given wavelength is determined by the equation:

$$E = h\upsilon = h\frac{c}{\lambda}$$

where: h – Planck's constant, h = 6.623 x 10^{-27} erg s; ν – frequency of radiation (1/s); c – speed of light, c = 3 x 10^{10} cm/s; λ – wavelength.

The photodegradation reaction takes place when the energy absorbed by polymer molecule is at least equal to the dissociation energy E_d.

Sometimes, however, the absorbed energy can be converted into heat or emitted radiation in form of light quanta.

In order to quantify the photodegradation process one introduces the concept of quantum efficiency Φ of the process:

$$\Phi = \frac{\text{number of times a specific event occurs}}{\text{photon absorbed by the system}}$$

To halt this process one applies usually appropriate photostabilizers. However, in the case when a change of structure of photodegradable polymers is needed appropriate photosensitizers are then used, which facilitate this process.

Similarly as the photolysis, but much easier, proceeds the radiation degradation of polymers. A characteristic feature of this process is the nearly complete lack of depolymerization, leading to the formation of monomers.

Different groups present in polymer chains exhibit different sensitivities to radiation. If the presence of certain groups, such as, for example, phenyl radical or double bonds increases the resistance of polymer to the ionizing radiation, then one speaks about the internal protective effect. Its action is not limited to the nearest mer only, but extends to the whole macromolecule.

10.7.3.2 Chemical Degradation

The degradation-taking place under the influence of chemical agents causes ruptures of chains as a result of chemical reaction. This concerns mainly the hydrolytic reactions, running principally in the case of condensation polymers and containing easy to break chain elements, such as, for example, ester or amide groups. Very important chemical degradation reactions are the oxidation reactions and attachment of ozone.

The molecular oxygen present in air facilitates the process of degradation of polymers and makes it possible at the relatively low temperatures, even at room temperature after a sufficiently long period of time. This process takes place more easily when performed in the presence of singlet oxygen or ozone.

The reaction of degradation in the presence of oxygen proceeds, with its participation, through the stage of formation of hydrogen peroxides. The macroradical, formed by thermal dissociation or other radical reactions, reacts with oxygen, and then with another polymer molecule in transfer reaction associated with the formation of a hydrogen peroxide:

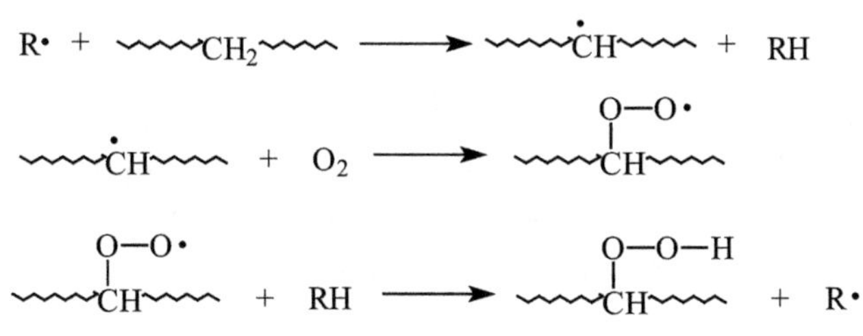

The formed hydrogen peroxides decompose easily by heating or by light illumination according to the scheme:

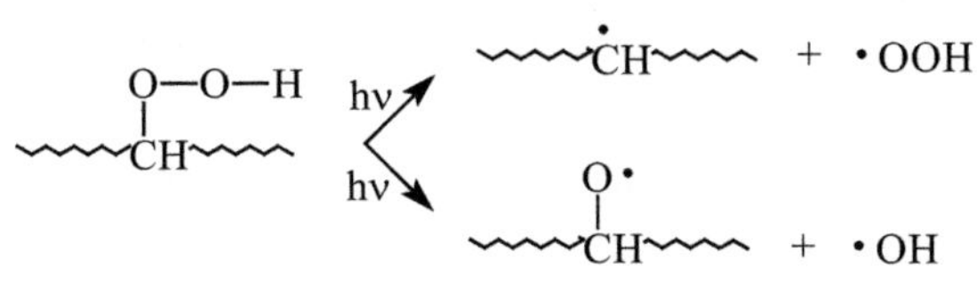

This causes a series of follow-up free radical reactions. The oxy-macroradical is easily disrupted in β position. During the disrupture an aldehyde group and a new macroradical are formed at the chain end:

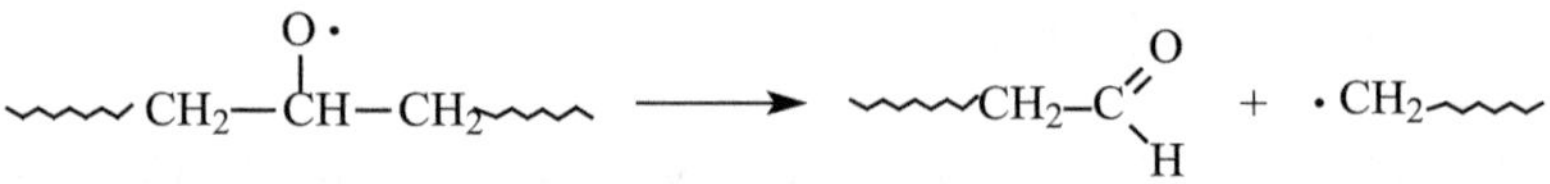

Even easier can be attached to the polymer the ozone molecule. As a result of this reaction a peroxide macroradical and a hydroxyl radical are formed.

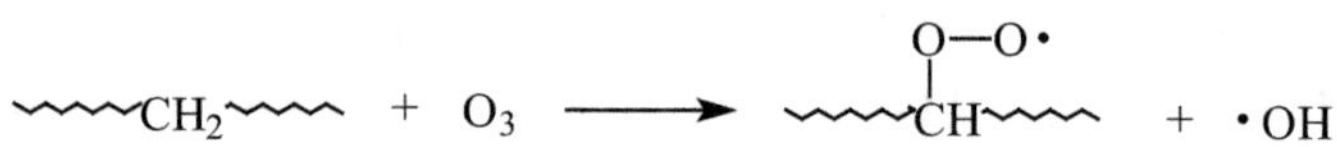

If the polymer molecule contains double bonds, then the ozonides are formed, which subsequently decompose into aldehydes and acids.

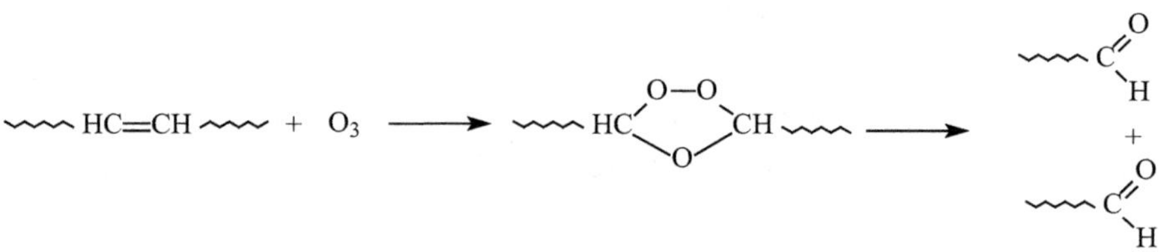

This phenomenon has been used recently for the fabrication of degradable polymers. The derived from them plastics decompose under natural conditions and after use may be disposed of in waste dumps.

The currently used degradable polymers can be divided into:

- photodegradable;
- biodegradable;
- composites obtained from biodegradable polymers.

The photodegradable polymers, containing sensitizers, undergo the degradation process by exposure to ultraviolet radiation in the presence of oxygen.

The photodegradable polymers, among which the most important are the polyolefins, are intended mainly for the manufacture of disposable packaging.

An interesting application of the photodegradable polyethylene film is the protection of young plants growth in countries with dry climates.

Crops are covered with foil containing a precise amount of sensitizer. The foil prevents the heat and the moisture loss. It prevents also the access of pests and weed growth and ensure a proper growth and development of crop root. After reaching a certain growth the intense sunlight causes degradation of the film and its disintegration into small, degraded fragments. Breakthrough time of the films depends on the amount of added sensitizer, assuming a cloudless sky. The crushed film residue does not need to be collected and can be mixed with the soil during the plowing.

The biodegradable polymers are natural polymers, and polymers obtained by the modification of natural as well as those manufactured by various methods of chemical synthesis and biotechnological processes. The world's first biodegradable polymer is produced using the biotechnology method is Bipol, produced by British company Zeneca. Bipol is a polyester derived from 3-hydroxybutyric and 3-hydroxypentanoic acids. The production process involves the fermentation of sugars with the use of *Alcaligenes entropus* bacteria.

During the last twenty years explored were the properties of many other, similar polyesters. As result, the fatty polyhydroxy acids are now one of the larger group of thermoplastic polymers (more than 100 different types). In addition to the biodegradability, an important feature of these products is that they can be processed by the same methods as other thermoplastics

The microbiological processes are also used for the production of poly(lactic acid) (PLA), which is produced by the *Lactobacillus* bacteria in the process of fermentation of sugars. It is used for the manufacture of packaging materials and fibers for surgical applications as well as, among the others, for carpet production.

The poly(aspartic acid) is a completely biodegradable polyamide. It is manufactured by Bayer AG company since 1997 by reacting the maleic anhydride with ammonia. The polymer is used to soften water and to prevent the formation of deposits in pipes and in reservoirs.

An enormous advantage of biodegradable polymers is the possibility of recycling of used products and other waste by composting.

KEYWORDS

- **chemical technology**
- **chemistry**
- **elastomers**
- **macromolecules**
- **polycondensation**
- **polymerization**

REFERENCES

1. Abusleme, J., Giannetti, E., Emulsion Polymerization, Macromolecules 1991, 24.
2. Adesamya, I., Classification of Ionic Polymers, Polym. Eng. Sci., 1988, 28, 1473.
3. Akbulut, U., Toppare, L., Usanmaz, A., Onal, A., Electroinitiated Cationic Polymerization of Some Epoksides, Makromol. Chem. Rapid. Commun., 1983, 4, 259.
4. Andreas, F., Grobe, K., Chemia propylenu, WNT 1974, Warszawa.
5. Barg, E. I., Technologia tworzyw sztucznych, PWT 1957, Warszawa.
6. Billmeyer, F. W., Textbook of Polymer Science, John Wiley and Sons, New York 1984.
7. Borsig, E., Lazar, M., Chemical Modification of Polyalkanes, Chem. Listy 1991, 85, 30.
8. Braun, D., Czerwiński, W., Disselhoff, G., Tudos, F., Analysis of the Linear Methods for Determining Copolymerization Reactivity Ratios, Angew. Makromol. Chem. 1984, 125, 161.
9. Burnett, G. M., Cameron, G. G., Gordon, G., Joiner, S. N., Solvent Effects on the Free Radical Polymerization of Styrene, J. Chem. Soc. Faraday Trans., 1–1973, 69, 322.
10. Challa, R. R., Drew, J. H. Stannet, V. T., Stahel, E. P., Radiation Induced Emulsion Polymerization of Vinyl Acetate, J. Appl. Polym. Sci., 1985, 30, 4261.
11. Candau, F., Mechanism and Kinetics of the Radical Polymerization of Acrylamide In Inverse Micelles, Makromol. Chem., Makromol Symp., 1990, 31, 27.
12. Ceresa, R. J., Kopolimery blokowe i szczepione, WNT 1965, Warszawa.
13. Charlesby, A., Chemia radiacyjna polimerów, WNT 1962, Warszawa.
14. Chern C.-S, Poehlein, G. W., Kinetics of Cross-linking Vinyl Polymerization, Polym. Plast. Technol. Eng., 1990, 29, 577.
15. Chiang W.-Y, Hu C.-M., Studies of Reactions with Polymers, J. Appl. Polym. Sci., 1985, 30, 3895 i 4045.
16. Chien, J. C., Salajka, Z., Syndiospecific Polymerization of Styrene, J. Polym. Sci. A, 1991, 29, 1243.

17. Chojnowski, J., Stańczyk WE., Procesy chlorowania polietylenu, Polimery, 1972, 17, 597.
18. Collomb, J., Morin, B., Gandini, A., Cheradame, H., Cationic Polymerization Induced by Metal Salts, Europ. Polym. J., 1980, 16, 1135.
19. Czerwiński, W. K., Solvent Effects on Free – Radical Polymerization, Makromol. Chem., 1991, 192, 1285 i 1297.
20. Dainton, F. S., Reakcje łańcuchowe, PWN, 1963 Warszawa.
21. Ding, J., Price, C., Booth, C., Use of Crown Ether in the Anionic Polymerization of Propylene Oxide, Eur. Polym. J., 1991, 27, 895 i 901.
22. Dogatkin, B. A., Chemia elastomerów, WNT 1976, Warszawa.
23. Duda, A., Penczek, S., Polimeryzacja jonowa i pseudojonowa, Polimery 1989, 34, 429.
24. Dumitrin, S., Shaikk, A. S., Comanita, E., Simionescu, C., Bifunctional Initiators, Eur. Polym. J., 1983, 19, 263.
25. Dunn, A. S., Problems of Emulsion Polymerization, Surf. Coat. Int., 1991, 74, 50.
26. Eliseeva, W. I., Asłamazova, T. R., Emulsionnaja polimeryzacja v otsustivie emulgatora, Uspiechi Chimii, 1991, 60, 398.
27. Erusalimski, B. L., Mechanism dezaktivaci rastuszczich ciepiej v procesach ionnoj polimeryzacji vinylovych polimerov, Vysokomol. Soed. A., 1985, 27, 1571.
28. Fedtke, M., Chimiczieskije reakcji polimerrov, Chimica, Moskwa 1990.
29. Fernandez-Moreno, D., Fernandez-Sanchez, C., Rodriguez, J. G., Alberola, A., Polymerization of Styrene and 2-vinylpyridyne Using AlEt-VCl Heteroaromatic Bases, Eur. Polym. J., 1984, 20, 109.
30. Fieser, L. F., Fieser, M., Chemia organiczna, PWN 1962, Warszawa.
31. Flory, P. J., Nauka o makrocząsteczkach, Polimery 1987, 32, 346.
32. Frejdlina, R. Ch., Karapetian, A. S., Telomeryzacja, WNT, Warszawa 1962.
33. Funt, B. L., Tan, S. R., The Fotoelectrochemical Initiation of Polymerization of Styrene, J. Polym. Sci. A, 1984, 22, 605.
34. Gerner, F. J., Hocker, H., Muller, A. H., Shulz, G. V., On the Termination Reaction In The Anionic Polymerization of Methyl Methacrylate in Polar Solvents, Eur. Polym. J., 1984, 20, 349.
35. Ghose, A., Bhadani, S. N., Electrochemical Polymerization of Trioxane in Chlorinated Solvents, Indian, J. Technol., 1974, 12, 443.
36. Giannetti, E., Nicoletti, G. M., Mazzocchi, R., Homogenous Ziegler-Natta Catalysis, J. Polym. Sci. A, 1985, 23, 2117.
37. Goplan, A., Paulrajan, S., Venkatarao, K., Subbaratnam, N. R., Polymerization of N, N'-methylene Bisacrylamide Initiated by Two New Redox Systems Involving Acidic Permanganate, Eur. Polym. J., 1983, 19, 817.
38. Gózdz, A. S., Rapak, A., Phase Transfer Catalyzed Reactions on Polymers, Makromol. Chem. Rapid Commun., 1981, 2, 359.
39. Greenley, R. Z., An Expanded Listing of Revised Q and e Values, J. Macromol. Sci. Chem. A, 1980, 14, 427.
40. Greenley, R. Z., Recalculation of Some Reactivity Ratios, J. Macromol. Sci. Chem. A, 1980, 14, 445.
41. Guhaniyogi, S. C., Sharma, Y. N., Free-radical Modification of Polipropylene, J. Macromol. Sci. Chem. A, 1985, 22, 1601.

42. Guo, J. S., El Aasser, M. S., Vanderhoff, J. W., Microemulsion polymerization of Styrene, J. Polym. Sci. A, 1989, 27, 691.
43. Gurruchaga, M., Goni, I., Vasguez, M. B., Valero, M., Guzman, G. M., An Approach to the Knowledge of The Graft Polymerization of Acrylic Monomers Onto Polysaccharides Using Ce (IV) as Initiator, J. Polym. Sci. Part C, 1989, 27, 149.
44. Hasenbein, N., Bandermann, F., Polymerization of Isobutene With VOCl3 in Heptane, Makromol. Chem., 1987, 188, 83.
45. Hatakeyama, H. et al., Biodegradable Polyurethanes from Plant Components, J. Macromol. Sci. A., 1995, 32, 743.
46. Higashimura, T., Law Y.-M., Sawamoto, M., Living Cationic Polymerization, Polym. J., 1984, 16, 401.
47. Hirota, M., Fukuda, H., Polymerization of Tetrahydrofuran, Nagoya-ski Kogyo Kenkynsho Kentyn Hokoku 1984, 51.
48. Hodge, P., Sherrington, D. C., Polymer-supported Reactions in Organic Syntheses, John Wiley and Sons, New York 1980, tłum. na rosyjski Mir, Moskwa 1983.
49. Inoue, S., Novel Zine Carboxylates as Catalysts for the Copolymerization of CO_2 With Epoxides, Makromol. Chem. Rapid Commun., 1980, 1, 775.
50. Janović, Z., Polimerizacije I polimery, Izdanja Kemije Ind., Zagrzeb 1997.
51. Jedliński, Z. J., Współczesne kierunki w dziedzinie syntezy polimerów, Wiadomości Chem., 1977, 31, 607.
52. Jenkins, A. D., Ledwith, A., ed., Reactivity, Mechanism and Structure in Polymer Chemistry, John Wiley and sons, London 1974, tłum.na rosyjski, Mir, Moskwa 1977.
53. Joshi, S. G., Natu, A. A., Chlorocarboxylation of Polyethylene, Angew. Makromol. Chem. 1986, 140, 99–115.
54. Kaczmarek, H., polimery, a środowisko, Polimery 1997, 42, 521.
55. Kang, E. T., Neoh, K. G., Halogen Induced Polymerization of Furan, Eur. Polym. J. 1987, 23, 719.
56. Kang, E. T., Neoh, K. G., Tan, T. C., Ti, H. C., Iodine Induced Polymerization and Oxidation of Pryridazine, Mol. Cryst. Lig. Cryst., 1987, 147, 199.
57. Karnojitzki, V., Organiczieskije pierekisy, I. I. L., 1961 Moskwa.
58. Karpiński, K., Prot, T., Inicjatory fotopolimeryzacji i fotosieciowania, Polimery 1987, 32, 129.
59. Keii, T., Propene Polymerization with MgCl2 – Supported Ziegler Catalysts, Makromol. Chem. 1984, 185, 1537.
60. Khanna, S. N., Levy, M., Szwarc, M., Complexes formed by anthracene with 'living' polystyrene, *Dormant Polymers. Trans Faraday Soc,* 1962, 58, 747–761.
61. Koinuma, H., Naito, K., Hirai, H. Anionic Polymerization of Oxiranes., Makromol. Chem., 1982, 183, 1383.
62. Korniejev, N. N., Popov, A. F., Krencel, B. A. Komplieksnyje metalloorganiczeskije katalizatory, Chimia, Leningrad 1969.
63. Korszak, W. W. Niekotoryje problemy polikondensacji, Vysokomol. Soed. A, 1979, 21, 3.
64. Korszak, W. W. Technologia tworzyw sztucznych, WNT, Warszawa 1981.
65. KowalskaE., Pełka, J., Modyfikacja tworzyw termoplastycznych włóknami celulozowymi, Polimery 2001, 46, 268.
66. Kozakiewicz, J., Penczek, P., Polimery jonowe, Wiadomości Chem. 1976, 30, 477.

67. Kubisa, P., Polimeryzacja żyjąca, Polimery 1990, 35, 61.
68. Kubisa, P., Neeld, K., Starr, J., Vogl, O., Polymerization of Higher Aldehydes, Polymer 1980, 21, 1433.
69. Kucharski, M., Nitrowe i aminowe pochodne polistirenu, Polimery 1966, 11, 253.
70. Kunicka, M. Choć, H. J., Hydrolitic degradation and mechanical properties of hydrogels, Polym. Degrad. Stab. 1998, 59, 33.
71. Lebduska, J., Dlaskova, M., Roztokova kopolymerace, Chemicky Prumysl 1990, 40, 419 i 480.
72. Leclerc, M., Guay, J., Dao, L. W., Synthesis and Characterization of Poly (alkylanilines), Macromolecules 1989, 22, 649.
73. Lee, D. P., Dreyfuss, P., Triphenylphesphine Termination of Tetrahydrofuran Polymerization, J. Polym. Sci., Polym. Chem. Ed. 1980, 18, 1627.
74. Leza, M. L., Casinos, I., Guzman, G. M., Bello, A., Graft Polymerization of 4-vinylpyridine Onto Cellulosic Fibers by the Ceric on Method, Angew. Makromol. Chem. 1990, 178, 109 i 119.
75. Lopez, F., Calgano, M. P., Contrieras, J. M., Torrellas, Z., Felisola, K., Polymerization of Some Oxiranes, Polymer International 1991, 24, 105.
76. Lopyrev, W. A., Miaczina, G. F., Szevaleevskii, O. I., Hidekel, M. L., Poliacetylen, Vysokomol. Soed. 1988, 30, 2019.
77. Mano, E. B., Calafate, B. A. L., Electrolytically Initiated Polymerization of N-vinylcarbazole, J. Polym. Sci., Polym. Chem. Ed. 1983, 21, 829.
78. Mark, H. F., Encyclopedia of Polymer Science and Technology, Concise 3rd ed., Wiley-Interscience, Hoboken, New York 2007.
79. Mark, H., Tobolsky, A. V., Chemia fizyczna polimerów, PWN 1957, Warszawa.
80. Matyjaszewski, K., Davis, T. P., Handbook of Radical Polymerization, Wiley-Interscience, New York 2002.
81. Matyjaszewski, K., Mülle, A. H. E., 50 years of living polymerization, Progress in Polymer Science, 2006, 31, 1039–1040.
82. Mehrotra, R., Ranby, B., Graft Copolymerization Onto Starch, J. Appl. Polym. Sci. 1977, 21, 1647, 3407 i 1978, 22, 2991.
83. Morrison, R. T., Boyd, N., Chemia organiczna, PWN 1985, Warszawa.
84. Morton, M., Anionic Polymerization, Academic Press, New York 1983.
85. Munk, P., Introduction to Macromolecular Science, J. Wiley and Sons, New York 1989.
86. Naraniecki, B., Ciecze mikroemulsyjne, Przem. Chem. 1999, 78(4), 127.
87. Neoh, K. G., Kang, E. T., Tan, K. L., Chemical Copolymerization of Aniline with Halogen-Substituted Anilines, Eur. Polym. J. 1990, 26, 403.
88. Nowakowska, M., Pacha, J., Polimeryzacja olefin wobec kompleksów metaloorganicznych, Polimery 1978, 23, 290.
89. Ogata, N., Sanui, K., Tan, S., Synthesis of Alifatic Polyamides by Direct Polycondensation with Triphenylphosphine, Polymer, J. 1984, 16, 569.
90. Onen, A., Yagci, Y., Bifunctional Initiators, J. Macromol. Sci. Chem. A, 1990, 27, 743, Angew. Makromol. Chem. 1990, 181, 191.
91. Osada, Y., Plazmiennaja polimerizacia i plazmiennaja obrabotka polimerov, Vysokomol. Soed. A, 1988, 30, 1815.
92. Pasika, W. M., Cationic Copolymerization of Epichlorohydrin With Styrene Oxide And Cyclohexene Oxide, J. Macromol. Sci. Chem. A, 1991, 28, 545.

93. Pasika, W. M., Copolymerization of Styrene Oxide and Cyklohexene Oxide, J. Polym. Sci. Part A, 1991, 29, 1475.
94. Pielichowski, J., Puszyński, A., Preparatyka monomerów, W. P.Kr. Kraków.
95. Pielichowski, J., Puszyński, A.,: Preparatyka polimerów, TEZA WNT 2005 Kraków.
96. Pielichowski, J., Puszyński, A., Technologia tworzyw sztucznych, WNT 2003, Warszawa.
97. Pielichowski, J., Puszyński, A., Wybrane działy z technologii chemicznej organicznej, W. P.Kr. Kraków.
98. Pistola, G., Bagnarelli, O., Electrochemical Bulk Polymerization of Methyl Methacrylate in the Presence of Nitric Acid, J.Polym.Sci., Polym. Chem. Ed. 1979, 17, 1002.
99. Pistola, G., Bagnarelli, O., Maiocco, M., Evaluation of Factors Affecting The Radical Electropolymerization of Methylmethacrylate in the Presence of HNO_3, J. Appl. Electrochem. 1979, 9, 343.
100. Połowoński, S., Techniki pomiarowo-badawcze w chemii fizycznej polimerów, W. P. L., Łódź 1975.
101. Porejko, S., Fejgin, J., Zakrzewski, L., Chemia związków wielkocząsteczkowych, WNT, Warszawa 1974.
102. Praca zbiorowa: Chemia plazmy niskotemperaturowej, WNT, Warszawa 1983.
103. Puszyński, A., Chlorination of Polyethylene in Suspension, Pol. J. Chem. 1981, 55, 2143.
104. Puszyński, A., Godniak, E., Chlorination of Polyethylene in Suspension in the Presence of Heavy Metal Salts, Macromol. Chem. Rapid Commun, 1980, 1, 617.
105. Puszyński, A., Dwornicka, J., Chlorination of Polypropylene in Suspension, Angew. Macromol. Chem. 1986, 139, 123.
106. Puszyński, J. A., Miao, S., Kinetic Study of Synthesis of SiC Powders and Whiskers In the Presence $KClO_3$ and Teflon, Int. J. SHS 1999, 8(8), 265.
107. Rabagliati, F. M., Bradley, C. A., Epoxide Polymerization, Eur. Polym. J. 1984, 20, 571.
108. Rabek, J. F., Podstawy fizykochemii polimerów, W. P.Wr., Wrocław 1977.
109. Rabek, T. I., Teoretyczne podstawy syntezy polielektrolitów i wymieniaczy jonowych, PWN, Warszawa 1962.
110. Regas, F. P., Papadoyannis, C. J., Suspension Cross-linking of Polystyrene With Friedel-Crafts Catalysts, Polym. Bull. 1980, 3, 279.
111. Rodriguez, M., Figueruelo, J. E., On the Anionic Polymerization Mechanism of Epoxides, Makromol. Chem. 1975, 176, 3107.
112. Roudet, J., Gandini, A., Cationic Polymerization Induced by Arydiazonium Salts, Makromol. Chem. Rapid Commun. 1989, 10, 277.
113. Sahu, U. S., Bhadam, S. N., Triphenylphosphine Catalyzed Polymerization of Maleic Anhydride, Makromol. Chem. 1982, 183, 1653.
114. Schidknecht, C. E., Polimery winylowe, PWT, Warszawa 1956.
115. Szwarc, M., Levy, M. Milkovich, R., Polymerization initiated by electron transfer to monomer. A new method of formation of block copolymers, *J Am Chem Soc,* 1956, 78, 2656–2657.
116. Szwarc, M., 'Living' polymers, *Nature,* 1956, 176, 1168–1169.
117. Sen, S., Kisakurek, D., Turker, L., Toppare, L., Akbulut, U., Electroinitiated Polymerization of 4-Chloro-2,6-Dibromofenol, New Polymeric Material, 1989, 1, 177.

118. Sikorski, R. T., Chemiczna modyfikacja polimerów, Prace Nauk. Inst. Technol. Org. Tw. Szt. P. Wr. 1974, 16, 33.
119. Sikorski, R. T., Podstawy chemii i technologii polimerów, PWN, Warszawa 1984.
120. Sikorski, R. T., Rykowski, Z., Puszyński, A., Elektronenempfindliche Derivate von Poly-methylmethacrylat, Plaste und Kautschuk 1984, 31, 250.
121. Simionescu, C. I., Geta, D., Grigoras, M., Ring-Opening Isomerization Polymerization of 2-Methyl-2-Oxazoline Initiated by Charge Transfer Complexes, Eur. Polym. J. 1987, 23, 689.
122. Sheinina, L. S., Vengerovskaya Sh.G., Khramova, T. S., Filipowich, A. Y., Reakcji piridinov i ich czietvierticznych soliej s epoksidnymi soedineniami, Ukr. Khim. Zh. 1990, 56, 642.
123. Soga, K., Toshida, Y., Hosoda, S., Ikeda, S., A Convenient Synthesis of a Polycarbonate, Makromol. Chem. 1977, 178, 2747.
124. Soga, K., Hosoda, S., Ikeda, S., A New Synthetic Route to Polycarbonate, J. Polym. Sci., Polym. Lett. Ed. 1977, 15, 611.
125. Soga, K., Uozumi, T., Yanagihara, H., Siono, T., Polymerization of Styrene with Heterogeneous Ziegler-Natta Catalysts, Makromol. Chem. Rapid. Commun., 1990, 11, 229.
126. Soga, K., Shiono, T., Wu, Y., Ishii, K., Nogami, A., Doi, Y., Preparation of a Stable Supported Catalyst for Propene Polymerization, Makromol. Chem. Rapid Commun. 1985, 6, 537.
127. Sokołov, L. B., Logunova, W. I., Otnositielnaja reakcionnosposobnost monomerov i prognozirovanic polikondensacji v emulsionnych sistemach, Vysokomol. Soed. A., 1979, 21, 1075.
128. Soler, H., Cadiz, V., Serra, A., Oxirane Ring Opening with Imide Acids, Angew. Makromol. Chem. 1987, 152, 55.
129. Spasówka, E., Rudnik, E., Możliwości wykorzystania węglowodanów w produkcji biodegradowalnych tworzyw sztucznych, Przemysł Chem. 1999, 78, 243.
130. Stevens, M. P., Wprowadzenie do chemii polimerów, PWN, Warszawa 1983.
131. Stępniak, I., Lewandowski, A., Elektrolity polimerowe, Przemysł Chem. 2001, 80(9), 395.
132. Strohriegl, P., Heitz, W., Weber, G., Polycondensation using silicon tetrachloride, Makromol. Chem. Rapid Commun. 1985, 6, 111.
133. Szur, A. M., Vysokomolekularnyje soedinenia, Vysszaja Szkoła, Moskwa 1966.
134. Tagle, L. H., Diaz, F. R., Riveros, P. E., Polymerization by Phase Transfer Catalysis, Polym. J. 1986, 18, 501.
135. Takagi, A., Ikada, E., Watanabe, T., Dielectric Properties of Chlorinated Polyethylene and Reduced Polyvinylchloride, Memoirs of the Faculty of Eng., Kobe Univ. 1979, 25, 239.
136. Tani, H., Stereospecific Polymerization of Aldehydes and Epoxides, Advances in Polym. Sci. 1973, 11, 57.
137. Vandenberg, E. J., Coordination Copolymerization of Tetrahydrofuran and Oxepane with Oxetanes and Epoxides, J. Polym. Sci. Part A, 1991, 29, 1421.
138. Veruovic, B., Navody pro laboratorni cviceni z chemie polymeru, SNTL, Praha 1977.

139. Vogdanis, L., Heitz, W., Carbon Dioxide as a Monomer, Makromol. Chem. Rapid Commun. 1986, 7, 543.
140. Vollmert, B., Grundriss der Makromolekularen Chemic, Springer-Verlag, Berlin 1962.
141. Wei, Y., Tang, X., Sun, Y., A study of the Mechanism of Aniline Polymerization, J.Polym.Sci. Part A, 1989, 27, 2385.
142. Wei, Y., Sun, Y., Jang G-W., Tang, X., Effects P-Aminodiphenylamine on Electrochemical Polymerization of Aniline, J. Polym. Sci., Part C, 1990, 28, 81.
143. Wei, Y., Jang G-W., Polymerization of Aniline and Alkyl Ring substituted Anilines in the Presence of Aromatic Additives, J. Phys. Chem. 1990, 94, 7716.
144. Wiles, D. M., Photooxidative Reactions of Polymers, J. Appl. Polym. Sci., Appl. Polym. Symp. 1979, 35, 235.
145. Wilk, K. A., Burczyk, B., Wilk, T., Oil in Water Microemulsions as Reaction Media, 7th International Conference Surface and Colloid Science, Compiegne 1991.
146. Winogradova, C. B., Novoe v obłasti polikondensacji, Vysokomol. Soed. A, 1985, 27, 2243.
147. Wirpsza, Z., Poliuretany, WNT, Warszawa 1991.
148. Wojtala, A., Wpływ właściwości oraz otoczenia poliolefin na przebieg ich fotodegradacji, Polimery 2001, 46, 120.
149. Xue, G., Polymerization of Styrene Oxide with Pyridine, Makromol. Chem. Rapid Commun. 1986, 7, 37.
150. Yamazaki, N., Imai, Y., Phase Transfer Catalyzed Polycondensation of Bishalomethyl Aromatic compounds, Polym. J. 1983, 15, 905.
151. Yasuda, H., Plasma Polymerization, Academic Press, Inc. Orlando 1985.
152. Yuan, H. G., Kalfas, G., Ray, W. H., Suspension polymerization, J. Macromol. Sci. Rev. Macromol. Chem. Phys., 1991, C 31, 215.
153. Zubakowa, L. B., Tevlina, A. C., Davankov, A. B., Sinteticzeskije ionoobmiennyje materiały, Chimia, Moskwa 1978.
154. Zubanov, B. A., Viedienie v chimiu polikondensacionnych procesov, Nauka, Ałma Ata 1974.
155. Żuchowska, D., Polimery konstrukcyjne, WNT, Warszawa 2000.
156. Żuchowska, D., Struktura i właściwości polimerów jako materiałów konstrukcyjnych, W.P.Wr., Wrocław 1986.
157. Żuchowska, D., Steller, R., Meisser, W., Structure and properties of degradable polyolefin – starch blends, Polym. Degrad. Stab. 1998, 60, 471.

CHAPTER 11

PREPARATION OF MICRO AND NANO-SIZED POLYMER-COLLOID COMPLEXES

E. I. KULISH,[1] M. V. BAZUNOVA,[1] E. I. KULISH,[1] L. A. SHARAFUTDINOVA,[1] V. G. SHAMRATOVA,[1] and G. E. ZAIKOV[2]

[1]*Bashkir State University, 32 Zaki Validi Street, 450076 Ufa, Republic of Bashkortostan, Russia*

[2]*Institute of Biochemical Physics named N.M. Emanuel of Russian Academy of Sciences, 4 Kosygina Street, 119334, Moscow, Russia*

CONTENTS

ABSTRACT

This chapter presents the results of the development process for the preparation of micro and nano-sized polymer-colloid complexes (PCC) on the basis of the chitosan (CTZ) and the sodium salt of chitosan succinamid (SCTZ) with silver halide sols in aqueous media. Results of research of CTZ, sodium salt of SCTZ solutions and PCC of CTZ and SCTZ with colloidal particles of silver iodide influence on structurally-functional properties of erythrocytes' membranes on model of acidic hemolysis are presented. The comparative analysis of results convinces that CTZ, SCTZ solutions and disperse systems on the basis of PCC of CTZ and SCTZ with colloidal particles of the silver iodide are capable to modulate variously matrix properties of erythrocytes of blood.

11.1 INTRODUCTION

The problem of creating of stable disperse systems based on lyophobic sols stabilized by non-covalent interaction of colloidal particles with macromolecules of natural and synthetic polymers is topical in the following task:

- synthesis of metal nanoparticles in polymer-protectors solutions;
- creating of metal-polymer nanocomposites with unique catalytic, electrical, optical and magnetic properties [1];
- obtaining of nano- and micro-sized polymeric containers for targeted delivery of slightly soluble drugs [2, 3].

Using of ultrafine systems is especially effective while introducing new ways to target delivery of drugs to the affected area of the body and new methods of prolonging the therapeutic effect of drugs. As a means of target delivery of drug with prolonged action nano-sized systems based on natural and synthetic biodegradable biocompatible polymers, including chitosan (CTZ) and its derivatives are often used.

The positive charge of CTZ macromolecules promotes its penetration through cell membranes and dense layers of the epithelium, provides good adhesion to mucosa and determines its bacteriostatic properties.

Spontaneously formed aggregates of macromolecules of CTZ and its derivatives are arousing particular interest during the creating of drug carriers based on these polysaccharides. It's possible to obtain polymeric carriers with desired size directly knowing the laws of aggregates of CTZ and its derivatives formation, and the factors that determine their size. The theory describing the behavior of polyelectrolytes being associated in dilute aqueous solutions predicts that the stabilization of intermolecular aggregates is determined by the competition between attraction and repulsion (caused by the charged groups on the polymer chain) of groups being associated and the osmotic pressure of counter ions, which prevents the aggregation. The aggregates sizes depend on the content of associating and charged groups. However, the available experimental data are contradictory. Also, there is evidence of instability of associated systems based on CTZ and its derivatives [4].

The size of polymeric nanoparticles for targeted drug delivery is equally important. It is known that polymeric carriers with sizes of 100–200 nm may provide directional transportation of antituberculosis drugs directly into macrophages, as macrophages are capable of absorbing foreign objects exactly with such sizes [5]. Particles of smaller size (30–40 nm to 200 nm) can accumulate passively in antitumor foci by a mechanism known as increased permeability and retention. This happens because of the increased blood supply and lower lymphatic drainage in tumor [6].

One of the approaches for establishing stable nanostructured systems with adjustable sizes may be using of the ability of macromolecules to self-assembly by intermolecular association via non-covalent bonds – on the example of a CTZ and its derivatives polymer-colloid complexes (PCCs), such as sodium salt of chitosan succinylamide (SCTZ) with inorganic colloidal particles of lyophobic sols, for example, sol of silver iodide. Dispersed system on the basis of sols in the presence of the polymer solution is essentially organic-inorganic nanocomposite in which the polysaccharide macromolecules form shields around inorganic nanoparticles.

The proposed approach is interesting because of:

- the resulting complexes may retain aggregate stability for a long time, that allows to predict the possibility of practical application of composites;
- inorganic colloidal particles of lyophobic sols can act as nucleus for encapsulation of slightly soluble drugs by using nano- and micro-sized containers [7].

An additional advantage of developing methods of targeted delivery of drugs based on the PCC of CTZ and SCTZ with colloidal particles of silver iodide is antiseptic properties of AgI.

Colloidal systems of drugs' delivery are usually designed both for oral and intravenous administration of drugs [8]. Therefore, the study of the biocompatibility of initial polysaccharides CTZ and SCTZ solutions and aqueous dispersions of PCCs of CTZ with blood cells gains special importance. Blood as one of the body fluids serves as a kind of informative biomarker for organism resistance to the action of various factors, allowing to establish the critical points of transition from a pronounced toxic to the stimulating effect. Erythrocytes, like other blood cells respond to changes in the external and internal environment, but due to the morphological simplicity their pathological reactions are slightly informative. Due to this, the action of certain chemicals in subtoxic doses, usually does not cause microscopically observed changes. However, one of the manifestations of their influence may be changes in cell resistance to the effects of hemolytic agents. Dysfunction of biological membranes under the influence of various exogenous substances is in many cases not only the result, but the cause of pathological changes in the cell and the organism as a whole [9].

11.2 AIM OF THE WORK

In connection with the foregoing, it is appropriate to obtain the PCC on the basis of water-soluble natural CTZ and SCTZ polymers with silver halide sols in aqueous media and to study the influence of the sample of CTZ, SCTZ solutions and the resulting dispersion PCC on the structural and functional properties of erythrocyte membranes. Taking into account that the main reaction of erythrocytes in contact with a foreign surface is lysis [10], as a model for evaluating the effects of these drugs on cells acid hemolysis selected.

11.3 EXPERIMENTAL PART

We used CTZ samples with deacetylation degree of 82% and M_w = 80,000 a.m.u. and SCTZ with Mw = 207000 a.m.u. with the substitution degree of 75% (produced by "Bioprogress" Russia, Shchelkovo).

The crossover concentration of CTZ and SCTZ determined from the dependence of logarithm of polysaccharides solutions dynamic viscosity on the concentration logarithm was 0.4 wt. % and 0.7 wt. %, respectively.

AgI sols prepared by standard methods from 7 mL of 0.01 N silver nitrate solution and 10 mL of 0.01 N potassium iodide.

The initial AgI micelle particles' size, defined by two independent methods – turbidimetry at a wavelength of 440 and 490 nm by FEC-56 instrument, and by particle size analyzer "Shimadzu Salid – 7101" – was between 95 and 120 nm.

Disperse systems on the basis of SCTZ PCC with colloidal particles of silver iodide are obtained by mixing of an aqueous solution of SCTZ with 0.4% concentration with freshly prepared AgI sol in volumetric ratio of 1:1 and 2:1 (hereinafter – 1:1 SCTZ:AgI sol and 2:1 SCTZ:AgI sol).

Disperse systems on the basis of CTZ PCCs with colloidal particles of silver iodide are obtained by mixing of CTZ dissolved in 1% acetic acid with concentration of 1 wt. % with freshly prepared AgI sol in volumetric ratio of 1:1 (hereinafter – 1:1 CTZ:AgI sol).

The 1:1 SCTZ:AgI sol, 2:1 SCTZ:AgI sol and 1:1 CTZ:AgI sol dispersions particles' size defined by the particle size analyzer "Shimadzu Salid – 7101" is between 100 and 375 nm.

Quantitative studies of CTZ and SCTZ complexes with AgI micelles are carried out by turbidimetric titration of AgI sols by polymers' solutions so that the $[Ag^+]$: unit of the polymer ratio is from 0.175:1 to 2:1.

Rheological measurements of CTZ, SCTZ solutions and PCC dispersion conducted on a modular dynamic rheometer Haake Mars III at 25°C. Flow curves and viscosity curves were obtained in a constant shear stress mode at a shear rate of 0.1 to 100 s^{-1}.

Influence of solutions CTZ, SCTZ and disperse systems based on CTZ and SCTZ PCCs with colloidal particles of silver iodide on the structural and functional properties of erythrocyte membranes was estimated by recording the kinetics of acid hemolysis [2]. In the in vitro experiment, used blood in an amount of 0.5 mL obtained from the tail vein of Wistar (n = 20) rats weighing 200–250 g in compliance with the European Convention requirements for the Protection of Vertebrate Animals used for experimental and other scientific purposes (Strasbourg, 1986) and the Russian Federal Law "On Protection of Animals on Cruelty" from 01.01.1997. Heparin is used as anticoagulant. Study of the influence degree of ultrafine systems on the basis of chitosan and its derivatives on the physicochemical properties of the lipid-protein complexes of the plasma membranes was performed on the model system CTZ sample-erythrocytes-HCl due to erythrocyte resistance change because of influence of hydrochloric acid (0.08 mL) in iso-osmotic solution of NaCl (0.85%). Hemolysis was carried out in quartz cuvettes with exterior dimensions of 20*40*10 mm and a volumetric displacement of 5 mL. Initial AgI sol, aqueous solution of SCTZ with concentration of 0.2 wt. %, CTZ dissolved in 1% acetic acid with concentration of 0.5 wt.%, aqueous dispersion of 1:1 SCTZ:AgI sol and 2:1 SCTZ: AgI sol, 1:1 CTZ: AgI sol are used as active agents. Hemolysis intensity in time recorded photometrically using photoelectrocolorimeter KFK – 3. Determination of extinction was performed every 15 seconds until no further changes. Reducing of extinction – result of the gradual destruction of red blood cells, and those cells are first broken, which have weaker resistance to hydrochloric acid. According to the obtained values acidic erythrograms depicting the distribution of erythrocytes by resistance were built: on the horizontal axis – hemolysis time, which is a measure of the resistance, delayed, the vertical axis – the percentage of erythrocytes. In the control, the differential distribution curve of erythrocytes by resistance takes the form of unimodal curve with steeply descending branch. Analysis of the kinetics of acidic erythrograms performed using the following parameters: start time, end time and the peak of hemolysis, the width of the interval of the dominant group of erythrocytes in the population. The proportion of cells with different resistance in the general population of erythrocytes was calculated.

Mathematical and statistical analysis of the results was carried out in the application package STATISTICA v. 7.0 ("StatSoftInc," USA). Comparison of acidic erythrograms was performed using the nonparametric Mann-Whitney test, the null hypothesis of no difference was rejected at $p < 0.05$.

11.4 RESULTS AND DISCUSSION

The particle sizes of initial AgI micelles, defined according to the turbidity data and measurement using a particle size analyzer "Shimadzu Salid – 7101" are, on average, 95–120 nm. Consequently, one can expect the formation of polycomplexes on the basis of AgI sols with water-soluble polymers having a particle size close to the nanometer range.

To find the critical molar ratio of [Ag^+]: polysaccharide unit, leading to the formation of an insoluble complex, the turbidimetric method was used at polymer concentrations below the crossover point to avoid the mutual influence of macromolecular coils. The results are shown in Figure 11.1.

It has been established that for PCC on the basis of SCTZ and AgI sol range of molar ratios wherein aggregate stability of the particles observed is larger than for systems based on CTZ– AgI sol and even particularly neutralized CTZ- AgI sol. This can probably be attributed to the higher density of the CTZ charge distribution on the polysaccharide macromolecules and high ionic strength of solution.

In any case, it was found that the addition of a weak polyelectrolyte increases sedimentation stability of AgI micelles.

The particle sizes of obtained PCC on the basis of SCTZ solution mixtures with AgI sol, ranging from 100 to 375 nm.

Significant increase in the dynamic viscosity SCTZ:AgI sol system at middle concentrations that are lower in comparison with SCTZ's (Figure 11.2) indicates the formation of the PCC and the structuring effect of AgI particles in SCTZ: AgI systems.

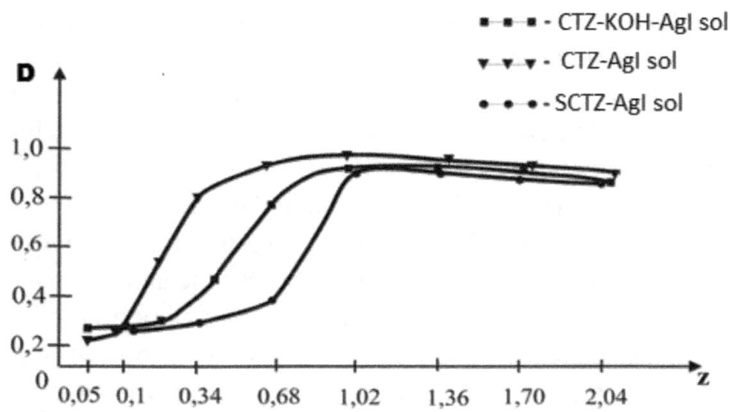

FIGURE 11.1 The dependence of the optical density of mixtures of CTZ solutions with AgI sol, of CTZ solutions in the presence of 0.01 mol/L KOH with AgI sol, of SCTZ solutions with AgI sol on composition of the mixture Z = [Ag^+]/[polymer unit] at t = 20°C, λ = 490 nm.

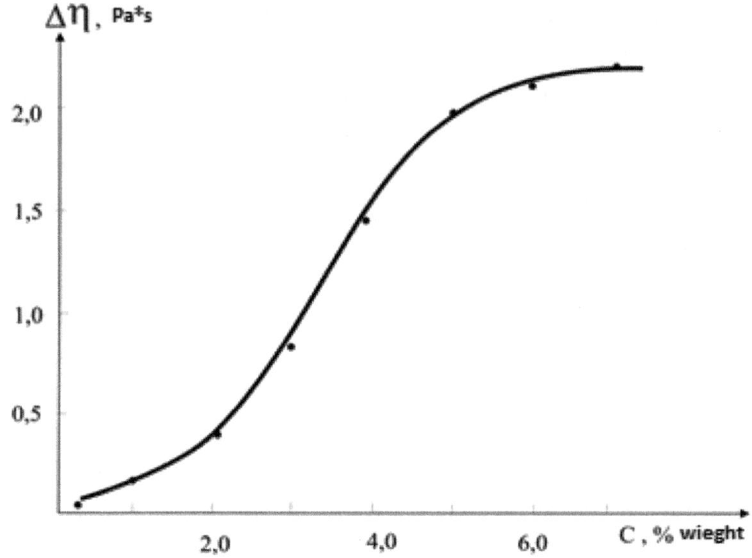

FIGURE 11.2 Increase of the dynamic viscosity of the SCTZ: AgI sol system in comparison with solution SCTZ with the polymer concentration increasing, 20°C, AgI particle concentration of 0.004 mol/L.

The results of studies on the hemocompatibility (Table 11.1) showed that the studied solutions of samples of CTZ, SCTZ and PCC dispersions of CTZ and SCTZ with colloidal particles of silver iodide have ambiguous effect on the functional state of the rat blood cell membranes. Comparative analysis confirms that the tested samples are capable to change the character of the distribution of red blood cells in terms of their stability and kinetic parameters of hemolysis in different ways. Particularly 0.2% SCTZ solution had no significant effect on the initial capacity of red blood cells to resist the action of hemolytic agent. At the same time, a sample of the disperse system of 2:1 SCTZ: AgI sol discovered the ability to change the functional state of erythrocyte membranes. During all periods of observation destruction intensity of red blood cells by the action of hydrochloric acid and the total lysis duration markedly differed from the original characteristics, increased for resistant erythrocytes. Apparently, this SCTZ composition has a stabilizing effect on the erythrocyte membranes.

Estimating the resistance degree after the incubation of red cells with dispersion of 1:1 HTZ: AgI sol, it should be noted that despite the slight

TABLE 11.1 Indicators of Acidic Resistance of Erythrocytes in the Control Group and After Exposure of Chitosan (CTZ) Solutions, Sodium Salt of Chitosan Succinylamide (SCTZ) and Polymer-Colloid Complexes (PCCs) of CTZ and SCTZ with Colloidal Particles of Silver Iodide (M ± m)

Indicators	**Control**	**0.2% SCTZ solution**	**0,5% CTZ solution in 1% acetic acid**	**The aqueous dispersion of 2:1 SCTZ: AgI sol**	**The aqueous dispersion of 1:1 SCTZ: AgI sol**	**The aqueous dispersion of 1:1 CTZ: AgI sol**
Time of the beginning of hemolysis, min	0.5±0.01	1.00±0.1	1.5±0.4	1.00±0.1	1.5±0.2	1.5±0.3
Time of the ending of hemolysis, min	5.5±0.1	6.5±0.2	6.5±0.4	6.5±0.3	6.5±0.2	6.5±0.4
Peak of the acidic erythrogram, min	2.00±0.3	2.00±0.1	2.00±0.1	3.00±0.2*	2.0±0.1	3.00±0.2*
Erythrograms base width, min	5.00±0.1	5.5±0.18	5.00±0.2	5.5±0.1	5.0±0.2	5.5±0.3
The percentage of highly resistant erythrocytes, %	12.9±1.6	11.5±1.4	13.16±2.1	34.1±4.7*	24.9±3.8*	24.6±2.9*
The percentage of erythrocytes with increased stability, %	4.78±0.31	8.6±1.1*	6.96±0.8	19.1±2.6*	13.00±2.1*	21.2±2.9*
The percentage of erythrocytes with average stability, %	73.19±5.86	63.3±4.7	72.5±5.2	35.8 ±3.7*	52.3±4.6*	41.6±4.1*
The percentage of erythrocytes with low stability, %	9.13±0.87	16.6±1.4*	7.38±0.6	11.0±1.8	11.29±2.3	12.6±2.5

* Statistically significant differences in comparison to the control group ($p<0.01$).

decrease erythrocyte fraction with average stability (41.6%), the number of red blood cells with a high and increased stability in comparison with the control (24–6 ± 2.9% and 21.2 ± 2.9% respectively, $p < 0.01$) significantly increases. Later more involvement of red blood cell in the process of hemolysis was found, that could indicate a protective effect of the compound.

Comparison of studies results of the effect of ultrafine systems based on chitosan and its derivatives on the integrity of red blood cells demonstrated the ability of some (2:1 SCTZ: AgI sol, 1:1 CTZ: AgI sol, 1:1 SCTZ: AgI sol) to provide a stabilizing effect on the membrane of red blood cells and others (0.2% aqueous SCTZ solution) to cause a decrease in resistance of erythrocytes. The observed effects may be determined by the variety of features of their impact on the nature of the bonds between the protein and lipid components of the membrane, the level of activity of cell enzyme systems, the value of the negative charge of the membrane surface [11]. Taking into account that the acid resistance of erythrocytes is mainly determined by erythrocyte membranes phospholipid bilayer state, it can be assumed that the observed effects are specified by reorganization of the lipid component of erythrocytes membranes under the exposure of ultrafine systems on the basis of chitosan and its derivatives.

11.5 CONCLUSION

Thus, the method of obtaining of stable nano- and micro-sized PCCs on the basis of water-soluble natural polymers CTZ and SCTZ with sols of silver halide in aqueous media was developed. It was found that CTZ and SCTZ solutions and dispersions of polymer-colloid complexes of CTZ and SCTZ with colloidal particles of silver iodide have a stabilizing or destabilizing effect on erythrocyte membranes depending on the composition of the sample, that is show their different hemocompatibility and points on the urgency of further search of chitosan and its derivatives complexes with inorganic colloidal particles having potentially protective properties. The results of these studies are promising at the obtaining of nano- and micro-sized polymeric containers for targeted delivery of slightly soluble drugs.

KEYWORDS

- **acid resistance of erythrocytes**
- **chitosan**
- **micelles of silver iodid**
- **polymer-colloidal complexes**

REFERENCES

1. Eliseeva, E. A., Litmanovich, E. A., Ostaeva, G. Yu., Chernikova, E. V., Papisov, I. M. Zoli medi, stabilizirovannye polietilenglikol'-600-monolauratom i ego kompleksami s poliakrilovoi kislotoi. Vysokormolekulyarnye soedinenieya. Seriya A, 2014, vol. 56, № 6, 631–637.
2. Bazunova, M. V., Kolesov, S. V., Hairullina, A. I., Zaikov, G. E. Podhod k sozdaniyu lekarstvennyh sredstv prolongirovannogo deistviya na osnove uglerod-polimernyh nositelei. Vestnik Kazanskogo tehnologicheskogo universiteta, 2014, vol.17, № 7, p. 145–147.
3. Bolotova, G. V. Polimernye nositeli dlya protivotuberkulyoznyh lekarstvennyh sredstv na osnove hitozana. Molodoi uchenyi, 2010, Vol.2, №5, p. 208–210.
4. Korchagina, E. V. Agregaciya hitozana I ego proizvodnyh v razbavlennyh vodnyh rastvorah: avtoreferat dissertacii kandidata fiziko-matematicheskih nauk: 02.00.06.: M.: MGU im. M. V. Lomonosova, 2012, p. 232.
5. Bolotova, G. V. Polimernye nositeli dlya protivotuberkulyoznyh lekartsvennyh sredstv na osnove hitozana. Molodoi uchyonyi, № 5 (16), vol. 2.
6. Nel, A. E., Madler, L., Velegol, D. Understanding biophysicochemical interactions at the nan-bio interface. Nature Materials. 2009, № 8, 543–557.
7. Inozemceva, O. A. Formirovovanie i fiziko-himicheskie svoistva polielektrolitnyh nanokompozitnyh mikrokapsul. Rossiiskie nanotehnologii, 2007, vol. 2, № 9, 68–80.
8. Lampreht Alf. Nanolekarstva. Koncepcii dostavki lekarstv v nanonauke. M.: Nauchnyi mir, 2001, p. 232.
9. Savlukov, A. I., V. M. Samsonov, R. F. Kamilov, E. D. Wakirova, R. N. Yapparov, D. F. Wakirov. Sostoyanie ustoichivosti eritrocitov kak zveno adaptacii organizma. Medicinskii vestnik Bawkortostana, 2011, vol. 6. №4, 13–17.
10. Sevast'yanov, V. I., Laksina, O. V., Novikova, S. P., Rozanova, I. B., Ceitlina, E. A., Wal'nev, B. I. Sovremennye gemosovmestimye materialy dlya serde4no-sosudistoi hirurgii, pod red. V. I. Wumakova, (medicina i zdravoohranenie, seriya hirurgiya, issue 2). M., VNIIMI, 1987.
11. Kalawnikova, I. V., Mehanizmy vzaimodeistviya antibiotikov penicilinovogo ryadas eritrocitami cheloveka. Byull. Eksp. Biologii i mediciny, 2008, vol. 146, № 10, 419–423.

CHAPTER 12

ENTROPIC NOMOGRAMS AND S-CURVES

G. A. KORABLEV,[1] N. G. PETROVA,[2] R. G. KORABLEV,[1]
P. L. MAKSIMOV,[1] and G. E. ZAIKOV[3]

[1]*Izhevsk State Agricultural Academy, Russian Federation*

[2]*Agency of Informatization and Communication, Udmurt Republic, Russian Federation, E-mail: biakaa@mail.ru*

[3]*Institute of Biochemical Physics, Russian Academy of Science, Russian Federation*

CONTENTS

ABSTRACT

The concept of the entropy of spatial-energy interactions is used similarly to the ideas of thermodynamics on the static entropy.

The idea of entropy appeared based on the second law of thermodynamics and ideas of the adduced quantity of heat.

These rules are general assertions independent of microscopic models. Therefore, their application and consideration can result in a large number of consequences, which are most fruitfully used in statistic thermodynamics.

In this research we are trying to apply the concept of entropy to assess the degree of spatial-energy interactions using their graphic dependence, and in other fields. The nomogram to assess the entropy of different processes is obtained. The variability of entropy demonstrations is discussed, in biochemical processes, economics and engineering systems, as well.

12.1 INTRODUCTION

In statistic thermodynamics the entropy of the closed and equilibrium system equals the logarithm of the probability of its definite macro state:

$$S = k, \tag{1}$$

where W – number of available states of the system or degree of the degradation of microstates; k – Boltzmann's constant.

Or:

$$W = e^{S/k} \tag{2}$$

These correlations are general assertions of macroscopic character, they do not contain any references to the structure elements of the systems considered and they are completely independent of microscopic models [1].

Therefore the application and consideration of these laws can result in a large number of consequences.

At the same time, the main characteristic of the process is the thermodynamic probability W. In actual processes in the isolated system the entropy growth is inevitable – disorder and chaos increase in the system, the quality of internal energy goes down.

The thermodynamic probability equals the number of microstates corresponding to the given macro-state.

Since the system degradation degree is not connected with the physical features of the systems, the entropy statistic concept can also have other applications and demonstrations (apart from statistic thermodynamics).

"It is clear that out of the two systems completely different by their physical content, the entropy can be the same if their number of possible microstates corresponding to one macroparameter (whatever parameter it is) coincide. Therefore, the idea of entropy can be used in various fields. The increasing self-organization of human society leads to the increase in entropy and disorder in the environment that is demonstrated, in particular, by a large number of disposal sites all over the earth" [2].

In this research, the concept of entropy is applied to assess the degree of spatial-energy interactions using their graphic dependence, and in other fields.

12.2 ENTROPIC NOMOGRAM OF THE DEGREE OF SPATIAL-ENERGY INTERACTIONS

The idea of spatial-energy parameter (P-parameter) which is the complex characteristic of the most important atomic values responsible for interatomic interactions and having the direct bond with the atom electron density is introduced based on the modified Lagrangian equation for the relative motion of two interacting material points [3].

The value of the relative difference of P-parameters of interacting atoms-components – the structural interaction coefficient α is used as the main numerical characteristic of structural interactions in condensed media:

$$\alpha = \frac{P_1 - P_2}{(P_1 + P_2)/2} 100\% \qquad (3)$$

Applying the reliable experimental data we obtain the nomogram of structural interaction degree dependence (ρ) on coefficient α, the same for a wide range of structures (Figure 12.1). This approach gives the possibility to evaluate the degree and direction of the structural interactions of phase formation, isomorphism and solubility processes in multiple systems, including molecular ones.

Such nomogram can be demonstrated [3] as a logarithmic dependence:

$$\alpha = \beta(\ln \rho)^{-1} \tag{4}$$

where coefficient β is the constant value for the given class of structures. β can structurally change mainly within ±5% from the average value. Thus coefficient α is reversely proportional to the logarithm of the degree of structural interactions and therefore can be characterized as the entropy of spatial-energy interactions of atomic-molecular structures.

Actually the more is ρ, the more probable is the formation of stable ordered structures (e.g., the formation of solid solutions), for example, the less is the process entropy. But also the less is coefficient α.

The Eq. (4) does not have the complete analogy with Boltzmann's equation (1) as in this case not absolute but only relative values of the corresponding characteristics of the interacting structures are compared which can be expressed in percent. This refers not only to coefficient α but also to the comparative evaluation of structural interaction degree (ρ), for example – the percent of atom content of the given element in the

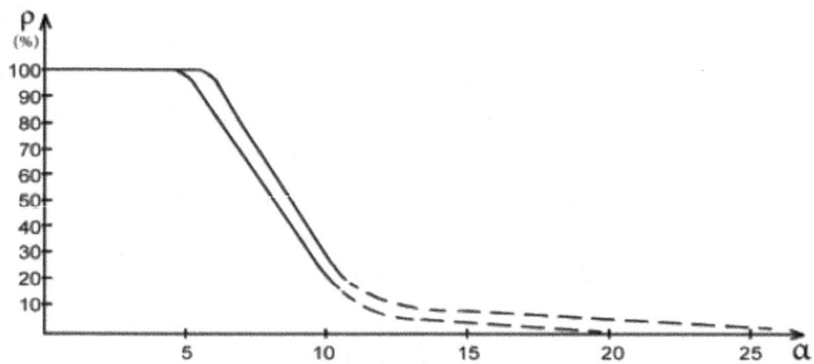

FIGURE 12.1 Nomogram of structural interaction degree dependence (ρ) on coefficient α.

solid solution relatively to the total number of atoms. Therefore, in Eq. (4) coefficient k = 1.

Thus, the relative difference of spatial-energy parameters of the interacting structures can be a quantitative characteristic of the inter action entropy: $\alpha \equiv S$.

12.3 ENTROPIC NOMOGRAM OF SURFACE-DIFFUSIVE PROCESSES

As an example, let us consider the process of carbonization and formation of nanostructures during the interactions in polyvinyl alcohol gels and metal phase in the form of copper oxides or chlorides. At the first stage, small clusters of inorganic phase are formed surrounded by carbon containing phase. In this period, the main character of atomic-molecular interactions needs to be assessed via the relative difference of P-parameters calculated through the radii of copper ions and covalent radii of carbon atoms.

In carbonization period the metal phase is being formed on the surface.

From this point, the binary matrix of the nanosystems C→Cu is being formed.

The values of the degree of structural interactions from coefficient α are calculated, for example, $\rho_1 = f\left(\frac{1}{\alpha_2}\right)$ – curve 2 given in Figure 12.2. Here, the graphical dependence of the degree of nano film formation (ω)

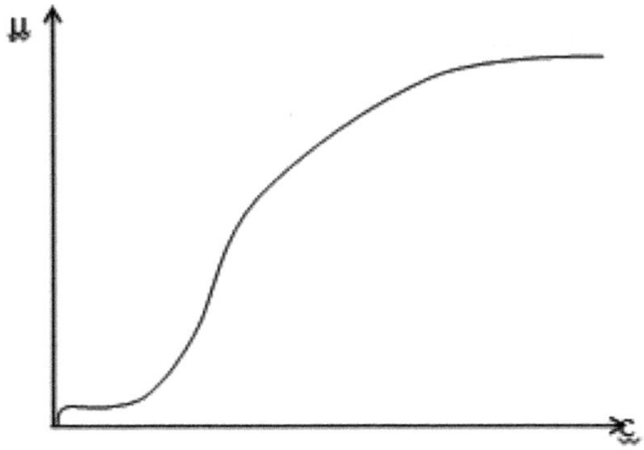

FIGURE 12.2.

on the process time is presented by the data from Ref. [4] – curve 1 and previously obtained nomogram in the form $\rho_1 = f\left(\frac{1}{\alpha_2}\right)$ – curve 3.

The analysis of all the graphical dependencies obtained demonstrates the practically complete graphical coincidence of all three graphs: $\omega = f(t)$, $\rho_2 = f\left(\frac{1}{\alpha_1}\right)$, $\rho_2 = f\left(\frac{1}{\alpha_2}\right)$, with slight deviations in the beginning and end of the process. Thus, the carbonization rate, as well as the functions of many other physical-chemical structural interactions, can be assessed via the values of the calculated coefficient α and entropic nomogram.

12.4 NOMOGRAMS OF BIOPHYSICAL PROCESSES

12.4.1 ON THE KINETICS OF FERMENTATIVE PROCESSES

"The formation of ferment-substrate complex is the necessary stage of fermentative catalysis. At the same time, n substrate molecules can join the ferment molecule" [6, p. 58].

For ferments with stoichiometric coefficient *n* not equal one, the type of graphical dependence of the reaction product performance rate (μ) depending on the substrate concentration (c) has [5] a sigmoid character with the specific bending point.

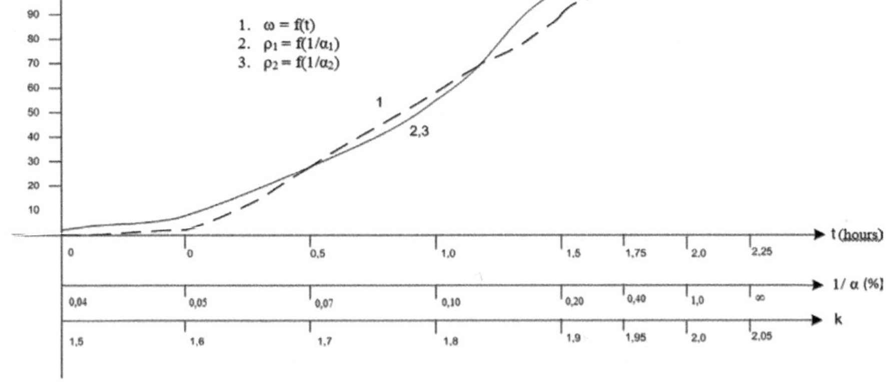

FIGURE 12.3.

The graph of the dependence of electron transport rate in bio-structures on the diffusion time period of ions is similar or dissimilar [5, p. 278].

In the procedure of assessing fermentative interactions (similarly to the previously applied in part 3 for surface-diffusive processes) the effective number of interacting molecules over 1 is applied.

In the methodology of P-parameter, a ferment as a limited isomorphic similarity with substrate molecules and does not form a stable compound with them, but, at the same time, such limited reconstruction of chemical bonds which "is tuned" to obtain the final product is possible.

12.4.2 DEPENDENCE OF BIOPHYSICAL CRITERIA ON THEIR FREQUENCY CHARACTERISTICS

a) The passing of alternating current through live tissues is characterized by the dispersive curve of electrical conductivity – this is the graphical dependence of the tissue total resistance (z-impedance) on the alternating current frequency logarithm (log ω). Normally, such curve, on which the impedance is plotted on the coordinate axis, and log ω – on the abscissa axis, formally, completely corresponds to the entropic nomogram (Figure 12.1).
b) The fluctuations of bio-membrane conductivity (conditioned by random processes) "have the form of Lorentz curve" [6, p. 99]. In this graph, the fluctuation spectral density (ρ) is plotted on the coordinate axis, and the frequency logarithm function log ω – on the abscissa axis.

The type of such curve also corresponds to the entropic nomogram in Figure 12.1.

12.5 LORENTZ CURVE OF SPATIAL-TIME DEPENDENCE

In Lorentz curve [7] the space-time graphic dependence (Figure 12.4) of the velocity parameter (θ) on the velocity itself (β) is given, which completely corresponds to the entropic nomogram in Figure 12.2.

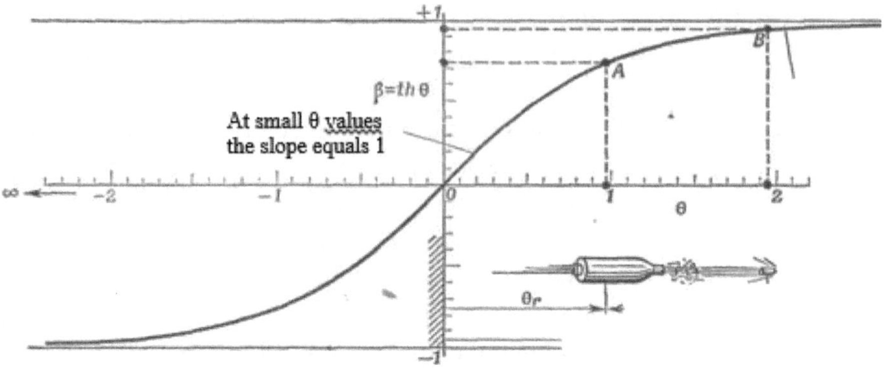

FIGURE 12.4 Connection between the velocity parameter Θ and velocity itself β=thΘ.

12.6 ENTROPIC CRITERIA IN BUSINESS AND NATURE

The main properties of free market providing its economic advantages are: (1) effective competition; and (2) maximal personal interest of each worker.

But on different economy concentration levels these ab initio features function and demonstrate themselves differently. Their greatest efficiency corresponds to small business – when the number of company staff is minimal, the personal interest is stronger and competitive struggle for survival is more active. With companies and productions increase the number of staff goes up, the role of each person gradually decreases, the competition slackens as new opportunities for coordinated actions of various business structures appear. The quality of economic relations in business goes down, for example, the entropy increases. Such process is mostly vivid in monostructures at the largest enterprises of large business (syndicates and cartels).

The concept of thermodynamic probability as a number of microstates corresponding to the given macrostate can be modified as applicable to the processes of economic interactions that directly depend on the parameters of business structures.

A separate business structure can be taken as the system macrostate, and as the number of microstates – number of its workers (N) which is the number of the available most probable states of the given business structure. Thus it is supposed that such number of workers of the business structure is the analog of thermodynamic probability as applicable to the processes of economic interactions in business.

Therefore it can be accepted that the total entropy of business quality consists of two entropies characterizing: (1) decrease in the competition efficiency (S_1), and (2) decrease in the personal interest of each worker (S_2), for example, $S = S_1 + S_2$. S_1 is proportional to the number of workers in the company: $S \sim N$, and S_2 has a complex dependence not only on the number of workers in the company but also on the efficiency of its management. It is inversely proportional to the personal interest of each worker. Therefore, it can be accepted that $S_2 = \frac{1}{\gamma}$, where γ – coefficient of personal interest of each worker.

By analogy with Boltzmann's equation (1) we have:

$$S = (S_1 + S_2) \sim \left[\ln N + \ln\left(\frac{1}{\gamma}\right)\right] \sim \ln\left(\frac{N}{\gamma}\right)$$

or $S = k\ln\left(\frac{N}{\gamma}\right)$,

where k – proportionality coefficient.

Here N shows how many times the given business structure is larger than the reference small business structure, at which N = 1, for example, this value does not have the name.

For non-thermodynamic systems when we consider not absolute but relative values, we take k = 1. Therefore:

$$S = \ln\left(\frac{N}{\gamma}\right) \tag{5}$$

In Table 12.1 you can see the approximate calculations of business entropy by the equation (5) for three main levels of business: small, medium and large. At the same time, it is supposed that number N corresponds to some average value from the most probable values.

When calculating the coefficient of personal interest, it is considered that it can change from 1 (oneself-employed worker) to zero (0), if such worker is a deprived slave, and for larger companies it is accepted as. $\gamma = 0.1 - 0.01$.

Despite of the rather approximate accuracy of such averaged calculations, we can make quite a reliable conclusion on the fact that business

TABLE 12.1 Entropy Growth

Structure parameters	Business		
	Small	**Average**	**Large**
$N_1 - N_2$	10–50	100–1000	10,000–100,000
γ	0.9–0.8	0.6–0.4	0.1–0.01
S	2.408–4.135	5.116–7.824	11.513–16.118
$\langle S \rangle$	3.271	6.470	13.816

entropy, with the aggregation of its structures, sharply increases during the transition from the medium to large business as the quality of business processes decreases [8].

In thermodynamics it is considered that the uncontrollable entropy growth results in the stop of any macro changes in the systems, for example, to their death. Therefore, the search of methods of increasing the uncontrollable growth of the entropy in large business is topical. At the same time, the entropy critical figures mainly refer to large business. A simple cut-down of the number of its employees cannot give an actual result of entropy decrease. Thus the decrease in the number of workers by 10% results in diminishing their entropy only by 0.6% and this is inevitably followed by the common negative unemployment phenomena.

Therefore for such super-monostructures controlled neither by the state nor by the society the demonopolization without optimization (i.e., without decreasing the total number of employees) is more actual to diminish the business entropy.

Comparing the nomogram (Figure 12.1) with the data from the Table 12.3, we can see the additivity of business entropy values (S) with the values of the coefficient of spatial-energy interactions (α), for example, $S = \alpha$.

It is known that the number of atoms in polymeric chain maximally acceptable for a stable system is about 100 units, which is 10^6 in the cubic volume. Then we again have $\lg 10^6 = 6$.

12.7 S-CURVES ("LIFELINES")

Already in the last century some general regularities in the development of some biological systems depending on time (growth in the number of bacteria colonies, population of insects, weight of the developing fetus, etc.) were found [9]. The curves reflecting this growth were similar, first of all, by the fact that three successive stages could be rather vividly emphasized on each of them: slow increase, fast burst-type growth and stabilization (sometimes decrease) of number (or another characteristic). Later it was demonstrated that engineering systems go through similar stages during their development. The curves drawn up in coordinate system where the numerical values of one of the most important operational characteristics (for example, aircraft speed, electric generator power, etc.) were indicated along the vertical and the "age" of the engineering system or costs of its development along the horizontal, were called S-curves (by the curve appearance), and they are sometimes also called "lifelines."

As an example, the graph of the changes in steel specific strength with time (by years) is demonstrated [9] (Figure 12.5).

Thus, the similarity between S-curves and entropic nomogram in Figure 12.2 is observed.

And in this case, the same as before, the time dependence(t) is proportional to the entropy reverse value ($1/\alpha$). As applicable to system, such curves characterize the process intensity, for example, sale of the given products.

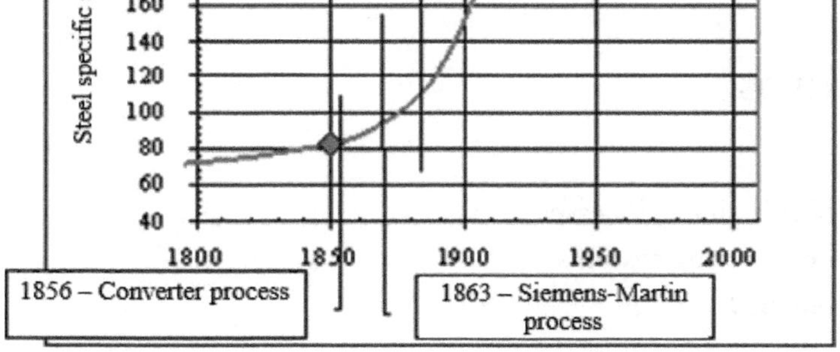

FIGURE 12.5 Dependence of steel specific strength on time.

At the same time, entropic nomograms in accordance with Figure 12.1 assess the business quality (ordinate in such graphs).

It is known that the entropy of isolated systems decreases. The entropy growth in open systems is compensated by the negative entropy due to the interaction with the environment.

All the above systems can be considered as open ones. This also refers to spatial-energy processes, when any changes in quantitative energy characteristics are conditioned by the interaction with external systems.

It is obviously observed in engineering and technological systems, the development of which is followed by additional innovations, modifications and financial investments.

The entropy in thermodynamics is considered as the measure of nonreversible energy dissipation. From the point of technological and economic principles, the entropy is mainly the measure of irrational energy resource utilization. With the time dependence increase, such processes stabilize in accordance with the nomogram to more optimal values – together with the growth of anti-entropy, for example, the value $1/\alpha = 1/\rho$.

The similar growth with time of rationality of technological, economic and physical and chemical parameters proves that such nomograms are universal for the majority of main processes in nature, technology and economy.

12.8 CONCLUSION

The idea of entropy is diversified in physical and chemical, economic, engineering and other natural processes that are confirmed by their nomograms.

KEYWORDS

- **biophysical processes**
- **engineering systems**
- **entropy**

- **nomogram**
- **spatial-energy parameter**
- **system**

REFERENCES

1. Reif, F. Statistic physics. M.: Nauka, 1972, 352 p.
2. Gribov, L. A., Prokofyeva, N. I. Basics of Physics. M.: Vysshaya shkola, 1992, 430 p.
3. Korablev, G. A. Spatial-Energy Principles of Complex Structures Formation. Brill Academic Publishers and VSP, Netherlands, 2005, 426pp. (Monograph).
4. Kodolov, V. I., Khokhriakov, N. V., Trineeva, V. V., Blagodatskikh, I. I. Activity of nanostructures and its manifestation in nanoreactors of polymeric matrixes and active media. Chemical Physics and Mesoscopy, 2008, V. 10, №4, p. 448–460.
5. Rubin, A. B. Biophysics. Book 1. Theoreticalbiophysics. M.: Vysshayashkola, 1987, 319 p.
6. Rubin, A. B. Biophysics. Book 2. Biophysics of Cell Processes. M.: Vysshaya shkola, 1987, 303 p.
7. Taylor, E., Wheeler, J. Spacetimephysics. Mir Publishers. M., 1987, 320 p.
8. Korablev, G. A., Petrova, N. G., Korablev, R. G., Osipov, A. K., Lekomtsev, P. L. Business entropy. Vestnik of Izhevsk State Agricultural Academy, 2013, №1, p. 76–79.
9. Kynin, A. T., Lenyashin, V. A. Assessment of the parameters of engineering systems using the growth curves. http://www.metodolog.ru/01428/01428.html.

CHAPTER 13

OXIDATION 2-(N-ACETYLAMINO)-2-(3,5-DI-TERT-BUTYL-4-HYDROXYPHENYL)-PROPIONIC ACID BY OXYGEN IN AN ALKALINE CONDITION

A. A. VOLODKIN,[1] G. E. ZAIKOV,[1] L. N. KURKOVSKAJA,[1] S. M. LOMAKIN,[1] I. M. LEVINA,[1] and E. V. KOVERZANOVA[2]

[1]*Federal State Budgetary Establishment of a Science of Institute of Biochemical Physics of N. M. Emanuelja of Russian Academy of Sciences, Russian Federation, E-mail: chembio@sky.chph.ras.ru*

[2]*Federal State Budgetary Establishment of a Science of Institute of Chemical Physics of N. N. Semenov of Russian Academy of Sciences, Russian Federation*

CONTENTS

ABSTRACT

At oxidation 2-(N-acetylamino)-2-(3,5-di-*tert*-butyl-4-hydroxyphenyl)-propionic acid by oxygen at 55–60°C it is formed 2,4-di-*tert*-butyl-bicyclo(4,3,1)deca-4,6-dien-8-(N-acetylamino)-3,9-dion-1-oxa, which constitution is based on the data of spectrums of NMR ^{1}H ^{13}C.

13.1 INTRODUCTION

Results of early works have shown perspectivity of analogs of tyrosine in biology and specificity of properties in the conditions of reactions with sour agents. For example, in reactions with thionyl chloride instead of acid chloride 2-(N-acetylamino)-2-(3,5-di-*tert*-butyl-4-hydroxy-phenyl)-propionic acid products of oxidative dimerization are formed [1]. The direction oxidation by oxygen 4-replaced 2,6-di-*tert*-butylphenols depends on conditions and a substituent constitution, and in each specific case results, as a rule, are ambiguous [2]. Thereupon oxidation 2-(N-acetylamino)-2-(3,5-di-*tert*-butyl-4-hydroxyphenyl)-propionic acid it is represented to one of the stages bound to an establishment of a constitution and properties of products inhibitor of oxidation, especially in the conditions of biological researches.

13.2 EXPERIMENTAL PART

NMR spectrums registered on the device “Avance-500 Bruker” rather TMS.

2-(N-acetylamino)-2-(3,5-di-*tert*-butyl-4-hydroxyphenyl)-propionic acid (1) it is synthesized on a method [3], m.p. 204–206°C. According to [4]: m.p. 204–206°C.

Oxidation 2-(N-acetylamino)-2-(3,5-di-*tert*-butyl-4-hydroxyphenyl)-propionic acid. In solution of 3.35 g (0.01 mol) compound **1,** 2.0 g (0.04 mol) NaOH in 60 mL of ethanol at 55–60°C passed oxygen within 6 h.

In a day to solution added HCl prior to the beginning of breaking (pH ≈6). Organic part have separated and after solvent and crystallization evaporation received 0.6 g (≈18%) 2,4-di-*tert*-butylbicyclo(4,3,1)deca-4,6-dien-8-(N-acetylamino)-3,9-dion-1-oxa **(2)**; m.p. 101–105°C. Found %: C 68.32 H 8.34. $C_{19}H_{27}NO_4$. Calculated %: C 68.44 H 8.16.

Spectrum NMR ^{1}H (DMSO-d_6, δ, ppm, *J*/Hz): 1.20 (s 18H, ^{t}Bu); 1.89 (s.3H, CH_3CO); 2.40 (t. 2H, $\underline{CH_2}$CH, *J* = 12.9); 4.86 (m. 1H, $CH_2\underline{CH}$,); 6.69 (d. 1H, *J* = 2.9); 6.95 (d. 1H, *J* = 2.9); 8.56 (d.1H, NH, *J* = 7.8). Spectrum NMR ^{13}C (DMSO-d6, δ, ppm): 22.18 ($\underline{C}H_3$CO); 29.09 (C–$\underline{C}H_3$); 36.84 ($\underline{C}H_2$); 48.56 ($\underline{C}$H); 76.28 ($\underline{C}$); 138.9 (C-$\underline{C}$-H) 139.6 (C-$\underline{C}$-H); 145.3 (C=$\underline{C}$); 146.0 ($\underline{C}$=C) 169.3 ($\underline{C}$ONH); 174.0 ($\underline{C}$OOH); 185.4 ($\underline{C}$=O).

2-(N-acetylamino)-2-(3,5-di-*tert*-butyl-4-hydroxyphenyl)-propionate sodium (3) Ad mixture of 3.35 g (0.01 mol) compound **1**, 0.5 g (0.01 mol) NaOH in 20 mL of ethanol maintain ≈15 mines, solvent evaporated, the residual crystallization from EtOH – H_2O (1:1). A yield≈3 g, m.p. > 250°C. Spectrum NMR ^{1}H (DMSO – d_6, δ, ppm, *J*/ Hz): 1.35 (s., 18 H, ^{t}Bu); 1.77 (s., 3H, $COCH_3$); 2.73 (dd., 1H, CH-CH$\underline{H}$, *J* = 6.8); 2.95 (dd, 1H, CH-C$\underline{H}$H, *J* = 4.6); 3,97 (m. 1H, C$\underline{H}$—CH_2); 6.6 (s., 1H, OH); 6.89 (s. 2H, Ar); 7.23 (d., 1 H, NH, *J* = 7.4). Spectrum NMR^{13}C (DMSO – d6, δ, ppm): 22.93 ($\underline{C}H_3$CO); 30.48 (C–$\underline{C}H_3$); 34.27 ($\underline{C}$); 37.45 ($\underline{C}H_2$); 55.73 ($\underline{C}$H); 125.5 (C=$\underline{C}$–H); 130.46 ($\underline{C}$–C=C); 138.09 (C–C=$\underline{C}$); 151.58 (C-$\underline{C}$-OH); 167.63 ($\underline{C}ONHCH_3$); 173.93 ($\underline{C}$OOH).

13.3 RESULTS AND DISCUSSION

In the presence of NaOH and 2-(N-acetylamine)-2-(3,5-di-*tert*-butyl-4-hydroxyphenyl)-propionic acid **(1)** at temperature of 55–60°C process of oxidation 2-(N-acetylamino)-2-(3,5-di-*tert*-butyl-4-hydroxyphenyl)-propionic sodium **(3)**. For 6 h. reactions in oxygen atmosphere ≈18% are formed 2,4-di-*tert*-butylbicyclo(4,3,1)deca-4,6-dien-8-(N-acetylamino)-3, 9-dion-1-oxa **(2)**, schema 1. At ambient temperature on air salt **3** does not react with oxygen. Composition of reactionary masses and a yield of compound **2** supervised from the data of a spectrum of NMR ^{1}H (Figure 13.1).

THE SCHEMA 1

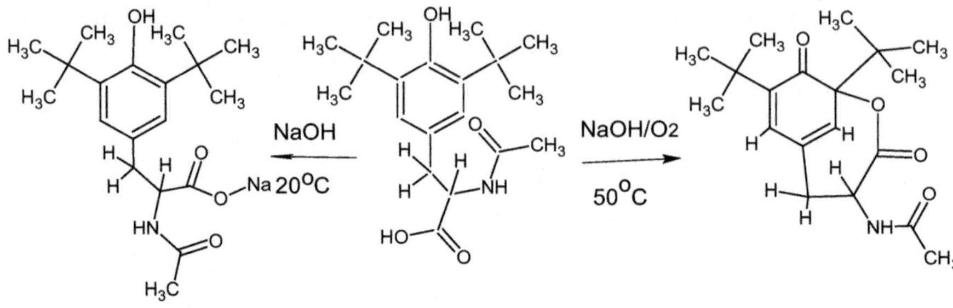

3 **1** **2**

From the data of a spectrum of NMR ^{1}H follows that signals of 6.69 and 6.95 ppm (J=2.9 Hz) to belong to two various protons in a hexatomic cycle of structure **2**. In a spectrum NMR ^{13}C signals of 138.92 and 139.59 ppm from carboneum of atoms correspond to the data which analysis in a format *debt* (Figure 13.2) specifies in communication of these atoms with hydrogen.

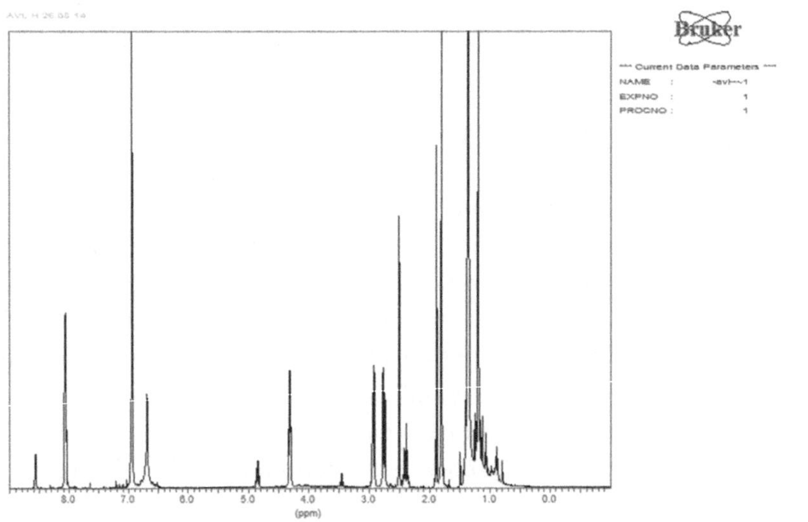

FIGURE 13.1 Spectrum of NMR ^{1}H reactionary mass of reaction of compound 1 with oxygen in an alkaline condition at 55–60°C for 6 h.

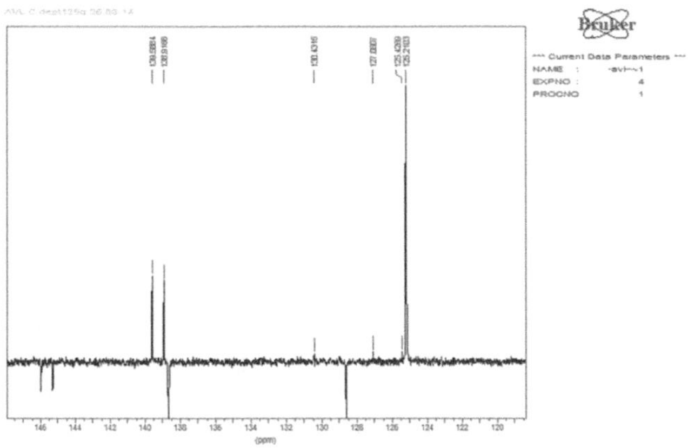

FIGURE 13.2 Fragment of a spectrum of NMR ^{13}C compound 2 in a format *dept.* (reactionary mass).

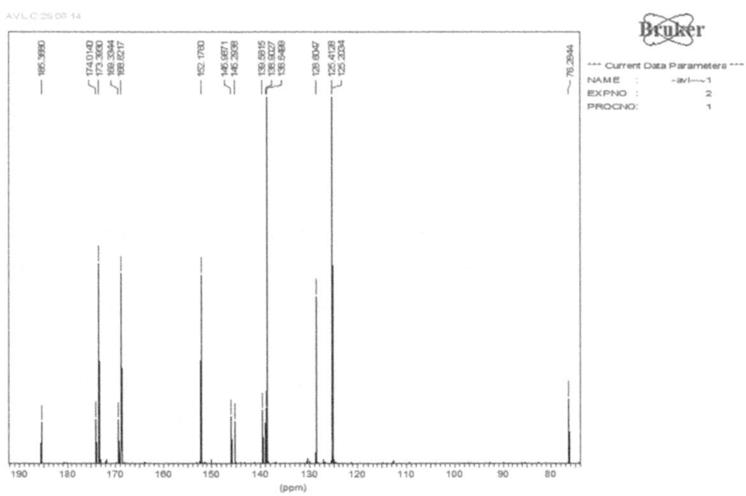

FIGURE 13.3 Spectrum of NMR ^{13}C from reactionary mass of reaction of compound 1 with oxygen in an alkaline condition at 55–60°C for 6 h.

Signal of 185.4 ppm in a spectrum NMR^{13}C (Figure 13.3) confirms presence of a carbonyl group at a hexatomic cycle of structure **2**.

At last, a signal of 76.28 ppm belongs to tetrahedral atom of carboneum, forming communications with cycle on carboneums atom, *tert*-butyl group and atom of oxygen (Figure 13.4).

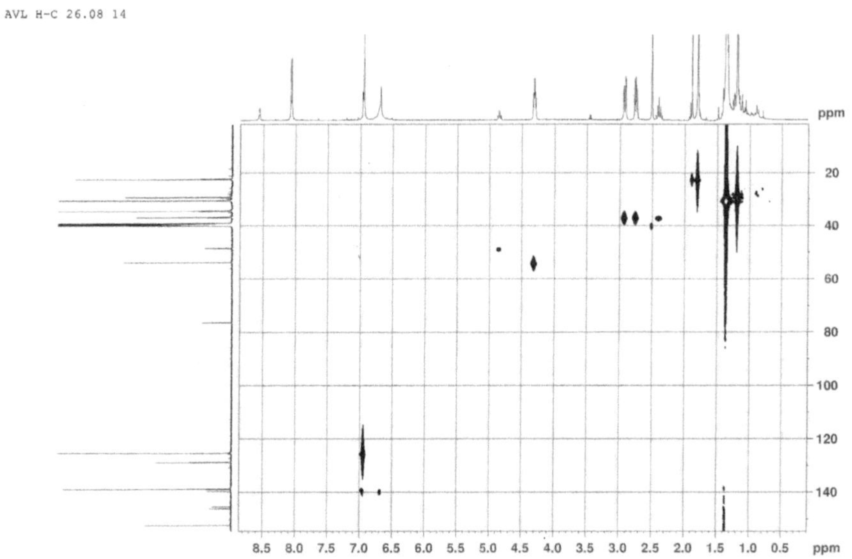

FIGURE 13.4 Correlation of data spectrums NMR ^1H and ^{13}C in system C–H.

According to Ref. [5], position of a signal from tetrahedral atom of carboneum in spiran to system with atom of nitrogen is in area of 75 ppm.

Data of spectrums of NMR is sufficient for conclusions about structure of oxidate 2-(N-acetylamino)-2-(3,5-di-*tert*-butyl-4-hydroxyphenyl)-propionic acid by O_2 at temperature of 50–60°C in solution of ethanol which confirm bonding between atom of oxygen of a carboxyl group and of carboneum atom from a hexatomic cycle.

13.4 CONCLUSION

The Process of oxidation 2-(N-acetylamino)-2-(3,5-di-*tert*-butyl-4-hydroxyphenyl)-propionicacidinanalkalineconditionwithallocation2,4-di-*tert*-butylbicyclo(4,3,1)deca-4,6-dien-8-(N-acetylamino)-3,9-dion-1-oxa.

KEYWORDS

- **2-(N-acetylamino)-2-(3,5-di-*tert*-butyl-4-hydroxyphenyl)-propionic acid**
- **2,4-di-*tert*-butylbicyclo(4,3,1)deca-4,6-dien-8-(N-acetylamino)-3,9-dion-1-oxa**
- **NMR-spectroscopy**
- **oxidation**
- **oxygen**

REFERENCES

1. Volodkin, A. A., Kurkovskaja, L. N., Zaikov, G. E., Lomakin, S. M.: Izv. Akad. Nauk, Ser. *Khim.*, 2013, 2265 [in Russian].
2. Volodkin, A. A., Malysheva, R. D., Ershov, V. V.: Izv. Akad. Nauk, Ser. *Khim.*, 1982, 1594 [in Russian].
3. Volodkin, A. A., Lomakin, S. M., Zaikov, G. E., Evteeva, N. M.: Izv. Akad. Nauk, Ser. *Khim.*, 2009, 900 [in Russian].
4. Teuder, H. J., Rause, H., Berariu, V.: Lieb. Ann. 1978, 757.
5. Vystorop, I. V., Konovalova, N. P., Neljubina, J. V., Varfolomeev, V. N., Fedorov, B. S., Sashenkova, T. E., Berseneva, E. N., Lysenko, K. A., Kostjanovsky, R. G.: Izv. Akad. Nauk, Ser. *Khim* 2010, 127–134 [in Russian].

CHAPTER 14

FORMATION 2-(3′, 5′-DI-TERT-BUTYL-4′-HYDROXYPHENYL) ETHYLOXY-*P*-CRESOL IN REACTION OF BASIC HYDROLYSIS DIETHYL N-ACETYLAMINO-(3,5-DI-TERT-BUTYL-4-HYDROXYBENZYL)-MALONATE

A. A. VOLODKIN,[1] G. E. ZAIKOV,[1] L. N. KURKOVSKAJA,[1] S. M. LOMAKIN,[1] I. M. LEVINA,[1] and E. V. KOVERZANOVA[2]

[1]*Federal State Budgetary Establishment of a Science of Institute of Biochemical Physics of N. M. Emanuelja of Russian Academy of Sciences, Russian Federation, E-mail: chembio@sky.chph.ras.ru*

[2]*Federal State Budgetary Establishment of a Science of Institute of Chemical Physics of N. N. Semenov of Russian Academy of Sciences, Russian Federation*

CONTENTS

ABSTRACT

New of condensation with simultaneous elimination *tertiary* butyl groups it is found in reaction of basic hydrolysis of diethyl N-acetylamino-(3,5-di-*tert*-butyl-4-hydroxybenzyl)-malonate in which result it is formed 2-(3,'5'-di-*tert*-butyl-4'-hydroxy-phenyl)ethyloxy-*p*-cresol. The compound constitution is based on the data NMR ^{1}H and ^{13}C, IR-spectra and synthesis from 5,7-di-*tert*-butylspiro(2,5)octa-4,7-diene-6-one.

14.1 INTRODUCTION

2-(N-Acetylamino)-2 (3,5-di-*tert*-butyl-4-hydroxyphenyl)-propionic acid is one of products of basic hydrolysis of diethyl N-acetylamino-(3,5-di-*tert*-butyl-4-hydroxybenzyl)-malonate, which development of chemistry are bound to synthesis of perspective antioxidants, analogs of tyrosine. Presence of two *is tertiary* butyl group in a benzene ring affects on reactivity of functional substituents that defines interest to researches of this bunch of compounds. For example, in classical reaction with thionyl chloride instead of acid chloride 2-(N-acetylamino)-2 (3,5-di-*tert*-butyl-4-hydroxyphenyl)-propionic acid products of oxidative dimerization [1] are formed. In reaction with hydrogen chloride products elimination of alkyl groups are formed [2].

In the present works researches with use of alkaline agents are continued. Throughout research of basic hydrolysis of diethyl N-acetylamino-(3,5-di-*tert*-butyl-4-hydroxybenzyl)-malonate **(1)**, it is positioned that in the course of reaction it is formed 2-(3,'5′-*di-tert-butyl*-4'-hydroxyphenyl)ethyloxy-*p*-cresol **(2)**. Formation of compound **(2)** is accompanied by elimination *tert*-butyl groups and loss acetylamine group, it is reactions uncharacteristic for earlier known derivatives acetylaminomolonic acids.

Compound **2** is synthesized from 5,7-di-*tert*-butylspiro(2,5)octa-4, 7-diene-6-one **(3)** and identity of the samples received by different methods is proved.

14.2 EXPERIMENTAL PART

NMR spectrums registered on the device "Avance-500 Bruker" rather TMS.

14.2.1 METHOD 1

Ad mixture of 43.6 g (0.1 mol) diethyl N-acetylamino-(3,5-di-*tert*-butyl-4-hydroxybenzyl)-malonate **(1)**, 4 g (0.1 mol) NaOH, 20 mL of water and 200 mL of iso-propyl alcohol heated at boiling under reflux. In a current of argon within 3 h. Further a reaction mixture cooled, the dropped out deposit separated and product isolated from aqueous ethanol (8:2). Received 23.4 g (62%) crystalohydrate sour 2-(N–acetylamino)-2-(3,5-di-*tert*-butyl-4-hydroxybenzile)-malonate sodium, m.p. > 258°C. According to Ref. [3], m.p. 258–260°C. Spectrum NMR ^1H (DMSO-d_6, δ, ppm): 1.26 (s, 18 H, ^tBu); 1.79 (s., 3 H, $COCH_3$); 3.37 (s, 2H, CH_2); 3.40–3.48 (with. уш, HOH);); 6.42–6.46 (s. уш., 1 H, OH); 6.79 (s, 2H, Ar); 7.20–7.27 (with. уш, s, 1 H, NH).

To a mother solution added 10% HC1 to pH ≈6, the dropped out deposit separated and isolated 2-(N-acetylamino)-2-(3,5-di-*tert*-butyl-4-hydroxyphenyl)-propionic acid from acetone. Yield of 5.37 g (17%); m.p. 205–206°C. According to Refs. [3, 4]: m.p. 205–206°C. Spectrum NMR ^1H (acetone – d_6, δ, ppm, *J*/Hz): 1.42 (c 18 H, ^tBu); 1.96 (s, 3 H, $COCH_3$); 2.94 (dd, 1H, CH-CH<u>H</u>, *J*=7.6); 3.11 (dd, 1H, CH-C<u>H</u>H), *J*=5.0); 4.63– 4.69 (m, 1H, C<u>H-</u>CHH); 5.90 (s, 1 H, OH); 7.05 (s, 2H, Ar); 7.45 (d, 1 H, NH, *J*=5.0).

Crystals dropped out of residual solution on the expiration of 30 days 2-(3,' 5′-di-*tert*-butyl-4'-hydroxyphenyl)ethyloxy-*p*-cresol (2) in quantity of 0.7 g; m.p. 95–96°C. Spectrum NMR ^1H ($CDCl_3$, δ, ppm., *J*/Hz): 1.43 (s., 18 H, ^tBu); 2.45 (s., 3H, CH_3); 2.90 (t., 2H, $CH_{2-}CH_{2-}$Ar, *J* = 7.4);

4.21 (t., 2H, CH_2-CH_{2-}Ar, J = 7.4); 5.13 (s., H, OH); 6.93 (s., 2H, Ar); 7.31 (d., 2H, Ar', J = 8.3); 7.74 (d.2H, Ar', J = 8.3). Spectrum NMR ^{13}C ($CDCl_3$, δ, ppm): 21.1 (CH_3); 29.7 (C–CH_3); 33.71 (C–H); 34.78 (CH_2), 70.60 (CH_2); 124.9 (C=C–H); 126.2 (C'= C-H); 127.3 (C' =C'–H); 129.2 (C=C-C=C-OH); 132.77 (C=C–C=C–OH); 135.6 (C' =C'–C' =C'–OH); 144.0 (HO–C=C); 152.19 (H-O–C' =C'). Spectrum NMR ^{17}O ($CDCl_3$, δ, ppm) 162.4. IR-Spectrum (KBr, ν/cm^{-1}): 3616 (OH), 1176 (C-O-C). Found %: C 81.25; H 9.44. $C_{23}H_{32}O_2$. Calculated %: C81.13; H 9.47.

14.2.2 METHOD 2

2-(3,' 5′-di-*tert*-butyl-4'-hydroxyphenyl)ethyloxy-*p*-cresol (2). Compound **3** in the form of a powder (4.5 g) seated in a weighing bottle and abandoned at ambient temperature ~ 6 months. The formed single crystal in number of 2.2 g separated, m.p. 95–96°C. Spectrum NMR ^{1}H ($CDCl_3$, δ, ppm., J/Hz): 1.43 (s., 18 H, ^{t}Bu); 2.45 (s., 3H, CH_3); 2.90 (t., 2H, CH_{2-}C$\underline{H}_{2-}$Ar, J = 7.4); 4.21 (t., 2H, C$\underline{H}_{2-}$$CH_{2-}$Ar, J = 7.4); 5.13 (s., H, OH); 6.93 (s., 2H, Ar); 7.31 (d., 2H, Ar', J = 8.3); 7.74 (d.2H, Ar', J = 8.3). Spectrum NMR ^{13}C ($CDCl_3$, δ, ppm): 21.1 (CH_3); 29.7 (C–$\underline{C}H_3$); 33.71 ((C–H)); 34.78 (CH_2), 70.60 (CH_2); 124.9 (C=$\underline{C}$–H); 126.2 (C'= $\underline{C}$-H); 127.3 ($\underline{C'}$ =C'–H); 129.2 ($\underline{C}$=C-C=C-OH); 132.77 (C=C–$\underline{C}$=C–OH); 135.6 ($\underline{C}$' =C'–C' =C'–OH); 144.0 (HO–$\underline{C}$=C); 152.19 (H-O–$\underline{C'}$=C'). Spectrum NMR ^{17}O ($CDCl_3$, δ, ppm) 162.4. IR-Spectrum (KBr, ν/ cm^{-1}): 3616 (OH), 1176 (C-O-C).

14.3 RESULTS AND DISCUSSION

Formation 2-(3,' 5′-di-*tert*-butyl-4'-hydroxyphenyl)ethyloxy-*p*-cresol **(2)** in the conditions of basic hydrolysis diethyl N-acetylamino-(3,5-di-*tert*-butyl-4-hydroxybenzyl)-malonate **(1)** is unexpected and is bound to dimerization and elimination coupled reactions. Unexpected formation of compound **2** in conditions solid state reactions is same 5,7-di-*tert*-butyl-spiro(2,5)octa-4,7-diene-6-one **(3)** (Schema 1).

THE SCHEMA 1

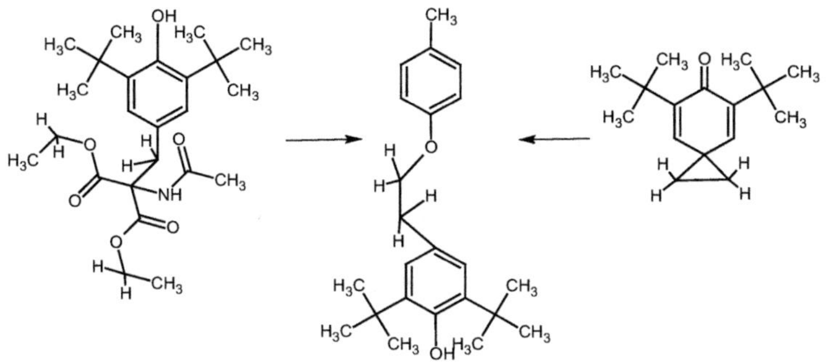

1 2 3

Identity of two samples of the same compound **2** is confirmed by comparison of their spectrums of NMR ^{1}H, [13]C (Figures 14.1 and 14.2) and IR-spectra (Figure 14.3).

Signals of 34.78 and 70.60 ppm fall into to carbon atoms of system group $CH_{2-}CH_2$ of compound **2**, that is confirmed by correlation in co-ordinates C–H. The data of a spectrum of NMR [13]C (21.06 ppm).

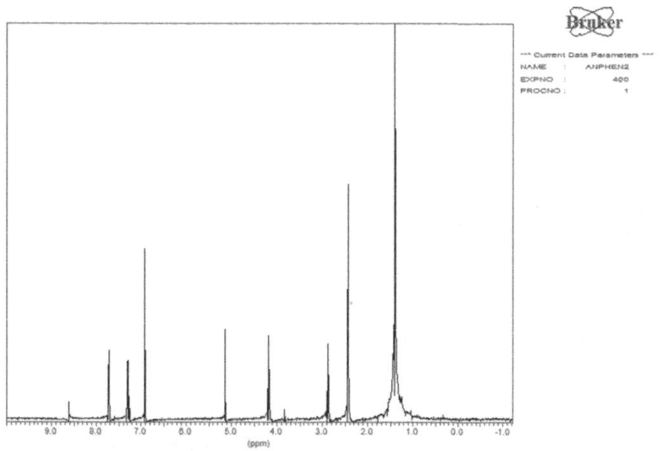

FIGURE 14.1 Spectrum of NMR [1]H 2-(3,'5'-di-*tert*-butyl-4'-hydroxyphenyl)ethyloxy-*p*-cresol, received on a method 1.

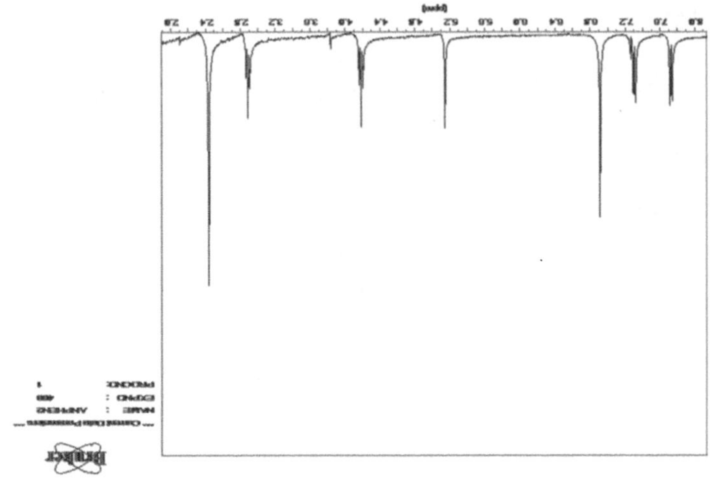

FIGURE 14.2 Spectrum of NMR [1]H 2-(3,'5′-di-*tert*-butyl-4'-hydroxyphenyl)ethyloxy-*p*-cresol, received on a method 2.

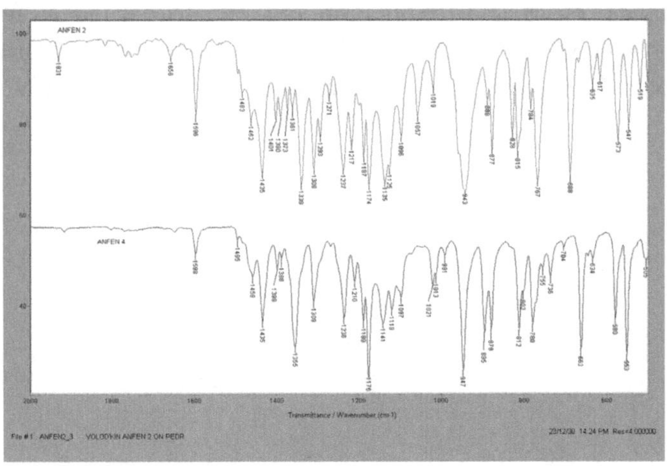

FIGURE 14.3 IR spectrums of samples 2-(3,'5′-di-*tert*-butyl-4'-hydroxyphenyl) ethyloxy-*p*-cresol.

NMR ^{1}H (2.45 ppm) specify on CH_3 group in pair position of a benzene ring of compound **2**. Spectrum NMR ^{1}H there are a signal (singlet) of 6.93 ppm. (meta-protons of an aromatic cycle), two doublet signals of 7.30 and 7.74 ppm. (4 protons of a fragment of frame p-kresol). This result confirms a stage loss of *tert*-butyl group in the course of formation of a single crystal **2**. There is opened a question of mechanism of formation of CH_3 group in the course of transformation of structure **1** in structure **2**.

The signal ^{1}H from phenolic hydroxyl is present at a spectrum NMR ^{1}H (5.1 ppm.) and IR – a spectrum (3616 cm^{-1}). Frequency of 1176 cm^{-1} is characteristic for C-O-C bunches. Presence at a molecular of atom of oxygen of bunch is confirmed by a NMR spectrum ^{17}O.

Taking into account formation of compound **2** in the course of spontaneous reaction 5,7-di-*tert*-butylspiro(2,5)octa-4,7-diene-6-one **(3)** it is possible to assume that in the conditions of diethyl N-acetylamino-(3,5-di-*tert*-butyl-4-hydroxybenzyl)-malonate **(1)** process of formation of compound **(3)**, for example, as a result of elimination acetylamine bunches is possible. The fact loss alkyl group proceeding is obvious at ambient temperature in a solid phase or neutral medium that earlier was not observed in practice of chemistry of the hindered phenols.

14.4 CONCLUSION

Condensation with simultaneous elimination alkyl groups in the course of basic hydrolysis diethyl N-acetylamino-(3,5-di-*tert*-butyl-4-hydroxybenzyl)-malonate, leads to formation 2-(3,’ 5’-di-*tert*-butyl-4’-hydroxyphenyl) ethyloxy-*p*-cresol.

KEYWORDS

- **2-(3,′ 5′-di-*tert*-butyl-4′-hydroxyphenyl)ethyloxy-*p*-cresol**
- **5,7-di-*tert*-butylspiro-(2,5)octa-4,7-diene-6-one**
- **diethyl N-acetylamino-(3,5-di-*tert*-butyl-4-hydroxybenzyl)-malonate**
- **IR spectroscopy**
- **NMR spectroscopy**

REFERENCES

1. Volodkin, A. A., Kurkovskaja, L. N., Zaikov, G. E., Lomakin, S. M.: Izv. Akad. Nauk, Ser. *Khim.*, 2013, 2265 [in Russian].
2. Volodskin, A. A., Zaikov, G. E., Kurkovskaja, L. N., Lomakin, S. M., Sofina, S. J., Vestn. Kasan. Technology University, 2013, v.16, №8, 18–21 [in Russian].
3. Volodkin, A. A., Lomakin, S. M., Zaikov, G. E., Evteeva, N. M. Izv. Akad. Nauk, Ser. *Khim.*, 2009, 900 [in Russian].
4. Teuder, H. J., Rause, H., Berariu, V. Lieb. Ann. 1978, 757.

INDEX

D

N

O

P

V

W

For Product Safety Concerns and Information please contact our EU representative GPSR@taylorandfrancis.com Taylor & Francis Verlag GmbH, Kaufingerstraße 24, 80331 München, Germany

Batch number: 10725935

Printed by Printforce, the Netherlands